T0403341

Functional Reverse Engineering of Machine Tools

Computers in Engineering Design and Manufacturing

Series Editor:
Wasim Ahmed Khan
Professor, Faculty of Mechanical Engineering, GIK (Ghulam Ishaq Khan)
Institute of Engineering Sciences and Technology, Pakistan

This new series aims to acquaint readers with methods of converting legacy software and hardware used in Computer Aided Design (CAD), Computer Aided Manufacturing (CAM), and Computer Aided Engineering (CAE), with or without Virtual Reality (VR) as well as current and future software modeling technology. CAD/CAM/CAE and VR technology, Design and Manufacturing, Thermofluids, System Dynamics, and Control Domains will be covered in the series. The series will also address monitoring of Product Life Cycle concepts applicable to discrete products and its integration in above-mentioned software packages on the World Wide Web. The concepts related to serial and parallel computing for the operation of CAD/CAM/CAE with or without VR technology at current atomic level Central Processing Unit (CPU) technology and future hardware technology based on graphene and quantum computing shall also be in the scope of the series. Current research in communication methods, both standard and proprietary at any given stage, shall also become the part of the book series at the appropriate level. The series will include books with an extensive background in mathematics, high-level software languages, software modeling techniques, computer hardware technology, communication methods and include mechanical engineering and industrial engineering domains.

Functional Reverse Engineering of Machine Tools
Edited by Wasim Ahmed Khan, Ghulam Abbas, Khalid Rahman,
Ghulam Hussain, and Cedric Aimal Edwin

For more information on this series, please visit: https://www.crcpress.com/Computers-in-Engineering-Design-and-Manufacturing/book-series/CRCCOMENGDES

Functional Reverse Engineering of Machine Tools

Edited by
Wasim Ahmed Khan, Ghulam Abbas,
Khalid Rahman, Ghulam Hussain, and
Cedric Aimal Edwin

CRC Press
Taylor & Francis Group
Boca Raton London New York

CRC Press is an imprint of the
Taylor & Francis Group, an **informa** business

CRC Press
Taylor & Francis Group
6000 Broken Sound Parkway NW, Suite 300
Boca Raton, FL 33487-2742

© 2020 by Taylor & Francis Group, LLC
CRC Press is an imprint of Taylor & Francis Group, an Informa business

No claim to original U.S. Government works

Printed on acid-free paper

International Standard Book Number-13: 978-0-367-07803-4 (Hardback)

Library of Congress Cataloging-in-Publication Data
Names: Khan, Wasim A., editor.
Title: Functional reverse engineering of machine tools / edited by Wasim Ahmed Khan [and 4 others].
Description: Boca Raton, FL : CRC Press/Taylor & Francis Group, 2019. | Series: Computers in engineering design and manufacturing | Includes bibliographical references.
Identifiers: LCCN 2019020278 | ISBN 9780367078034 (hardback : acid-free paper) | ISBN 9780429022876 (ebook)
Subjects: LCSH: Reverse engineering. | Machine theory. | Automatic control.
Classification: LCC TA168.5 .F86 2019 | DDC 621.9/02—dc23
LC record available at https://lccn.loc.gov/2019020278

Visit the Taylor & Francis Web site at
http://www.taylorandfrancis.com

and the CRC Press Web site at
http://www.crcpress.com

To late professor Dr. Abdul Raouf S.I.

Professor Dr. Abdul Raouf was a distinguish scholar of international ranking. He was a PhD in Industrial Engineering, and practiced teaching, research, and consultancy for 57 years. He was at the University of Windsor, Ontario, Canada, for more than two decades as Head of Industrial Engineering. He served King Fahd University of Petroleum and Minerals, Dahran, Saudi Arabia, as Professor in Systems Engineering Department for ten years. He was Rector of Ghulam Ishaq Khan (GIK) Institute of Engineering Sciences and Technology, Pakistan, for six years. He served as University Professor and Advisor of University of Management Technology, Lahore, Pakistan, and was Patron and Professor, Institute of Quality and Technology Management, University of Punjab until his retirement.

Dr. Raouf published extensively in the areas of performance evaluation that included Modeling and Optimization of Tasks involving Information Conservation, Information Reduction, Information Generation, and Production System Optimization in the areas of Quality, Safety, and Maintenance of Production Systems. He has authored/co-authored 10 books and contributed to more than 150 research papers to refereed journal and refereed conference proceedings.

Recognizing his scholarly pursuits, Dr. Abdul Raouf was bestowed upon the coveted title of "Sitare-e-Imtiaz (S.I.)" by the Government of Pakistan. He was declared "Best Scientist" in the field of engineering by the Pakistan Academy of Sciences. The Higher Education Commission of Pakistan conferred upon him the title of Distinguished National Professor.

Until retirement, besides being the Editor-in-Chief of the International Journal of Quality in Maintenance Engineering, *Dr. Abdul Raouf was on the editorial board of nine international research journals. He was Chairman of the National Quality Assurance Committee constituted by the Higher Education Commission of Pakistan. He was a member of the accreditation committee, Education Department, Government of Punjab. He was a member of governing bodies of a number of public and private sector universities.*

He was one of the honorary chairs of the First International Workshop on Functional Reverse Engineering of Machine Tools (WRE2018).

Contents

Foreword ix
Preface xi
Acknowledgments xvii
Disclaimer xix
Editors xxi
Contributors xxv

Part 1   The Cybernetics

1. **Microprocessors and Microcontrollers: Past, Present, and Future** 3
   Abdul Haseeb, Muhammad Faizan, Muhammad Faisal Khan, and Muhammad Talha Iqrar

2. **Selection of Sensors, Transducers, and Actuators** 29
   Memoon Sajid, Jahan Zeb Gul, and Kyung Hyun Choi

Part 2   Sensor Development

3. **Reverse Engineering of Titanium to Form TiO$_2$ for Humidity Sensor Applications** 55
   Azhar Ali Haidry and Linchao Sun

Part 3   Re-Manufacturing

4. **Manufacturing, Remanufacturing, and Surface Repairing of Various Machine Tool Components Using Laser-Assisted Directed Energy Deposition** 79
   Muhammad Iqbal, Asif Iqbal, Malik M. Nauman, Quentin Cheok, and Emeroylariffion Abas

Part 4   Metrology

5. **Investigation on Dynamical Prediction of Tool Wear Based on Machine Vision and Support Vector Machine** 91
   Chen Zhang and Jilin Zhang

6. **In-Process Measurement in Manufacturing Processes** 105
   Ahmad Junaid, Muftooh Ur Rehman Siddiqi, Riaz Mohammad, and
   Muhammad Usman Abbasi

Part 5   Design of Mechanisms and Machines

7. **The Convex Hull of Two Closed Implicit Surfaces** 137
   Xiaoping Wang

8. **Design of Multi-DOF Micro-Feed Platform Based on Hybrid**
   Compliant Mechanism 153
   Jianqiang Huo, Chen Zhang, and Yun Song

9. **Metal Forming Presses: Technology, Structure, and**
   Engineering Design 167
   Volkan Esat

Part 6   Decision Making

10. **Modular Approach of Writing MATLAB Code for a Markov**
    Decision Process 183
    Ali Nasir

11. **A Smart Microfactory Design: An Integrated Approach** 215
    Syed Osama bin Islam, Liaquat Ali Khan, Azfar Khalid, and
    Waqas Akbar Lughmani

12. **Reverse Engineering the Organizational Processes:**
    A Multiformalism Approach 255
    Abbas K. Zaidi and Edward Huang

Part 7   Next-Generation Communication

13. **Next-Generation National Communication Infrastructure**
    (NCI): Emerging Future Technologies—Challenges and
    Opportunities 277
    Javed I. Khan

Index 309

Foreword

Introduction of secrets of technology to less developed nations forces the modern world to find ways to advance further in order to keep its dominance. Efforts by machine tool research center at the Ghulam Ishaq Khan (GIK) Institute of Engineering Sciences and Technology, Pakistan, are admirable as the team is forging ahead with system design and manufacturing of the machine tools. In line with its vision to assist the government of Khyber Pakhtunkhwa in making a shift to knowledge-based economy, the Directorate of Science and Technology (DoST), Government of Khyber Pakhtunkhwa, Pakistan, is proudly co-pioneering and financing the commendable effort. The initiative, which carries a great promise, will enable Khyber Pakhtunkhwa and GIK Institute to develop the capability of functional reverse engineering – the first of its kind in Pakistan.

The disruption is making the youth of the Next Eleven Nations think about promoting enterprises of functional reverse engineering for self-employment, job creation, financial freedom and contribution to socio economic uplift of Pakistan. It will surely fuel the passion for further technological development and advancement in the field by DoST, Government of Khyber Pakhtunkhwa, Pakistan, and GIK Institute.

Hopefully, in next 2 years, the machine tool research center will be able to formalize the codes and procedures for functional reverse engineering for use by various sectors in Next Eleven Countries.

Dr. Khalid Khan
Director
Directorate of Science and Technology
Department of Science and Technology & Information Technology
Government of Khyber Pakhtunkhwa
Pakistan

Preface

Functional reverse engineering is a concept whereby machinery and equipment have their system design performed and the object manufactured using available production facility based on understanding of the functionality of the object. The object may have the same functionality or may have enhanced innovative and multidisciplinary features.

Functional reverse engineering involves detailed system design with respect to mechanical, electrical, electronics, computer software, computer hardware, and control engineering features of the object. Ergonomics plays an important role as well. The object may be an existing product and/or an existing product with additional features.

The system design does not violate any intellectual property rights and is developed using licensed tools. Similarly, the manufacturing of the object is performed using system approach to manufacturing with available resources for the manufacturing and assembly. The production parameters are indigenously developed, and the production does not violate any intellectual property right or any international regulation when it is done using legally imported or indigenously developed machinery.

The basic objective of the engineering design and manufacturing today is to produce innovative and multidisciplinary products for the global market that meet highest quality standards and allow customization. The basic philosophy of performing engineering design and manufacturing in capitalist society is to maximize the profits, generate highly skilled jobs, and develop technology that is unique in nature in order to command the market for a significant time.

The need for functional reverse engineering arises when an independent company, a new additional company in a conglomerate of companies, or a completely new group of people develops interest in producing an object or a range of objects that are in short supply or can be produced at a lower cost locally. Such a need also arises when a nation endeavors to reduce their imports and increase the exports.

In the prevailing economic horizon, the developed world (North America and Western Europe) is leading in product development through technology that produces novel, innovative, high quality, and customized products for local and international markets. The developed nations are followed by the BRICS nations attempting to develop novel products for local and international consumption thus trying to optimize their import–export portfolio. Then comes the Next Eleven Countries that are striving for a balance of payment leading to significant net profit that gives a satisfactory socio-economic outlay.

Manufacturing of finished goods is the key area that the developed nations adopt to produce products for local use and to export to other nations. BRICS and Next Eleven Countries lag behind in manufacturing of finished goods at a competitive price through Small to Medium Enterprises (SMEs). Developed nations rely on SMEs working with appropriate production philosophy individually or in amalgamation such as Lean Manufacturing, Agile Manufacturing, Just in Time, Theory of Constraint (ToC), Group Technology, MRP, MRPII, and Supply Chain; manage the SMEs through Computer Integrated Manufacturing (CIM); produce products at Six Sigma Quality standard; and continuously improve in various aspects through KAIZAN. The main feature of these SMEs is high level of automation in both tangible and intangible aspects of production (manufacturing and manufacturing services). These SMEs are capable of performing management control and client services through Cloud, organizational control through Industrial Ethernet, and shop floor control through SERCOSE III.

Industry 4.0 with its holistic approach is the futuristic view of SMEs in developed world. Industry 4.0 is also the key turning point for SMEs in BRICS and Next Eleven Countries to perform as per the characteristics of SMEs in the developed nations.

A discrete product is an object that is countable. The type and number of such products is enormous. It ranges from a pen to write to an aeroplane to travel in. To produce a discrete product or range of discrete products, hundreds of manufacturing processes are used independently or in combination. An ideal classification of manufacturing processes is provided by Kalpakjian and Schmidt [1]. A manufacturing system comprises a processing area where manufacturing processes are founded as per a predetermined scheme. To produce a discrete product or range of discrete products, the manufacturing system also has an assembly area. There are special arrangements of methods of assembly. The processing area and the assembly area have defined inputs of information, energy, and material. The output from the manufacturing system (enterprise) is a discrete product or range of discrete products. There are types of manufacturing systems for the production of discrete product or range of discrete products. Such manufacturing enterprises are ideally classified by Chryssoulouris [2].

The discrete products chosen for product development are machine tools. The machine tools are categorized as strategic machine tools and non-strategic machine tools. The strategic machine tools can produce any discrete component, structure, assembly, mechanism, or machine as well as non-strategic machine tools. The non-strategic machine tools produce the basic necessities of mankind, i.e., food, apparel, and shelter. Next Eleven Countries rely highly on import of strategic machine tools and non-strategic machine tools as well as products produced by these machine tools thus spending very significant amount of prime foreign exchange on these items. Next Eleven Countries often face embargo and poor after sales service even on legally imported machinery and other products often compromising on sovereignty of a nation. The scenario limits Next Eleven Countries to manufacture with competitive production facility. In order to enhance the productivity of the nation, the Next Eleven Countries do not have any other choice but to start developing capacity in the production of strategic and non-strategic machine tools so as to minimize dependency on developed world and the BRICS nations thus taking a step toward a better balance of payment of their import–export bill.

In order to introduce an intervention in the economic horizon of Next Eleven Countries, the introduction of SMEs in the area of development of strategic and non-strategic machine tools for various sectors such as discrete product manufacturing for home appliances, automotive, agricultural machinery, bio-medical engineering, etc. is going to play an important role over a relatively short period of time. These developments shall effectively reduce the burden of imports for Next Eleven Countries and shall improve the availability of skilled jobs in these nations that will also align with the higher education targets of these nations.

The strategic and non-strategic machine tools' design is dependent on many fields of engineering and technology. The manufacturing of these products is equally a difficult task. Today's machine tool development relies on use of number of materials (metals, alloys, etc.), machine elements (gears, bearings, etc.), computer hardware (microprocessors, microcontrollers), control elements (sensors, transducers, and actuators), computer software (operating systems, compilers, high- and low-level languages, etc.), communication protocols (serial and parallel data transmission protocols), and digital data storage mechanisms (on local RAM, on Cloud, etc.). Craftsmanship is also an equally important factor. In order to develop state-of-the-art machine tool, the relevant human resources have to align themselves with the global practices for machine tool development.

Computer Numerical Control (CNC) machine tools are designed and manufactured for various manufacturing processes. They vary in size, capacity, features, and special assemblies attached to such equipment.

The development of machine tool design for discrete manufacturing processes is a multidisciplinary exercise undertaken to cover various requirements conveyed as "needs" by the user.

In all these multidisciplinary aspects of machine tool design and manufacturing, a level of maturity is required which may not be the benchmark of a single organization. For example, an organization excels in the mechanical design and manufacturing of the equipment but at the same time is not capable of producing a controller for the machine tool. It is very common practice in today's machine tool development to use knowledge of different vendors to produce a state-of-the-art machine tool.

Such an exercise shall involve vendors with high-level design, manufacturing, and assembly maturity in following areas:

i. Structural and thermal analysis of the machine tool

ii. Vibration analysis of the machine tool

iii. Design and production of machine tool base assembly

iv. Design and production of machine tool clamping devices

v. Design and production of machine tool assemblies such as scrap removal mechanism, automatic tool changer, automatic pallet changer, etc.

vi. Design and production of machine elements such as gears, bearings, shafts, etc.

vii. Design and production of coolant circulation pumps, lubrication pumps, etc.

viii. Design and development of CNC controller for specific manufacturing processes including Human Computer Interface (HCI) development, provision for use of a standard language for part programming, and availability of part programming features

ix. Selection of computer hardware for CNC controller including microprocessors, microcontroller, graphic adopter, internal memory device, external backup devices, keyboard layout, provision of internet connectivity, etc.

x. Selection of appropriate sensors and transducers for closed-loop control of the machine tool

xi. Design and Development of control characteristics of the CNC controller such as development of transfer function and adjusting the transfer function to produce optimal values for applied voltage versus lead screw rotation

xii. Foundation design for the machine tool.

In such a scenario, it becomes a concurrent design and manufacturing exercise that results in a high-value product – a modern machine tool. Such a strategic or non-strategic machine tool is capable of attracting a large sector of the market thus leading to increased profitability for the machine tool designer and manufacturer.

It also means using indigenous resources where soft knowledge or hard technology is available, and for rest relying on bilateral agreements with modern world and BRICS nations with a deletion policy in mind.

The call for proposal for the First International Workshop on Functional Reverse Engineering of Machine Tools is broad to attract engineers and technologists from various disciplines (giki.edu.pk/wre2018). The submissions, as expected, are from various categories. It is expected that during the next two calls for proposal, each area presented in this book shall be strengthened.

This first book of three that will be written on this topic is divided into seven parts. The first part of this book covers state of the art related to microprocessors, microcontrollers, sensors, transducers, and actuators. This part comprises two chapters. The second part relates to sensor development. The third part comprises a chapter on laser cladding (remanufacturing) as it is a deposition technique similar to deposition on substrate as in sensor development. The fourth part addresses "measurement". Online tool wear and embedded quality control through machine vision are addressed in this part. Part 5 relates to design of mechanisms and machine. The chapters presented here entail knowledge related to surface manipulation, mechanism for texture development, and development of a strategic machine tool. Part 6 includes chapters on technology that contribute to decision making in operations management. The last part gives details of fifth-generation communication technology.

Editors
March 2019

References

1. S. Kalpakjian, and S. R. Schmid, *Manufacturing Processes for Engineering Materials*, New York, NY: Prentice Hall, 2008.
2. G. Chryssolouris, *Manufacturing Systems: Theory and Practice*, New York, NY: Springer-Verlag, 2010.

Acknowledgments

Editors are deeply indebted to Directorate of Science and Technology (DoST), Government of Khyber Pakhtunkhwa, Pakistan, for their all-out financial and administrative support. Partial financial support by Higher Education Commission of Pakistan is acknowledged. Technical support of Institution of Mechanical Engineers (IMechE), UK and Institute of Electrical and Electronics Engineers (IEEE), USA needs recognition.

Editors are indebted to the organizing committee of First International Workshop on Functional Reverse Engineering of Machine Tools (WRE2018) for their contribution. Editors owe special thanks to all our overseas guests who travelled from USA, China, Turkey, and Brunei and all other overseas guests who contributed from overseas through digital media. Efforts of contributors from within the country are also appreciated.

Voluntary work done by the IMechE GIK (Ghulam Ishaq Khan) Institute chapter needs a mention. Graduate students in Faculty of Mechanical Engineering, namely Mr. Faraz Ahmed and Mr. Sarmad Ishfaq, contributed to the compilation of this book.

Disclaimer

Every effort has been made to keep the contents of this book accurate in terms of description, examples, intellectual rights of others, and contents of websites at the time of browsing. The editors, authors, and publishers are not responsible for injury, loss of life, or financial loss arising from use of material in this book.

Editors

Prof. Dr. Wasim Ahmed Khan holds the first degree in Mechanical Engineering from NED University of Engineering and Technology, Karachi, Pakistan. He later obtained a PhD degree in the area of Operations Research from the Department of Mechanical Engineering, University of Sheffield, England, UK. He is a life member of Pakistan Engineering Council. He is also a chartered engineer (CEng) of the Engineering Council, UK, and a fellow of the Institution of Mechanical Engineers (FIMechE), UK. He was recently elected as a senior member of the Institution of Electrical and Electronics Engineers (IEEE), USA. He is also a member of IEEE Computer Society. He has diverse work experience including working with manufacturing industry; software development for local and overseas clients; and teaching production engineering, business, and computer science students. He has two books with international publishers to his credit. He has authored more than 30 research papers in journals and conferences of international repute. He is also Principal Investigator of two major research grants. Prof. Dr. Wasim A. Khan is currently working as a Professor in the Faculty of Mechanical Engineering, Ghulam Ishaq Khan (GIK) Institute of Engineering Sciences and Technology. He is also acting as the Director, Office of Research, Innovation and Commercialization (ORIC); Senior Advisor, The Catalyst – GIK Incubator; and Coordinator, GIK Institute Professional Education Program.

Dr. Ghulam Abbas completed his BS (1999–2002) in Computer Science at Edwardes College, Peshawar, and received a MS (2003–2005) in Distributed Systems from University of Liverpool (QS World Rank # 173), England, with distinction in research dissertation. The award of distinction rendered him recipient of a bursary from Liverpool Hope University (Times UK Rank # 49), England, to study for his Doctorate. He earned his PhD (2006–2010) from the Department of Electrical Engineering and Electronics, University of Liverpool. Dr. Abbas joined Ghulam Ishaq Khan (GIK) Institute of Engineering Sciences and Technology as Assistant Professor (2011–2016) in the Faculty of Computer Sciences and Engineering. He has also served the Institute as Director IT Department (2012–2017). Currently, he is serving as Associate Professor, Director (Huawei Authorized Information and Network Academy), and Advisor of IT Department.

Prior to joining GIK, Dr. Abbas was with Liverpool Hope University, where he served as Research Assistant (2006–2009) in the Intelligent and Distributed Systems (IDS) Laboratory, and later as Project Coordinator (2009–2010) at the IDS Laboratory on a knowledge transfer project funded by the Higher Education Funding Council for England. After the completion of his Doctorate, Dr. Abbas served as Seminar Tutor (2010–2011) at the Department of Mathematics and Computer Science, Liverpool Hope University. He also contributed as Mentor (2003–2011) for many Edwardian students at Liverpool Hope University during his stay in Liverpool. Before embarking on his PhD, Dr. Abbas served briefly as Lecturer (2005–2006) at Edwardes College, Peshawar, and formerly as Teaching Assistant (2002–2003) at the same college. In recognition of his professional standing, Dr. Abbas was elected as Fellow of the Institute of Science & Technology, England, in 2011, was elected to the grade of the Senior Member of IEEE in 2015, and was elected as Fellow of the British Computer Society in 2016.

Dr. Khalid Rahman is an Associate Professor in the Faculty of Mechanical Engineering at Ghulam Ishaq Khan (GIK) Institute of Engineering Sciences and Technology where he has been a faculty member since 2012. He received his BS in Mechanical Engineering from GIK Institute of Engineering Sciences and Technology, MS and PhD degree from Jeju National University, South Korea in 2012. He also has industrial experience of 7 years in design and manufacturing. His research interests include direct write technology for electronic devices and sensor fabrication and applications.

Dr. Ghulam Hussain has a PhD in Mechanical Engineering from the Nanjing University of Aeronautics & Astronautics, PR China and is currently working as an Associate Professor in the Ghulam Ishaq Khan (GIK) Institute of Engineering Sciences and Technology, Pakistan. He is working on advanced manufacturing processes and is the author of more than 70 publications. He stands among pioneers and leading researchers of die-less incremental forming processes. He is actively involved in doing research with renowned international universities. Based on his scientific contributions, he is listed in "Who is Who" and "Productive Scientists of Pakistan". He has also been selected as a Foreign Expert on Manufacturing in China and a Foreign Research Member of King Abdulaziz University. Moreover, he is a reviewer and editorial board member of several reputed international journals.

Dr. Cedric Aimal Edwin is a Business Adviser, Strategist, Management Consultant, and Start-up Mentor. Currently, he is the Associate Project Director of Khyber Pakhtunkhwa Impact Challenge (KPIC), an innovative project of LUMS-SOE (Lahore University of Management Sciences, School of Education) and Directorate of Youth Affairs (DYA), Government of Khyber Pakhtunkhwa (KP). Dr. Cedric has a PhD in Business Management from University of Liverpool, UK, and extensive experience of commercializing innovative products. Under the KPIC project, he trained more than 330 members of KP youth in entrepreneurial skills training, and more than 200 have received government funding to start their own business. Dr. Cedric has supported a total of 150 start-ups (with over 150 products and services) through his high-tempo product experimentation techniques. Dr. Cedric is the lead trainer and associate director of this project, responsible for managing and executing the project on the ground.

Previously, he was Head of Incubation Center, Coordinator Industrial Linkages and Assistant Professor of Business Management at Ghulam Ishaq Khan Institute of Engineering Sciences and Technology from 2014 to 2018. Dr. Cedric, under the leadership of Prof. Dr. Wasim A. Khan, secured a grant of PKR 127 Million and PKR 105 Million for Establishment of Technology Incubation Center and Promotion of Enterprises of Reverse Engineering, respectively, from Directorate of Science and Technology (DoST), Government of Khyber Pakhtunkhwa.

Over the years, Dr. Cedric has helped many academic institutions in establishing innovation center and departments. He specializes in launching new products and services. During his tenure as the Head of Incubation Center, he helped launch six innovative products and four services. While he was at the core of the management team behind Enterprises of Reverse Engineering, they produced ten state-of-the-art production machines. Dr. Cedric has profoundly lead big R&D teams to launch innovative products and services that fit the present market needs. His innovative new product development methodology is a mix of lean philosophy, design thinking, design sprints, blue ocean strategy, hacking growth, and outcome-based learning.

Contributors

Emeroylariffion Abas
Faculty of Integrated Technologies
Universiti Brunei Darussalam
Bandar Seri Begawan, Brunei

Muhammad Usman Abbasi
PhD student (Institute of Biomedical
    Engineering)
Department of Engineering Science
University of Oxford
Oxford, United Kingdom

Quentin Cheok
Faculty of Integrated Technologies
Universiti Brunei Darussalam
Bandar Seri Begawan, Brunei

Kyung Hyun Choi
Department of Mechatronics
    Engineering
Jeju National University
Jeju, South Korea

Volkan Esat
Mechanical Engineering Program
Middle East Technical University -
    Northern Cyprus Campus
Mersin, Turkey

Muhammad Faizan
Department of Electrical
    Engineering
Hamdard University
Karachi, Pakistan

Jahan Zeb Gul
Department of Mechatronics
    Engineering
Jeju National University
Jeju, South Korea

Azhar Ali Haidry
College of Materials Science and
    Technology
Nanjing University of Aeronautics
    and Astronautics
Nanjing, China

Abdul Haseeb
Department of Electrical
    Engineering
Hamdard University
Karachi, Pakistan

Edward Huang
Systems Engineering & Operations
    Research, Volgenau School of
    Engineering
George Mason University
Fairfax, Virginia

Jianqiang Huo
College of Mechanical and Electrical
    Engineering
Nanjing University of Aeronautics
    and Astronautics
Nanjing, China

Asif Iqbal
Faculty of Integrated
    Technologies
Universiti Brunei Darussalam
Bandar Seri Begawan, Brunei

Muhammad Iqbal
Faculty of Integrated
    Technologies
Universiti Brunei Darussalam
Bandar Seri Begawan, Brunei

Muhammad Talha Iqrar
Department of Electrical
    Engineering
Hamdard University
Karachi, Pakistan

Syed Osama bin Islam
Department of Mechanical
    Engineering
Capital University of Science and
    Technology (CUST)
Islamabad, Pakistan

Ahmad Junaid
Department of Mechanical
    Engineering
CECOS University
Peshawar, Pakistan

Azfar Khalid
Department of Mechanical
    Engineering
Capital University of Science and
    Technology (CUST)
Islamabad, Pakistan

Javed I. Khan
Networking and Media
    Communications Laboratory,
    Department of Computer
    Science
Kent State University
Kent, Ohio

Liaquat Ali Khan
Department of Mechanical
    Engineering
Capital University of Science and
    Technology (CUST)
Islamabad, Pakistan

Muhammad Faisal Khan
Department of Electrical
    Engineering
Hamdard University
Karachi, Pakistan

Waqas Akbar Lughmani
Department of Mechanical
    Engineering
Capital University of Science and
    Technology (CUST)
Islamabad, Pakistan

Riaz Mohammad
Department of Mechanical
    Engineering
CECOS University
Peshawar, Pakistan

Ali Nasir
Department of Electrical
    Engineering
University of Central Punjab
Lahore, Pakistan

Malik M. Nauman
Faculty of Integrated Technologies,
Universiti Brunei Darussalam
Bandar Seri Begawan, Brunei

Memoon Sajid
Department of Mechatronics
    Engineering
Jeju National University
Jeju, South Korea
and
Faculty of Electrical Engineering
GIK Institute of Engineering
    Sciences and Technology
Topi, Pakistan

Muftooh Ur Rehman Siddiqi
Department of Mechanical
    Engineering
CECOS University
Peshawar, Pakistan

Yun Song
College of Mechanical and Electrical
    Engineering
Nanjing University of Aeronautics
    and Astronautics
Nanjing, China

Linchao Sun
College of Materials Science and
    Technology
Nanjing University of Aeronautics
    and Astronautics
Nanjing, China
and
Key Laboratory of Materials
    Preparation and Protection for
    Harsh Environment
Ministry of Industry and
    Information Technology
Nanjing, China

Xiaoping Wang
Research Center of CAD/CAM
    Engineering
Nanjing University of Aeronautics
    and Astronautics
Nanjing, China

Abbas K. Zaidi
Systems Engineering & Operations
    Research, Volgenau School of
    Engineering
George Mason University
Fairfax, VA, USA

Chen Zhang
College of Mechanical and Electrical
    Engineering
Nanjing University of Aeronautics
    and Astronautics
Nanjing, China

Jilin Zhang
College of Mechanical and Electrical
    Engineering
Nanjing University of Aeronautics
    and Astronautics
Nanjing, China

Part 1

The Cybernetics

This part deals with the history and working of the controller of machine tools. A detailed survey of these controllers is provided in Khan et al. (2011). Next Eleven Countries do not produce microprocessors and microcontrollers and are dependent on Modern World for off the shelf purchase of microprocessors, microcontrollers, associated electronic integrated circuits, sensors, transducers, and actuators. Engineers in Next Eleven Countries are capable of producing circuit design using proprietary circuit design software such as Proteus, Multisim, or any open source circuit design software. The printed circuit boards (PCBs) can be developed. Embedded systems with Wi-Fi control for Industrial Internet of Things and e-Commerce capability is state of the art in Next Eleven Countries.

Reference

W. A. Khan, A. Raouf, K. Cheng, 2011. *Virtual Manufacturing*, Springer Series in Advanced Manufacturing. London, UK: Springer Verlag.

1

Microprocessors and Microcontrollers: Past, Present, and Future

Abdul Haseeb, Muhammad Faizan, Muhammad Faisal Khan, and Muhammad Talha Iqrar

Hamdard University

CONTENTS

1.1 Introduction 4
    1.1.1 Terminologies Related to Microprocessor and
        Microcontroller 4
    1.1.2 Difference between Microprocessor and Microcontroller 5
1.2 Popular Brands of Microprocessors 7
1.3 Classification of Microcontrollers 20
    1.3.1 According to Bits 20
        1.3.1.1 8-Bit Microcontroller 20
        1.3.1.2 16-Bit Microcontroller 20
        1.3.1.3 32-Bit Microcontroller 20
    1.3.2 According to Memory 21
        1.3.2.1 Embedded Memory Microcontrollers 21
        1.3.2.2 External Memory Microcontrollers 21
    1.3.3 According to Instruction Set 21
        1.3.3.1 Reduced Instruction Set Computer (RISC)
                Microcontroller 21
        1.3.3.2 Complex Instruction Set Computer (CISC)
                Microcontroller 21
    1.3.4 According to Memory Architecture 21
        1.3.4.1 Harvard Architecture Microcontroller 21
        1.3.4.2 Princeton Architecture Microcontroller 22
1.4 History of Microprocessors and Microcontrollers 22
1.5 Comparison of Various Microcontrollers of Modern Age 25
1.6 Future Trends 26
1.7 Conclusions 27
References 27

1.1 Introduction

Microprocessor is a single chip with built-in components such as an Arithmetic Logic Unit (ALU), a control unit, and a number of registers. The ALU performs arithmetic and logical operations, while the registers hold the data being processed by the microprocessor. The control unit directs and coordinates all the processing. Advance microprocessors have additional components, e.g. memory cache that holds additional data related to the instruction being processed by the microprocessor. Microprocessor is a general-purpose device, contains many functional blocks, and is designed to support multitasking. Figure 1.1 presents the block diagram of a typical microprocessor.

When processor unit, memories, peripherals, and other devices are fabricated in a single chip, it is called a microcontroller. Initially, when microprocessors and microcontrollers were launched, both these devices did not have a lot of features, but with the passage of time and advancement of technology, a number of characteristics have been added to them. A new generation microcontroller contains advanced peripherals and communication blocks like Digital to Analog Converter (DAC), Analog to Digital Converter (ADC), I2C, SPI, Wi-Fi, Bluetooth, etc.

This chapter presents basic theory related to microprocessors and microcontrollers along with a brief history and features of the latest microprocessors and microcontrollers. It is to be understood that the microprocessors were invented first, and later on due to the need of a special chip for input–output intensive actions, microcontrollers were devised. This chapter also provides a comparison of some recent microcontroller boards. Future trends are also part of this chapter. Figure 1.2 gives a schematic of a typical microcontroller.

1.1.1 Terminologies Related to Microprocessor and Microcontroller

There are some important terms related to microprocessor and microcontroller which are commonly used. These terms are described in this section:

System-in-a-Package (SiP) is a number of integrated circuits (ICs) embedded in a module. SiP is used in digital music players, mobile phones, etc.

System on Chip (SoC) is an IC that integrates all components of a computer (present on motherboard) in a single chip. This is just like a microcontroller which is a single-chip central processing unit (CPU), but SoC contains more powerful devices like advanced RISC machines (ARM) microcontroller, graphics processing unit (GPU), etc. The advantage of SoC is the reduction in per-unit cost and the size of chip, in addition to higher speed that can be achieved because no external buses are attached to the system.

Computer-on-Module (CoM) is similar to a microcontroller board like Arduino or Raspberry Pi. It is a complete single-board computer (SBC) based on microcontroller but has less performance as compared to a desktop computer. Its benefits are time-to-market product with less cost and reduction of risk.

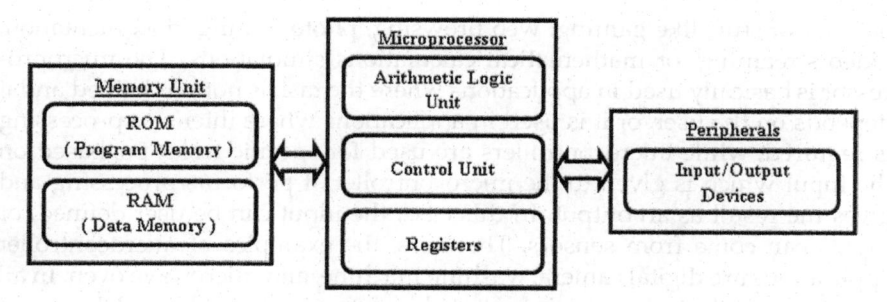

FIGURE 1.1
Central processing unit (CPU) with block diagram of microprocessor.

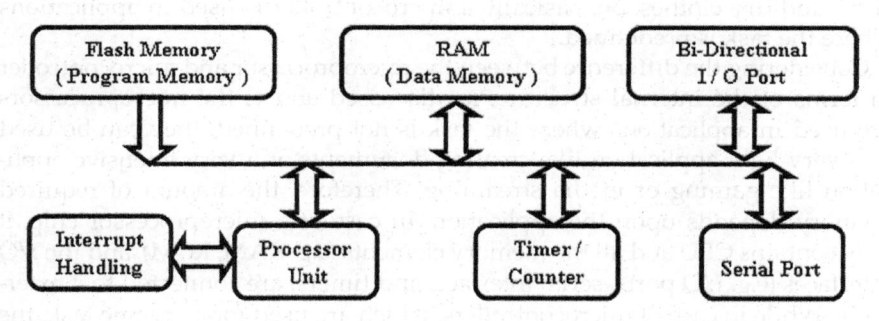

FIGURE 1.2
Block diagram of microcontroller.

SBC is a small portable motherboard having all the functionalities of motherboard like microprocessor, input/output devices, memory unit all embedded on a single printed circuit board (PCB). The best example of SBC is Raspberry Pi.

System on Module (SoM) is a type of SBC with a board-level circuit that integrates several system functions in a single module. The main advantage of SoM is its reuse design because it can be integrated by many embedded system components like microprocessor/microcontroller, digital signal processor (DSP), memories, universal serial bus (USB), Ethernet, serial peripheral interface (SPI), ADC, DAC, etc.

1.1.2 Difference between Microprocessor and Microcontroller

Although microprocessors and microcontrollers look very identical to each other, there are many differences between them in many aspects, such as in terms of application in which they are used, in terms of costs, processing power which they possess, and power consumption [1,2]. First, consider the difference between them in terms of applications. A classic example of microprocessor application is a personal computer or laptop through which one can do a lot of stuff like gaming, web browsing, photo editing, documentation, video streaming, or mathematical calculations/simulations. The microprocessor is basically used in applications where the task is not predefined and it depends on the user, or it is used in applications where intensive processing is required, while microcontrollers are used for specific tasks. So, based on the input which is given to the microcontroller, it performs processing and gives the result as an output. In this case, the input can be user defined, or inputs can come from sensors. Therefore, the examples of microcontroller applications are digital camera, washing machine, and microwave oven. In all these devices, the task which is going to be performed is predefined. In case of microwave oven, once user sets the power and timings, it gives cooked food. In case of washing machine, once user sets the parameters of machine, it gives clean and dry clothes. So, basically a microcontroller is used in applications where the task is predefined.

Considering the difference between the microprocessor and microcontroller in terms of the internal structure, as discussed above, the microprocessors are used in applications where the task is not predefined; they can be used in a very light application like creating documents or a very intensive application like gaming or media streaming. Therefore, the amount of required memory depends upon the application. In case of a microprocessor chip, it only contains CPU and all the memory elements (e.g. RAM, ROM), and the I/O interfaces (e.g. I/O ports, serial interface, and timers) are connected to it externally. While in case of microcontrollers, which are used for a specific task, the amount of memory and the required I/O ports which are required are limited. So, in case of microcontroller, as all the memory elements and I/O ports are integrated (along with the CPU) inside a single chip, the size of an overall system is much smaller, while in case of microprocessor, as all memory elements and I/O ports are connected externally, the overall size of the system is larger than a microcontroller.

Considering the difference between microprocessors and microcontrollers in terms of processing power and memory, microprocessor is operated at much higher speed, so its clock speed is in gigahertz (GHz) range which varies from 1 to 4 GHz for the high-end processors. As a microprocessor has to run an operating system, the amount of memory is quite high, i.e. RAM which ranges from 512 MB to 32 GB and ROM ranges from 128 GB to 2 TB. The common peripheral interfaces which are seen in microprocessors are like USB, high-speed Ethernet, and universal asynchronous receiver/transmitter (UART). While in case of microcontroller, the clock speed is in megahertz (MHz) range which varies from 1 to 300 MHz in high-end microcontrollers. As these microcontrollers are defined for a specific task, the amount of memory that is required by them is quite less. Its range of RAM is from 2 to 256 KB and its range of flash memory or program memory is 32 KB to 2 MB. The common peripheral interfaces which one can find inside the microcontroller are like inter-integrated circuit (I2C), SPI, and UART. Basically, all these are the serial interfaces which user finds in modern-day

TABLE 1.1

Comparison of Microprocessor and Microcontroller

Parameters	Microprocessor	Microcontroller
Designed for	Multitasking	Single tasking
Applications	General purpose, servers, data centers, gamming, application designing, etc.	Embedded system design like microwave oven, washing machine, camera, smart watches, etc.
Internal structure	Contains functional block for data processing and execution like ALU, etc.	Processor, memory and I/O interfacing are fabricated in a Single chip
Cost	High	Low
Power consumption	High	Low
Cooling system required	Heat sink and cooling fan	Air cooled for normal operation
Memory	Contains registers and cache memory	Contains RAM, ROM, and flash memory
Vendors	Intel and AMD	ARM, ATMEL, Microchip, ST Microelectronics, Toshiba, etc.

microcontrollers. In comparison to microcontrollers, the modern-day microprocessors are either 32-bit or 64-bit microprocessors. A 32-bit microprocessor means a microprocessor that can handle 32 bits of binary data at the same time. Similarly, a 64-bit microprocessor can handle 64 bits of data at the same time. In case of 64-bit microprocessor, the address bus and data buses are of 64 bits. Similarly, in the case of 32-bit microprocessor, the address and data buses are of 32 bits. In comparison to microprocessors, modern-day microcontrollers are either 8-bit, 16-bit, or 32-bit microcontrollers; the amount of data which can be handled by a microprocessor in a single cycle is higher than that handled by the microcontrollers.

Considering the difference between microprocessors and microcontrollers in terms of power consumption and cost, in microprocessors, as memory elements and I/O ports are connected externally, the overall cost of the system, as well as power consumption, is higher than that of the microcontrollers.

A summary of differences between microprocessors and microcontrollers based on various parameters is given in Table 1.1.

1.2 Popular Brands of Microprocessors

Various popular brands of microprocessors are used to power calculators, process controllers, and Computer Numerical Controller inference engines.

They include ranges of microprocessors from Integrated Electronics (Intel), Motorola, Advanced Micro Devices (AMD), and others. A brief history of some of the popular microprocessors is provided below.

Integrated Electronics (Intel):

S. No.	Brand	Configuration	Year of Introduction	Features
1	Intel 4004	4-bit microprocessor	1971	• 740 KHz clock • Metal oxide silicon technology • 1 K data memory • 4 K program memory • 16 registers of 4 bits each • 46 instructions • Instruction cycle time 10.8 µs • 16 pin DIP (dual inline package)
2	Intel 4040	4-bit microprocessor	1974	• 0.5–0.8 MHz clock • 8-bit data bus • 14-bit address bus • Can address 16 KB of memory (RAM & ROM) • 8 input ports • 24 output ports • Peripheral chips available • Interrupt handling • 18 pin DIP
3	Intel 8008	8-bit microprocessor	1972	• 500 KHz clock • 16 4-bit input ports • 16 4-bit output ports • 60 instructions • 8 KB program memory • 640-byte data memory • 24 4-bit registers • Interrupt handling • Executes 60,000 instructions per second • 24 pin DIP
4	Intel 8080	8-bit microprocessor	1974	• 2 MHz clock • 8-bit data bus • 16-bit address bus • Memory size 64 KB (RAM & ROM) • Supports 256 I/O ports • 112 instructions • Peripheral chips available • 40 pin DIP
5	Intel 8085	8-bit microprocessor	1977	• 3.5 & 6 MHz clock • 8-bit data bus • 16-bit address bus

(Continued)

S. No.	Brand	Configuration	Year of Introduction	Features
				• Memory size 64 KB (RAM & ROM) • Serial I/O ports • 246 instructions • Peripheral chips available • 40 pin DIP
6	Intel 8086	16-bit microprocessor	1978	• 5–10 MHz clock • 16-bit data bus • 20-bit address bus • 1 MB maximum addressable memory • 64 KB of I/O space • 2.5 million instructions per second • Multiplexed address and data bus • Complex instruction set computer (CISC) design methodology • High-performance n-channel Metal Oxide Semiconductor (HMOS/CHMOS) technologies • Instruction set: x86-16 • Recognizes floating point instructions • 40-pin DIP plastic package
7	Intel 8088	16-bit microprocessor	1979	• 5–10 MHz clock • 8-bit data bus • 20-bit address bus • 1 MB maximum addressable memory • 2.5 million instructions per second • HMOS/CHMOS technologies • Peripheral chips available • 40-pin DIP & 44-pin plastic leaded chip carrier (PLCC) package
8	Intel 80186 & Intel 80188	16-bit microprocessor	1982	• Clock rates 6–25 MHz • 16-bit data bus • 20-bit address bus • Clock rates 6–25 MHz • 16-bit data bus • 24-bit address bus • 4 million instructions per second • Instruction set: x86-16 • Allows multitasking

(Continued)

S. No.	Brand	Configuration	Year of Introduction	Features
				• 16 MB of addressable memory • Peripheral chips available • 68-pin pin grid array (PGA), ceramic leadless chip carrier (CLCC), and PLCC
9	Intel 80286	16-bit microprocessor	1982	• Different versions such as 80C186, 80186 EA, & 80186 EB with additional components such as interrupt controller, clock generator, local bus controller and counters, etc. • Peripheral chips available • 68-pin PLCC, 100-pin plastic quad flat pack (PQFP), & 68-pin PGA
10	Intel 80386	32-bit microprocessor	1985	• Clock rate 12–40 MHz • 32-bit data and address buses • 4 million instructions per second • 4 GB addressable memory • Instruction set: x86-32 • 64-terabyte virtual address space • Various versions such as 80386 Dx, 80386 Sx, and 803386 Se • Read protected and virtual modes supported • Peripheral chips available • 132-pin PGA
11	Intel 80486	32-bit microprocessor	1989	• Clock speed 16–100 MHz • 32-bit data and address buses • 4 GB maximum addressable memory • 8 KB cache memory • 50 MHz • 40 million instructions per second • Peripheral chips available • 168-pin PGA
12	Intel Pentium	32-bit microprocessor	1993	• Clock speed 66 MHz • 64-bit data bus • 32-bit address bus • 4 GB maximum addressable memory • Instruction set: x86 plus Pentium extension • 8 KB cache each for data and instructions

(Continued)

S. No.	Brand	Configuration	Year of Introduction	Features
				• Peripheral chips available • 110 million instructions per second • Dual processor configuration • 273-pin ceramic PGA
13	Intel Pentium Pro	32-bit microprocessor	1995	• Clock speed 150–200 MHz • 64-bit data bus • 32-bit address bus • 64 GB addressable memory • 256 KB of level 1 & level 2 cache • 8 KB cache memory each for data and instruction • Instruction set: x86 • Quad processor configuration • Mainly used at servers and work stations • Peripheral chips available • 387-pin staggered pin grid array (SPGA)
14	Intel Pentium II	32-bit microprocessor	1997	• Clock speed 233–450 MHz • 64-bit data bus • 32-bit address bus • 64 GB addressable memory • Instructions set: x86 • 333 million instructions per second • Supports MMx technology • Peripheral chips available • 242-pin single edge contact (SEC)
15	Intel Pentium II Xeon	32-bit microprocessor	1998	• Clock speed 400–450 MHz • Similar to Pentium II with the exception of 1.2 cache and packaging • 1.1 cache of 32 KB • 1.2 cache was of 512 KB, 1 MB, or 2 MB • Instruction set: x86-64 • Peripheral chips available • Used at workstations, servers, and embedded systems
16	Intel Pentium III	32-bit microprocessor	1999	• Clock speed 400 MHZ–1.4 GHz • Instruction set MMx, SSE • Katmai, Coppermine, Coppermine T, and Tualatin cores • Peripheral chips available • Family of processors

(Continued)

S. No.	Brand	Configuration	Year of Introduction	Features
17	Intel Pentium III Xeon	32-bit microprocessor	1999	• Clock speed up to 1 GHz • Same architecture as Pentium III desktop processor • Ability to run in four-way systems • Tanner core and cascade core • Peripheral chips available • Family of processors
18	Intel Pentium IV	32-bit microprocessor	2000	• Clock speed 1.3–3.8 GHz • NetBurst architecture • Instruction set: x86, MMx, SSE, SSE2, SSE3 • 32 KB L1 cache • 256 KB L2 cache • Peripheral chips available • Family of processors
19	Intel Celeron	32-bit microprocessor	1998	• Clock speed: 266 MHz–3.6 GHz • Instruction set: x86-64 • P6, NetBurst, core, Sandy Bridge technology • Slower bus speed • Slower cache speed • Celeron D, Celeron M, Celeron dual core, Mobile Celeron dual core versions • Peripheral chips available • Family of processors
20	Intel Dual Core	32- or 64-bit microprocessor	2006	• Clock speed: 1.3–2.6 GHz • Instruction set: MMx, SSE, SSE2, SSE3, SSSE3, x86-64 • Two cores • Positioned above Celeron in terms of price and performance • Peripheral chips available • Family of processors
21	Intel core 2 Duo	64-bit microprocessor	2006	• Clock speed: 1.6–3.5 GHz • Micro architecture: core • Single, dual, and quad core microprocessors • Instruction set: x86-64 • Up to 6 MB L2 cache • Supplemental streaming SIMD extensions 3 (SSSE3), Trusted Execution Technologies • Peripheral chips available • Family of processors

(Continued)

S. No.	Brand	Configuration	Year of Introduction	Features
22	Intel Core 2 Quad	64-bit microprocessor	2006	• Clock speed: 24 GHz • Core technologies • 8 MB L2 cache • Instruction set: x86-64 • Peripheral chips available • Family of processors
23	Intel core i3-2125 2nd generation	64-bit microprocessor	2011	• Clock speed: 3.30 GHz • No of cores: 2 • No. of threads: 4 • Intel smart cache: 3 MB • Supports virtualization & Hyper threading • Thermal design power (TDP): 65 W
24	Intel Core i3-3250 3rd generation	64-bit microprocessor	2013	• Clock speed: 3.50 GHz • No of cores: 2 • No. of threads: 4 • Intel smart cache: 3 MB • Supports virtualization & Hyper threading • Anti-theft technology • TDP: 55 W
25	Intel Core i3-4370 4th generation	64-bit microprocessor	2014	• Clock speed: 3.8 GHz • No of cores: 2 • No. of threads: 4 • Intel smart cache: 4 MB • Supports virtualization & Hyper threading • Advanced encryption standard (AES) new instructions • Secure key • TDP: 54 W
26	Intel Core i3-5157U 5th generation	64-bit microprocessor	2015	• Clock speed: 2.5 GHz • No of cores: 2 • No. of threads: 4 • Cache: 3 MB • Supports virtualization & Hyper threading • OS guard • TDP: 28 W
27	Intel Core i5-2550K 2nd generation	64-bit microprocessor	2012	• Clock speed: 3.4–3.8 GHz • No of cores: 4 • No. of threads: 4 • Intel smart cache: 6 MB • Supports virtualization technology • AES new instructions • TDP: 95 W

(Continued)

S. No.	Brand	Configuration	Year of Introduction	Features
28	Intel Core i5-3570K 3rd generation	64-bit microprocessor	2012	• Clock speed: 3.4–3.8 GHz • No of cores: 4 • No. of threads: 4 • Intel smart cache: 6 MB • Supports virtualization technology • Anti-theft technology • TDP: 77 W
29	Intel Core i5-4690K 4th generation	64-bit microprocessor	2014	• Clock speed: 3.5–3.9 GHz • No of cores: 4 • No. of threads: 4 • Intel smart cache: 6 MB • Supports virtualization technology • Intel My WiFi technology • TDP: 88 W
30	Intel Core i5-5287U 5th generation	64-bit microprocessor	2015	• Clock speed: 2.9 GHz • No of cores: 2 • No. of threads: 4 • Cache: 3 MB • Supports virtualization & Hyper threading • Intel Smart Response Technology • Intel Small Business Advantage • TDP: 28 W
31	Intel Core i7	64-bit microprocessor	2008	• Clock speed 2.66–3.33 GHz • Four physical cores • 64 KB of L1 cache per core • 256 KB of L2 cache • 8 MB of L3 cache • Instruction set: SSE4.2, SSSE3 • Supports virtualization, Hyper threading & Turbo Boost Technologies • Uses point to point protocol for I/O devices • 1366 LGA package • Peripheral chips available • Family of processors
32	Intel core i7-2700K 2nd generation	64-bit microprocessor	2011	• Clock speed: 3.5–3.9 GHz • No of cores: 4 • No. of threads: 8 • Intel smart cache: 8 MB • Supports virtualization, Hyper threading, & Execute Disable Bit • TDP: 95 W

(Continued)

S. No.	Brand	Configuration	Year of Introduction	Features
33	Intel Core i7-3770K 3rd generation	64-bit microprocessor	2012	• Clock speed: 3.5–3.9 GHz • No of cores: 4 • No. of threads: 8 • Intel smart cache: 8 MB • Supports virtualization & Hyper threading • Intel Turbo Boost Technology: 2.0 • TDP: 77 W
34	Intel Core i7-4790K 4th generation	64-bit microprocessor	2014	• Clock speed: 4–4.4 GHz • No of cores: 4 • No. of threads: 8 • Intel smart cache: 8 MB • Supports virtualization, Hyper threading, & OS guard • TDP: 88 W
35	Intel Core i7-5960X Extreme Edition	64-bit microprocessor	2014	• Clock speed: 3–3.5 GHz • No of cores: 8 • No. of threads: 16 • Intel smart cache: 20 MB • Supports virtualization & Hyper threading • TDP: 140 W
36	Intel Core i7-5557U 5th generation	64-bit microprocessor	2015	• Clock speed: 3.1–3.4 GHz • No of cores: 2 • No. of threads: 4 • Cache: 4 MB • Supports virtualization & Hyper threading • Intel Smart Response Technology • TDP: 23 W
37	Intel Core i9-9980XE	64-bit microprocessor	Q4/2018	• 18 core, 36 threads • 3 GHz • 4×Double data rate (DDR)4-2666 • L2 Cache 18×1024 KiB • L3 cache 24.75 MiB
38	Intel Core i9-9980XE	64-bit microprocessor	2019	• 14 core, 28 threads • 4 GHz • 4×DDR4-2666 • L2 cache 14×1024 KiB • L3 cache 19.25 MiB

Complete instruction set along with programming guide for all these Intel processors are available at the Internet.

Motorola:

S. No.	Brand	Configuration	Year of Introduction	Features
1	Motorola 6800	8-bit microprocessor	1974	• Clock speed: up to 1 MHz • 8-bit bi-directional data bus • 16-bit address bus • 64 KB addressable memory • I/O devices addressed as memory • 72 instructions, seven addressing modes with a total of 197 opcodes • Peripheral chips available • Family of processors
2	Motorola 6809	8-bit microprocessor	1978	• Clock speed: up to 2 MHz • Some 16-bit microprocessor features • Source compatible with 6800 • 64 KB addressing range to full 1 MB in 4 KB pages • Two 8-bit accumulators combined to a single 16-bit register • Peripheral chips available • Family of processors
3	Motorola 68000	32-bit microprocessors	1979	• Clock speed: up to 20 MHz • 32-bit data and address registers • 16 MB addressable memory • 8-bit external data bus • Peripheral chips available • Family of processors

Complete instruction set along with programming guide for all Motorola microprocessors is available at the Internet.

AMD:

S. No.	Brand	Configuration	Year of Introduction	Features
1	AM 286	16-bit microprocessor	1989	• Clock speed: 8–20 MHz • Instruction set: x86(IA-16) • Single core • L1 cache • Intel 80286 compatible • Desktop & embedded system application • 68-pin CLCC, PLCC, & PGA packages

(Continued)

S. No.	Brand	Configuration	Year of Introduction	Features
				• Peripheral chips available • Family of processors
2	AM 386	32-bit microprocessor	1991	• Clock speed: 20–40 MHz • 32-bit address bus • AM386 Dx-40 has 32-bit width on its data bus • Intel 80386 clone with performance touching Intel 80486 processor • Peripheral chips available • Family of chips • Various types of packaging
3	AM 486	32-bit microprocessor	1993	• Clock speed 25–120 MHz • 32-bit data bus • 32-bit address bus • Intel 80486 clone with performance touching Intel Pentium processor • Peripheral chips available • Family of chips • Various types of packaging
4	AM 586	32-bit microprocessor	1995	• Clock speed 133–150 MHz • Upgrade for AM486 system • Family of processors • Peripheral chips available • Various types of packaging
5	AMD K5	32-bit microprocessor	1996	• Clock speed 75–133 MHz • Upgrade for AM5x86 • Front-side bus (FSB) Speed 50–66 MHz
6	AMD K6	32-bit microprocessor	1997	• Clock speed 166–300 MHz • Upgrade for K5 • FSB speed 66 MHz
7	AMD K6-2	32-bit microprocessor	1998	• Clock speed 266–550 MHz • Upgrade for K6 • FSB speed 66–100 MHz
8	AMD K6-III	32-bit microprocessor	1999	• Clock speed 350–550 MHz • Upgrade for K6-2 • FSB speed 66–100 MHz
9	AMD Athlon K7	32-bit microprocessor	1999–2005	• Clock speed 500 MHz–2.3 GHz • Upgrade for K6-III • FSB speed 200–400 MT/s
10	AMD Athlon K8	64-bit microprocessor	2003–present	• Core clock 1,600–3,200 MHz • Upgrade for K7 • FSB speed 800–1,000 MHz

(Continued)

S. No.	Brand	Configuration	Year of Introduction	Features
11	AMD Athlon K10h	64-bit microprocessor	2007–present	• Core clock 1,700–3,700 MHz • Upgrade for K8 • FSB speed 1,000–2,000 MHz
12	AMD Bulldozer	64-bit microprocessor	2011–present	• Core clock 2,800–4,200 MHz • Upgrade for K10 • FSB speed 1,000–2,000 MHz
13	AMD Piledriver	64-bit microprocessor	2012–present	• Core clock 3,000–4,700 MHz • Upgrade for Bulldozer • FSB speed 1,000–2,000 MHz

Complete instruction set along with programming guide for all AMD microprocessors is available at the Internet.

Zilog:

S. No.	Brand	Configuration	Year of Introduction	Features
1	Z80	8-bit microprocessor	1976	• Clock speed: 2.5–4 MHz • 16-bit address bus • 64 KB of addressable memory • Number of instructions: 252 with 78 instructions as subset of Intel 8080 • Ease of design with dynamic random access memory (DRAM) • 44-pin surface-mount device (SMD) & quad flat package (QFP) packages
2	Z180	8-bit microprocessor	–	• Clock speed: 6–33 MHz • Code compatible with Z80 • High-performance peripheral integration • Peripheral chips available • Family of processors • Various types of packaging
3	Zilog Z8000	16-bit microprocessor	1979	• Clock speed: 10 MHz • 6 MB addressable space • 16 16-bit registers • 110 distinct instruction types • System and normal mode • 64 K I/O ports • Multi-processing capability
4	Zilog Z80000	32-bit microprocessor	1986	• Clock speed similar to Intel 386 • 256-byte cache memory • 4 GB addressable memory • 16 general purpose registers • Virtual memory • Multitasking

Complete instruction set along with programming guide for all Zilog microprocessors is available at the Internet.

Mostech:

S. No.	Brand	Configuration	Year of Introduction	Features
1	Mostech 6502	8-bit microprocessor	1975	• Clock speed: 1–2 MHz • 8-bit data bus • 16-bit address bus • 64 KB addressable memory

Complete instruction set along with programming guide for the Mostech microprocessor is available at the Internet.

IBM Power PC:

S. No.	Brand	Configuration	Year of Introduction	Features
1	Power 1 & Power 2	32-bit microprocessor	1990s	• Clock speed up to 62.5 MHz • 32-bit physical address • 52-bit virtual address • 8 KB instruction cache • 32 KB data cache
2	Power 3 & Power 4	64-bit microprocessor	1998	• Clock speed up to 1.9 GHz • 64-bit power PC instruction set • 32 KB instruction cache
3	Power 5	64-bit microprocessor	2004	• Clock speed up to 1.9 GHz • Dual core • 64 GB of DDR and DDR2 memory • 64-bit power PC architecture
4	Power 6	64-bit microprocessor	2007	• Clock speed up to 4.5 GHz • Dual core • Multiway threading • 64 KB instruction & data cache

Complete instruction set along with programming guide for all IBM Power PC microprocessors is available at the Internet.

Sun Scalable Processor Architecture (SPARC):

S. No.	Brand	Configuration	Year of Introduction	Features
1	Ultra SPARC T2	64-bit microprocessor	2007	• Clock speed up to 1.4 GHz • Multicore, multithreading CPU • 8 KB data cache • 16 KB instruction cache

Complete instruction set along with programming guide for Sun SPARC microprocessor is available at the Internet.

1.3 Classification of Microcontrollers

Microcontrollers can be classified according to the following schemes.

1.3.1 According to Bits

1.3.1.1 8-Bit Microcontroller

For 8-bit microcontroller, Arithmetic Logic Unit (ALU) performs the arithmetic and logic operations on a byte (8-bit instruction). Examples of 8-bit microcontroller (MCUs) are Intel 8031/8051, PIC1x, and Motorola MC68HC11 families.

1.3.1.2 16-Bit Microcontroller

In 16-bit microcontroller, ALU performs arithmetic and logical operations on a word (16 bits). The internal bus width of 16-bit microcontroller is of 16 bits. Examples of 16-bit microcontrollers are Intel 8096 family and Motorola MC68HC12 and MC68332 families. 16-bit microcontroller provides greater performance and computing capabilities with greater precision as compared to 8-bit microcontroller.

1.3.1.3 32-Bit Microcontroller

In 32-bit microcontroller, the ALU performs arithmetic and logical operations on a double word (32-bits). The internal bus width of 32-bit microcontroller is of 32 bits. Examples of 32-bit microcontrollers are Intel 80960 family, Motorola M683xx, and Intel/Atmel 251 family. The performance and computing capability of 32-bit microcontrollers are enhanced with greater precision as compared to 16-bit microcontrollers.

The bits of microprocessor/microcontroller define the internal data bus of microprocessor which means the size of data transfer and manipulation in one clock cycle. They also define the internal registers and memory storage. An 8-bit microcontroller is a low-cost controller, designed for projects having single operation, while a 16-bit controller is just like 8-bit controller but has more precision of arithmetic calculation. A 32-bit controller is a high-performance and relatively high-cost controller having high precision in arithmetic calculation and can interface large memories. 32-bit and 64-bit microcontrollers are generally used in multitasking applications and are capable of running operating system. With the increase in the bits of microcontroller, the performance enhances with increase in cost. Generally, a low-performance microcontroller is used for devices with less critical task and low-cost design. 32-bit and 64-bit microcontrollers are high-performance microcontrollers with relatively high cost and give performance like PCs such as multitasking, multimedia interfacing, and Internet connectivity.

1.3.2 According to Memory

1.3.2.1 Embedded Memory Microcontrollers

Embedded memory microcontrollers have all the functional blocks such as program and data memory, I/O ports, serial communication, counters, timers, interrupts, and control logic built on a chip. For example, Intel 8051 is an embedded microcontroller.

1.3.2.2 External Memory Microcontrollers

In an external memory microcontroller, the microcontroller does not have all hardware and software units present as single unit. It has all or part of the memory unit externally interfaced through a glued circuit. For example, in Intel 8031, program memory is externally interfaced. Intel 8051 has both internal and external memory units.

1.3.3 According to Instruction Set

1.3.3.1 Reduced Instruction Set Computer (RISC) Microcontroller

In RISC microcontroller architecture, the microcontroller has an instruction set that supports fewer addressing modes for the arithmetic and logical instructions and just a few (load, store, push, pop) instructions for data transfer. RISC implements each instruction in a single cycle using a distinct hardwired control. There is reduced instruction set and instructions are of fixed number of bytes and take a fixed amount of time for execution. An example of RISC architecture is the ARM processor family-based microcontrollers.

1.3.3.2 Complex Instruction Set Computer (CISC) Microcontroller

CISC microcontroller architecture provides a microcontroller that has an instruction set which supports many addressing modes for the arithmetic and logical instructions. The advantages of the CISC architecture are that many of the instructions are macro-like, allowing the programmer to use one instruction in place of many simpler instructions. An example of CISC microcontrollers is Intel 8096.

1.3.4 According to Memory Architecture

1.3.4.1 Harvard Architecture Microcontroller

Microcontrollers that are based on the Harvard architecture have separate data bus and an instruction bus. This allows execution to occur in parallel. When an instruction is being "pre-fetched", the current instruction is executing on the data bus. Once the current instruction is complete, the next instruction is ready to go. This pre-fetch theoretically allows for much faster execution than Von-Neumann architecture, on the expense of complexity.

1.3.4.2 Princeton Architecture Microcontroller

Princeton architecture microcontroller is based on Von-Neumann architecture. Microcontrollers that are based on this architecture have a single data bus that is used to fetch both instructions and data. Program instructions and data are stored in a common main memory. When a controller addresses main memory, it first fetches an instruction and then the data to support the instruction.

The two separate fetches slow up the controller's operation. Its main advantage is that it simplifies the microcontroller design because only one memory is accessed. In microcontrollers, the contents of RAM can be used for data storage and program instruction storage. The Motorola 68HC11 microcontroller uses Von-Neumann architecture.

1.4 History of Microprocessors and Microcontrollers

The first single-chip microprocessor was introduced by Intel as Intel 4004 in 1971 which has 4-bit data bus [3]. The Intel 4004 series named as MCS-4 chip set required 256 bytes of ROM. Microprocessor is a single chip in which all CPU functions of a computer are available. Microprocessors are programmable, meaning they can be given instructions and return results based on those instructions. Before their invention, multiple chips were required to do the same thing, often spread across numerous racks. Before the 4004, Intel was a memory chip company. The Intel 4001 and 4002 had 256 bytes of ROM and 40 bytes of RAM respectively. Also, the Intel 4003 had an optional 10-bit parallel output shift register for I/O functions for scanning keyboards, displays, printers, etc. [4].

The microprocessor was designed for general purpose operation and to perform multitasking which required a complete set of chips called motherboard to execute a program. For special purpose singe-task operation, a low-cost device was needed, which gave rise to the development of the microcontroller.

Intel's first microcontroller was Intel 8048/49, launched in 1976. Intel's first microcontroller was used in the Magnavox Odyssey video games in the original IBM PC keyboard. In 1980, Intel introduced MCS-51 microcontroller commonly known as 8051. The Intel 8051 became popular because of its low cost, improved instruction set, and availability of development tools. MCS-51 includes two timers, four ports, and 128 bytes or more of board RAM. The MCS-51 family is now made by several companies such as AMD, ATMEL 89x, Dallas Maxim, SiLabs, Philips, and many more with many different features.

With the passage of time, microcontroller design improved, and now it contains many features to meet modern design challenges. Today microcontrollers have almost every feature that is found in PCs or laptops.

The growth and development of microcontrollers has rapidly increased since the start of 1970s. Basically, a microcontroller is divided into categories as per memory, architecture type, bits, and instruction set. Major microcontrollers which are available in markets are 8051, Peripheral Interface Controller (PIC), AVR, and ARM microcontrollers.

Intel Corporation introduced 8051 microcontroller in 1980. It is an 8-bit microcontroller, having 4 KB ROM, 128 bytes RAM, 40-pin DIP, four parallel 8-bit ports, on-chip crystal oscillator, two timers of 16 bits, and five interrupt sources. 8052 and 8031 are other two members of the 8051 family, where 8052 microcontroller has all features of 8051 along with three additional timers and RAM of 256 bytes. 8031 microcontroller has all attributes of 8051 except ROM. A popular microcontroller of this family is AT89C51 [5].

PIC has built-in ADC as compared to 8051. It is based on RISC architecture, whereas 8051 is based on CISC.

Intel 8051 has 250 instructions whose execution consumes one to four machine cycles where each machine cycle takes 12 clock cycles, while PIC on average has one to two machine cycles where each machine cycle takes four clock cycles. Popular microcontrollers of this PIC family are PIC18fXX8, PIC16f88X, PIC32MXX, etc.

Atmel AVR microcontroller is also based on RISC architecture. It has an additional feature of electrically erasable programmable read-only memory (EEPROM) for storing data over a power off time. In order to handle hanging software states, there is also a watchdog feature. The instruction time is faster than 8051 due to reduction of divider between external crystal oscillator and machine cycle time, from 12 to 1 [6]. It is available in three categories: Tiny AVR for less memory, small size, and simpler application; Mega AVR for good memory up to 256 KB and higher number of inbuilt peripherals; and XMega AVR for commercial application, large program memory, and higher speed [7]. Popular microcontrollers of the AVR family are Atmega8, Atmega16, Atmega32, Arduino series.

ARM processor is a multicore CPU based on RISC architecture. It is available in 32-bit and 64-bit RISC multicore processor. It is smaller in size, operates at higher speed, and performs extra MIPS (million instructions per second). Due to its smaller size, it is usually found in ICs. Popular microcontrollers of ARM controller family are LPC2148, ARM Cortex-M0 to ARM Cortex-M7, etc.

The history and evolution of microcontroller is summarized in Table 1.2.

TABLE 1.2

Microprocessor and Microcontroller History [8]

Processor Name	Processor Size	Launch Date
Intel 4004	4-bit	1971
Intel 8085	8-bit	1974
Intel 8048	8-bit	1976
Microchip PIC	8-bit	1976
Intel 8086	16-bit	1978
Zilog Z8	8-bit	1979
Intel 8031	8-bit (ROM-less)	–
Intel 8051	8-bit	1980
Atmel At89C51	8-bit (flash memory)	1984
Microchip PIC16C64	8-bit	1985
Motorola 68HC11	8-bit (on-chip ADC)	1985
Hitachi H8	8-bit	1990
PIC16x84	8-bit	1993
Atmel AVR	8-bit RISC	1996
Microchip dsPIC	16-bit (digital signal controller)	2001
Atmel AVR32	32-bit	2006
Atmel AVR Butterfly	Atmega169PV	June 2006
Microchip PIC16Mx	32-bit	November 2007
Beagle Board	ARM + GPU	July 2008
Arduino Nano	8-bit	2008
Arduino Mega	8-bit	March 2009
Arduino UNO	8-bit	September 2010
BeagleBone	ARM + GPU	October 2011
Raspberry Pi	ARM + GPU	2012
BeagleBone Black	ARM + GPU	April 2013
Microchip PIC32MZ	MIPS M14K	November 2013
Raspberry Pi 2	ARM + GPU	February 2015
Raspberry Pi 3	ARM + GPU	February 2016
Microchip PIC32MM	Specialized for motor control, industrial control, industrial IoT and multi-channel controller area network (CAN) applications	June 2016
Microchip PIC32MK	32-bit	2017
Microchip PIC32MZ DA Family	Graphics controller, graphics processor, and 32 MB of DDR2 DRAM	2017
Raspberry Pi 3 B+	ARM + GPU	March 2018
Qualcomm Dragon Board	64-bit Qualcomm® Kryo™ quad-core CPU	2018

1.5 Comparison of Various Microcontrollers of Modern Age

Features available in new microcontroller series boards [9–12] are given in Table 1.3.

TABLE 1.3

List of Modern Microcontrollers

Development Board	Model	Communication						Memory		General Purpose Input Output (GPIO)
		Ethernet	Wi-Fi	Bluetooth	LoRaWAN	Sigfox	GSM-LTE	RAM	Flash Memory	
Raspberry Pi	Zero wireless		✓	✓				512 MB	SD card	17
	Model A 1							256 MB	SD card	8
	Model A 1+							512 MB	SD card	17
	Model B 1	✓						512 MB	SD card	8
	Model B 1+	✓						512 MB	SD card	17
	Model B 2	✓						1 GB	SD card	17
	Model B 1.2	✓						1 GB	SD card	17
	Model B 3	✓	✓	✓				1 GB	SD card	17
	Model B 3+	✓	✓	✓				1 GB	SD card	17
Arduino	Uno							2 KB	32 KB	14
	Nano							2 KB	32 KB	14
	Mega2560							8 KB	256 KB	54
	DUE							96 KB	512 KB	54
	Yun							1 KB	32 KB	14
	Ethernet	✓						2 KB	32 KB	14
	UNO Wi-Fi							2 KB	48 KB	14
	TIAN	✓	✓					32 KB	256 KB	8
	GENUINO 101							24 KB	196 KB	14
Arduino MKR	MKR ETHERNET SHIELD	✓						32 KB	256 KB	8
	MKR Wi-Fi 1010		✓	✓				32 KB	256 KB	8
	MKR WAN 1300				✓			32 KB	256 KB	8
	MKR FOX 1200					✓		32 KB	256 KB	8
	MKR GSM 1400						✓	32 KB	256 KB	8

(Continued)

TABLE 1.3 (*Continued*)

List of Modern Microcontrollers

Development Board	Model	Communication						Memory		
		Ethernet	Wi-Fi	Bluetooth	LoRaWAN	Sigfox	GSM-LTE	RAM	Flash Memory	General Purpose Input Output (GPIO)
Pycom	WiPy		✓	✓				4 MB	8 MB	24
	LoPy		✓	✓	✓			512 KB	4 MB	24
	SiPy		✓	✓				512 KB	32 MB	24
	LoPy4		✓	✓	✓	✓		4 MB	8 MB	24
	Gpy		✓	✓				4 MB	8 MB	22
	FiPy		✓	✓	✓	✓	✓	4 MB	8 MB	22

The Raspberry Pi (explained in Table 1.2) is simply a computer with privilege of I/O pins. It is an SBC containing Broadcom's SoC. It contains ARM microcontroller and GPU. Moreover, it has an Ethernet port and new series containing Wi-Fi and Bluetooth which enable it to connect to the external world.

Arduino is the most popular microcontroller board series, having a number of variants and software supports. It contains different microcontroller ranging from 8-bit AVR ATMega microcontroller to 32-bit ARM microcontroller.

Arduino MKR is a new series from Arduino mainly designed for Internet of Things (IoT) platform and supports Ethernet, Wi-Fi, Bluetooth, LORA, Sigfox, and long-term evolution (LTE) communication.

Pycom is the IoT-based microcontroller board that supports Wi-Fi, Bluetooth, LORA, Sigfox, and LTE communication.

Qualcom Dragon Board and National Instruments DACs have alternate futuristic trend for next-generation embedded systems.

1.6 Future Trends

There are many aspects of daily life where one can use microcontroller-based systems through which ease to human being can be provided in order to handle routine matters in a better way. In early stages, the major part of microcontrollers was RAM, ROM, and CPU. Later on, microchip introduced further improvements, and EEPROM was included for storing data in memory which is not volatile. Afterward, Atmel introduced ADC which helps to process analogue data, and timers and a large memory option along with serial communication are also available in their microcontrollers named ATmega series. In latest microcontrollers, there is also a Wi-Fi option available to transmit data wirelessly which is beneficial for two most popular technologies of today's world, i.e. IoT and Artificial Intelligence (AI).

1.7 Conclusions

Microprocessors and microcontrollers are very important parts of an electrical/electronic machine or device. A microcontroller is basically a computer that comes in variety of packages and sizes. It plays an important role behind most of the projects where one needs to automate any process, for conversion of data, for data logging purpose, and many other aspects. From early stages of microcontrollers, there are many improvements introduced by several manufacturers which are discussed in this chapter. Theory related to the subject matter has been thoroughly discussed, and the history of microcontrollers has also been explained. A comparison of latest microcontroller-based development boards including Raspberry Pi, Arduino MKR, Pycom, etc. has also been discussed to make the reader aware of different features of these boards. This chapter provides information to the reader on selecting a particular microcontroller for a particular purpose by comparing different microcontroller features.

References

1. Administrator, "Difference between Microprocessor and Microcontroller", Electronics Tutorial, *Electronics Hub website*, 29 May 2015, www.electronicshub.org/difference-between-microprocessor-and-microcontroller/. Accessed 7 Nov 2018.
2. H. Choudhary, "Difference Between Microprocessor and Microcontroller", *Engineersgarage.Com*, www.engineersgarage.com/tutorials/difference-between-microprocessor-and-microcontroller. Accessed 9 Nov 2018.
3. "The Story of the Intel® 4004", *Intel*, 2018, www.intel.com/content/www/us/en/history/museum-story-of-intel-4004.html.
4. "Intel® I/O Controller Hub 10 (ICH10) Family Datasheet", *Intel*, 2018, www.intel.com/content/www/us/en/io/io-controller-hub-10-family-datasheet.html.
5. *Electronics Hub*, 24 Nov 2018, www.electronicshub.org.
6. "Comparison AVR - 8051", *Shrubbery Networks*, www.shrubbery.net/~heas/willem/8051-avr2/compars.htm.

7. H. Dhorajiya, P. Babreeya and M. Patel, "RFID based toll tax collection system", Project report, October 2015, LDRP Institute of Technology and Research, Gandhinagar, India.

8. Microcontrollers Lab, et al. "Introduction to 8051 Microcontroller: Getting Started Tutorial", *Microcontrollers Lab*, 2018, http://microcontrollerslab.com/8051-microcontroller/.

9. "Arduino - Products", *Arduino.Cc*, 2018, www.arduino.cc/en/Main/Products.

10. "MKR Family - Arduino", Store.Arduino.Cc, 2018, https://store.arduino.cc/usa/arduino/arduino-mkr-family.

11. "Products Archive - Raspberry Pi". *Raspberry Pi*, 2018, www.raspberrypi.org/products/.

12. "Fipy·Gitbook", Docs.Pycom.Io, 2018, https://docs.pycom.io/datasheets/development/fipy.

2

Selection of Sensors, Transducers, and Actuators

Memoon Sajid

Jeju National University
GIK Institute of Engineering Sciences and Technology

Jahan Zeb Gul and Kyung Hyun Choi

Jeju National University

CONTENTS

2.1 Introduction...30
2.2 Transducers...31
    2.2.1 Optoelectronic Interface 31
    2.2.2 Electromechanical Interface 32
    2.2.3 Optomechanical Interface..33
    2.2.4 Electrochemical Interface 33
2.3 Sensors..33
    2.3.1 Characteristics of Sensors..34
        2.3.1.1 Linearity and Calibration 34
        2.3.1.2 Sensitivity...36
        2.3.1.3 Specificity...36
        2.3.1.4 Detection Range and Limit...36
        2.3.1.5 Stability and Reproducibility 37
        2.3.1.6 Response Time...37
    2.3.2 Passive Sensors..37
        2.3.2.1 Resistive Sensors 38
        2.3.2.2 Capacitive Sensors 39
        2.3.2.3 Inductive Sensors 40
        2.3.2.4 Electrochemical Sensors 40
        2.3.2.5 Optical Sensors..40
    2.3.3 Active Sensors 41
2.4 Actuators...41
    2.4.1 Hydraulic Actuators 42
    2.4.2 Pneumatic Actuators 43
    2.4.3 Electric Actuators..43
    2.4.4 Mechanical Actuators 43

    2.4.5   Smart Soft Actuators 43
              2.4.5.1   Electrochemical Actuators 44
              2.4.5.2   SMA/SMP Actuators 44
              2.4.5.3   Piezoelectric Actuators 46
              2.4.5.4   Magnetic Actuators 47
    2.5   Conclusions 47
    References 48

2.1 Introduction

In the current era of automation, humans are becoming increasingly dependent on machines to perform their routine tasks. Most modern machines
are based on electromechanical systems that can deliver performance
much better than their human counterparts. With advancements in design
and fabrication technologies, the systems are becoming better, smarter, and
more versatile day by day. Almost all advanced systems consist of complex
combination of electrical and mechanical parts working in harmony that
are linked together through three basic components. These crucial components are the sensors, transducers, and actuators. They equip the system
with awareness by sensing various surrounding parameters like temperature, humidity, gases, pressure, light, stress, strain, sound, etc. They
make the system intelligent by using their senses as feedback to control
its resulting action. And they make the system autonomous by enabling it
to respond to that feedback through physical actuation. Transducers play
the role of the bridge that converts one form of signal or energy to another.
Collectively, sensors, transducers, and actuators are the ears, eyes, nose,
spine, muscles, and limbs of a machine.

There are hundreds of types of sensors, transducers, and actuators based
on their design, materials, working principle, and application [1,2]. The
most common sensors that are used in our daily life gadgets and machines
include temperature [3], humidity [4,5], pressure [6], touch [7,8], strain [9],
light [10,11], and Hall Effect sensors. They are present in our devices ranging from a hand phone to a microwave oven, a washing machine, an air
conditioner, our automobiles, and even in our rice cookers. Transducers
convert the signals generated by the sensors into machine readable electrical signals that can be used to interact with a human or another machine
or a machine part [12]. Some commonly used transducers in our daily life
include microphone (sound to electricity), speaker (electricity to sound),
solar cell (light to electricity), light bulb (electricity to light), electromagnetic
generator (mechanical motion to electricity), motor (electricity to motion),
etc. Actuators on the other hand are less common, but with the advancement in micro-level fabrication technologies like micro-electromechanical
systems (MEMS), they are rapidly catching up. They can be described as a

subset of transducers that produce physical motion in response to any kind of input signal. There are actuators ranging from a few microns to a couple of feet in all our routinely used machinery. The common actuators include motors, automatic door locks, airplane fins, car wipers, robotic limbs, etc. This means that effectively everything of a modern machine other than the body, processors, electronic circuitry, and passive mechanical parts consists of sensors, transducers, and actuators.

2.2 Transducers

A transducer is a device that converts one form of energy to another. Itreceives an input signal and generates an output signal of different form in response. Transducers act as an interface or a bridge between various working environments. Signals from optical systems, mechanical systems, and electrical systems can be interconverted and fed to each other using transducers. They are also a crucial component of both sensors and actuators. They translate the response of the sensors to a user recognizable or a machine readable signal. They also translate the user input or machine signal to a compatible input for an actuator. Transducers can be classified based on the type of interface they are providing by converting one form of signal to another. Most commonly, in sensors and actuators research, transducers are the interface electrodes that provide a link between the signal produced by the active region of the sensors and the processing circuitry [12–14]. Some main types of transducers commonly used in our devices are shown in Figure 2.1.

2.2.1 Optoelectronic Interface

This type of transducers converts either an optical signal in to an electrical signal or vice versa. Their most common examples include solar cells and light emitting diodes [15,16] (LEDs). Optoelectronic transducers are commonly used in optical sensors [17], photodetectors [18,19], isolation circuits, user interfaces, displays, etc. It can be explained by an example of a handheld photodetector meter that provides the reading for the intensity of incident light. The photons in the light are absorbed by the active area of the photodetector which is then translated to a change in resistance of the active area as shown in Figure 2.1a. This change is converted into an electrical signal by the transducer electrodes, and that signal is processed by a microcontroller circuit and is converted into light intensity using the sensor's calibration curve. Finally, the value of the light intensity is displayed to the user through a liquid crystal display (LCD) that is a transducer converting the electrical signal to an optical signal. Optoelectronic transducers are usually used in

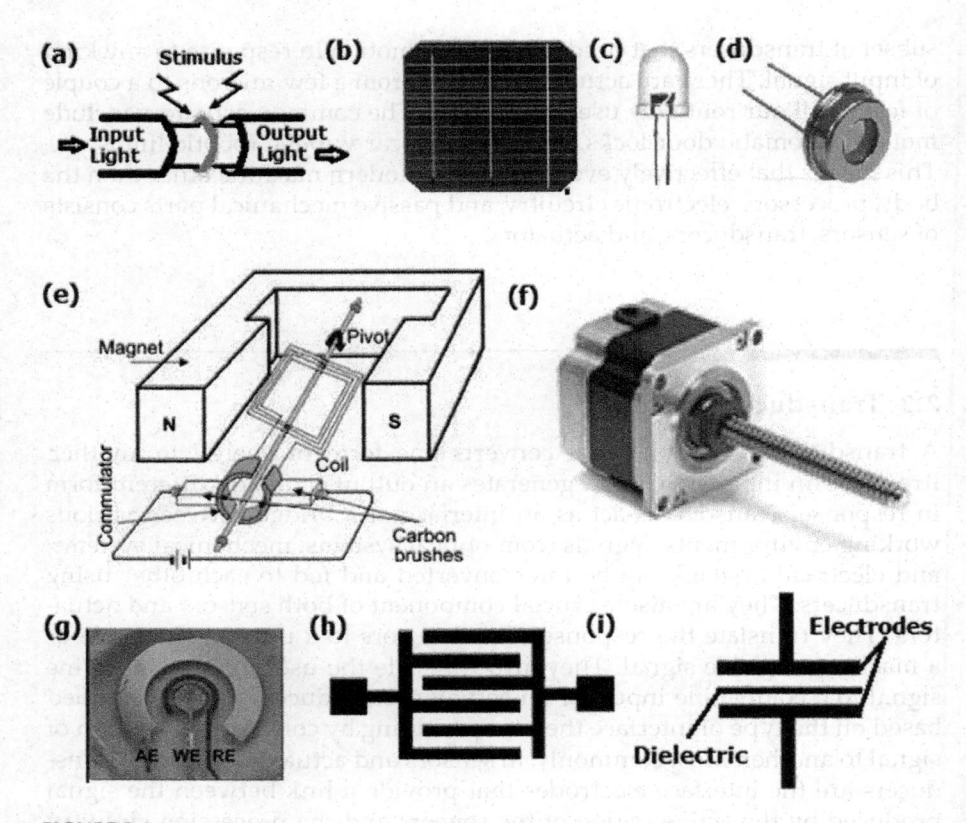

FIGURE 2.1
Different types of transducers showing (a) optical-fiber-based transducer, (b) solar cell, (c) LED, (d) photodetector, (e) motor, (f) linear stepping actuator, (g) three-electrode transducer for electrochemical cell, (h) interdigital transducer -type (IDT) transducer, and (i) sandwich-type transducer electrodes.

communication systems and applications involving non-physical, non-tangible stimuli like light, temperature, etc. They are generally more expensive and complex in operation when compared to simple electrode-based transducers and are usually employed in high-end systems.

2.2.2 Electromechanical Interface

This is the most common type of transducers that can be seen around in daily life. They provide an interface between electrical and mechanical systems [20,21]. The most common example of conversion of an electrical signal to a mechanical motion is a motor presented in Figure 2.1b and c. Then there comes a speaker that converts electrical signals into sound. A motor can also be considered as an actuator making actuators as a subset of the transducers category. The transducers that convert mechanical energy to electrical energy include generators and microphones (sound is a form of mechanical energy). Another type of electromechanical transducers quickly gaining interest are triboelectric, piezoelectric, and pyroelectric transducers that convert different types of mechanical inputs like pressure, stress, strain, or twisting into electrical signals [22]. They can be used as generators and can also be used in self-powered sensors that are explained later in sensors portion. An example of a system consisting of an electromechanical interface provided by transducers is similar to the one provided in optoelectronic interface. We can replace the input sensor with a microphone to measure the noise level and the LCD display with a speaker output to play the output to the user. Electromechanical transducers are generally cheaper and simpler in operation and are thus widely used in daily life machinery and gadgets. They are also easier to interface and are quite robust and reliable.

2.2.3 Optomechanical Interface

This is a rather less common category of transducers, but there are systems requiring direct conversion of optical signals to mechanical motion and vice versa. The most common example are optical actuators in which the material properties of the actuator change upon exposure to light resulting in expansion or contraction. The detailed mechanism will be discussed in the actuators section. They are commonly used in environments where direct supply of electricity is not feasible for mechanical motion and light is used as an alternative to induce mechanical motion. A research level example includes light guided mini-robots.

2.2.4 Electrochemical Interface

In case of chemical and biological sensing devices, these types of transducers are employed. They are effectively specifically designed electrodes as presented in Figure 2.1g that can collect the signal produced as a result of some electrochemical reaction or a bio-chemical instance. In case of electrochemical sensors, the target analyte comes in contact with the active region resulting in a redox reaction that produces ions and electrons [23,24]. These current carriers are then collected by the transducer electrodes, and the amount of analyte is quantified by the amount of flowing current.

2.3 Sensors

A sensor is a device that responds to an external stimulus or an analyte. The stimulus can be anything ranging from a physical entity like touch, pressure, strain, an electromagnetic signal, heat, light, chemical, gas, sound, water, particles, bio-reagents, etc. [1,25–27]. The sensing portion of the device is known as the active region that collects or receives the stimulus and then generates a signal of its own in response [2]. The response of the sensor can be in the form of a passive change in its intrinsic properties like color, electrical resistance, capacitance, and inductance or an active electrical signal in the form of a current or potential difference. We can categorize sensors based on the type of analyte they are detecting like environmental sensors, biosensors, etc., or we can categorize them based on their working mechanism like active sensors and passive sensors. Here, we will follow the later one due to its better versatility and easier understanding, but before that, we will develop understanding about some basic terms associated with sensor characteristics.

2.3.1 Characteristics of Sensors

There are many characteristics and evaluation parameters associated with sensing devices that can help determine their performance. These parameters include sensitivity, specificity, detection range, detection limit, reproducibility, stability, and response time. In design and selection of sensing devices, these parameters are optimized according to the application requirements. Below, we will discuss each of them briefly.

2.3.1.1 Linearity and Calibration

The output of a sensing device can change in any manner in response to the input target stimulus. The magnitude of the sensor's output can increase or decrease in linear, exponential, or even random pattern as shown in Figure 2.2a [1]. A known magnitude of target stimulus is applied to the sensor measured using a reference sensor, and then a calibration curve is established. A formula is calculated for that curve to find the exact relationship between the input and the output. This formula can be used to quantify the amount of change in sensor's output in response to the input and can also be used to predict the response. The easiest to interpret and the most stable relationship between the input and the output is a linear relationship as presented in Figure 2.2b [25]. It can be increasing or decreasing (having positive or negative slope) but can be processed by both the user and the machine without much trouble and with low processing power. It is also simpler to calculate other sensor characteristics for a linear sensor as compared to non-linear ones [2,28]. It is preferred while designing a sensing device to tune its response to a linear curve, but it is not possible in all cases without compromising other parameters of the sensor like its detection range and sensitivity. So, a trade-off is made based on prioritization of the parameters defined by the target application. Other commonly used calibration curves include exponential curves, Boltzmann

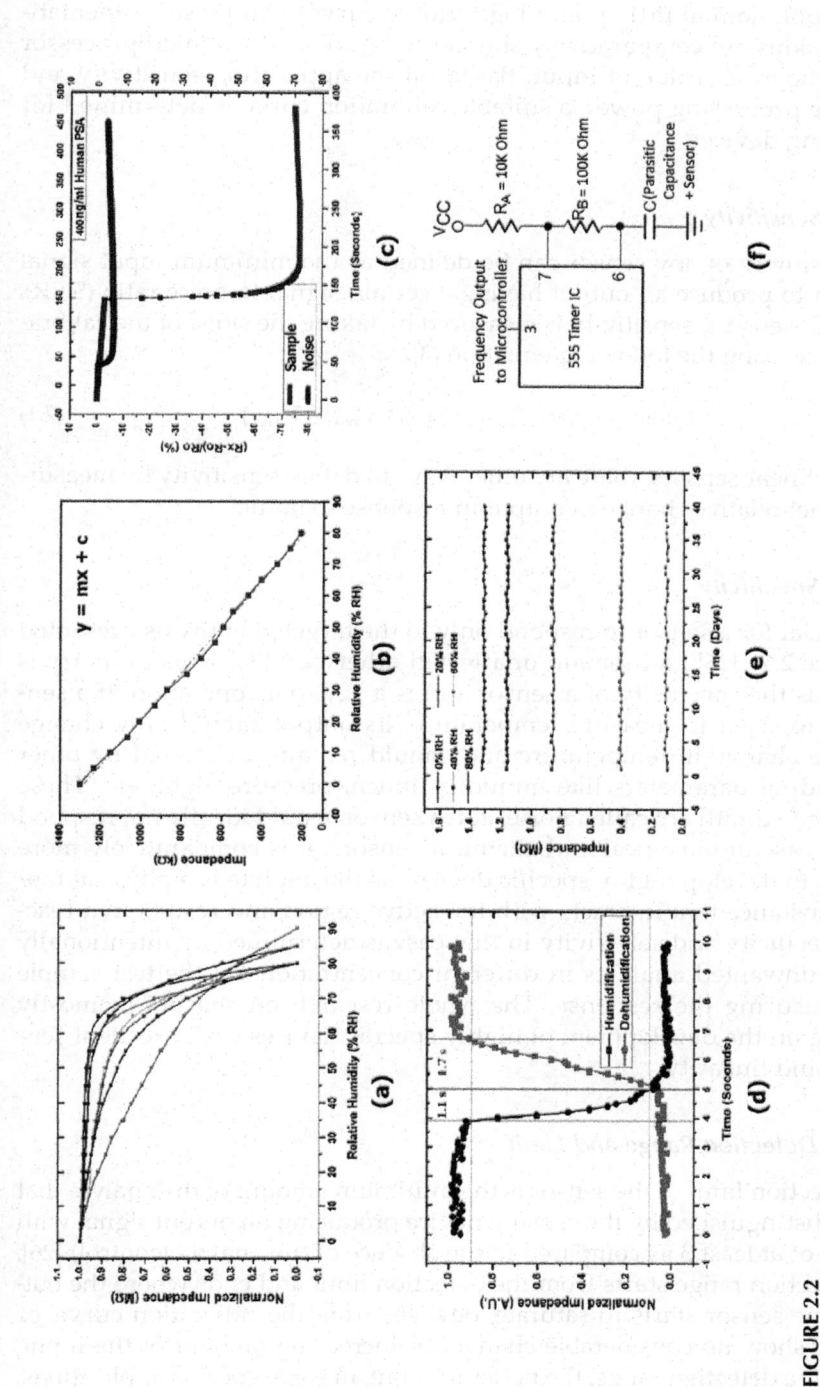

FIGURE 2.2

Sensor characteristics showing (a) various output response curve shapes for sensors [1], (b) calibration and linear curve fit [25], (c) specificity of a biosensor [26], (d) response and recovery times [2], (e) sensor stability over days [25], and (f) readout circuit for capacitive sensors [13].

curves, polynomial fitting, and logarithmic curves. All these mathematical equations are comparatively simpler to be solved by a microprocessor to find the exact value of input. Based on the application sensitivity and available processing power, a suitable calibration curve is determined for all sensing devices.

2.3.1.2 Sensitivity

The sensitivity of any sensor can be defined as the minimum input signal required to produce an output having a certain signal-to-noise ratio (SNR). For linear sensors, sensitivity is measured by taking the slope of the calibration curve using the following equation [3].

$$\text{Sensitivity} = (Y_{max} - Y_{min})/(X_{max} - X_{min}) \tag{2.1}$$

For non-linear sensors, there are other ways to define sensitivity by measuring the net relative change in output in response to input.

2.3.1.3 Specificity

It is crucial for a sensor to respond only to the targeted entity as presented in Figure 2.2c [26] and remain unaffected otherwise [26]. This property is known as the specificity of a sensor and is a key to its operation. If a sensor is supposed to measure temperature, its output should only change with the change in temperature and should remain unaffected by other surrounding parameters like humidity, touch, pressure, light, etc. These unwanted stimuli are called noise, and a sensor should ideally not respond to it. In case of biological and chemical sensors, it is comparatively more difficult to develop highly specific devices as the analyte is a physical tangible substance that interacts with the active region and results in a reaction. Specificity and selectivity in that case is determined by intentionally adding unwanted analytes in different concentrations in the test sample and measuring the response. The whole research on sensors in mostly focusing on the development of highly specific devices with excellent sensitivity and linearity.

2.3.1.4 Detection Range and Limit

The detection limit of the sensor is the minimum amount of the analyte that can be distinguished by the sensing device producing an output signal with an SNR of at least 3 as compared to the absence of the analyte (control) [26]. The detection range starts from the detection limit and ends where the output of the sensor starts to saturate, deviates from the calibration curve, or starts to show no considerable change for increasing amount of the input. Higher the detection range, the better it is. But, in some specific applications, the detection range can be smaller with more focus on improving the sensitivity, specificity, and response time.

2.3.1.5 Stability and Reproducibility

Ideal sensing devices should have a highly stable and reproducible response to the target analyte. A minimum of three full range curves are recorded, and a standard deviation is calculated. The average error should ideally not exceed from ±3%–4%. Also, over the time, the response of the sensor should remain stable and repeatable and should not decay as shown in Figure 2.2e [25]. The stability is usually measured by exposing the sensors to the maximum measurable concentration of the analyte for extended periods of time and then recording the deviation in its response curve [25]. It should also remain within ±3%–4%.

2.3.1.6 Response Time

The response time and recovery time of a sensor are very important in defining its performance and interfacing it to a system. The response time is the time taken by the output of the device to reach from 0% to 90% of its maximum value for an input changing from control to maximum detectable quantity [13]. The recovery time corresponds to the time taken by the output of the sensor to return from maximum value to 10% for an input changing from maximum to control. The sensors should have a fast response time and recovery time to ensure there is no hysteresis in the data or there is no lag in the overall system as presented in Figure 2.2d [2]. Again, sensors with fast response usually have compromised sensitivity or have high cost. If the target application is not using the sensors in real-time feedback but is logging the readings after certain intervals of time, a device with slower response and recovery time is also acceptable.

Now two basic categories of sensors based on their signal output and working principle will be discussed. Both active and passive sensors are used in all forms of monitoring, and their selection is purely based upon the requirements of the targeted application.

2.3.2 Passive Sensors

Passive sensors respond to an external target stimulus by changing their intrinsic properties like color, resistance, capacitance, or inductance [2,6,29,30]. They do not produce an active signal carrying energy by themselves, and therefore need external power sources for operation. Majority of sensing devices that currently exist are passive devices that need external power to perform their action or at least translate the detection event into a machine or user readable output. The most common types of passive sensors used in machinery are electrical sensors while color changing sensors are usually used in biology or chemistry to detect the presence or absence of a certain target analyte. Passive sensors are usually easier to fabricate and are less expensive when compared to active sensing devices. Passive sensors can be further divided into subcategories based on their working mechanism (Figure 2.3).

2.3.2.1 Resistive Sensors

Resistance is defined as the property of a material to hinder the flow of current through it as expressed by equation 1.2. Resistive sensors are the most common type of passive sensors in which the intrinsic resistance of the sensor's active region changes in response to the change in amount of target analyte. A number of categories of sensors are resistive type including humidity, temperature, flex, light, pressure, touch, and so on [29,32–34]. When the target analyte comes in contact with the sensing device, it effects the current flowing capability of the active layer by either hindering or enhancing the electron mobility. For instance, in case of temperature sensors, an increase in

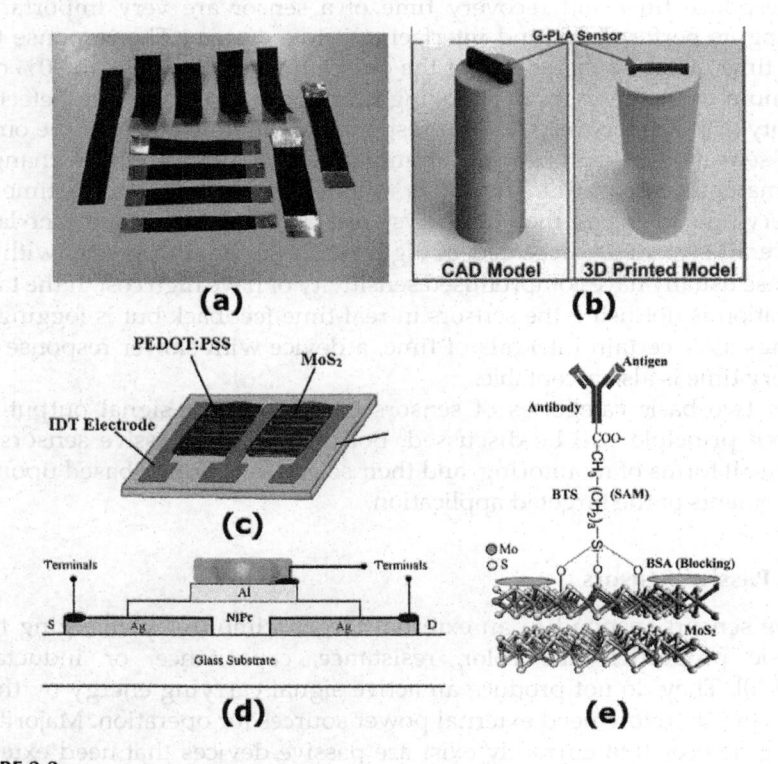

FIGURE 2.3
Different types of sensors showing (a) resistive flex sensors [9], (b) 3D printed resistive temperature sensor [3], (c) dual-element transducer-based resistive humidity sensor [26], (d) FET-type passive photodetector [31], and (e) active region of immunosensors showing binding event [26].

temperature results in increase of random motion of electrons in the active region and in return reduction in the current carrying capability, thus increasing the resistance. In contrast, in case of light dependent resistors (LDRs), the intrinsic resistance of the active region exposed to light reduces with increasing intensity of exposed light. In case of strain sensors, the resistance of the active region increases with increasing strain due to the formation of micro-level cracks and discontinuations in the current flow path.

$$\text{Resistance} = \text{voltage}/\text{current} \qquad (2.2)$$

Resistive sensors are usually cheap and are very simple and easy to interface with any type of electrical circuit as their transducers simply consist of an electrode pair. The sensor is connected as a variable resistor in a potential divider configuration with a known fixed resistor with value near to half of the range of sensor's output. The change in potential is measured that can be easily translated to the amount of analyte using a calibration curve.

2.3.2.2 Capacitive Sensors

Capacitance is defined as the ability of a material to store charge. Capacitive sensors are the second most common type of sensors and have the same reason behind their popularity: low cost and simple integration [4]. In case of capacitive sensors, the transducers are again a simple pair of electrodes that provide an interface between the sensing region and the electrical readout circuit. Capacitive sensors work on two basic principles: change in the dielectric constant of the active area material resulting in change of the capacitance and change in the physical distance between the two transducer electrodes that can be expressed using equation 1.3.

$$\text{Capacitance} = (\text{area} \times \varepsilon \times \text{dielectric coefficient})/\text{distance} \qquad (2.3)$$

Most of the pressure and touch sensors work on the principle of changing the distance between the transducer electrodes that result in the change in capacitance [35,36]. A deformable elastic material is used to fabricate the active region. In contrast, most of the gas sensors and humidity sensors work on the principle of change in capacitance based on the change in dielectric constant of the active area materials [4,29,37]. When the gas or water molecules are adsorbed inside the sensing layer, the cumulative effective dielectric constant of the material changes. Porous materials are usually selected to fabricate the sensing layer to allow easy adsorption and desorption of the target analyte. In most cases, the capacitance is converted into a friendlier machine readable quantity like frequency using simple a readout circuit as presented in Figure 2.2f because measurement of direct capacitance is complicated and requires a more expensive and bigger circuit.

2.3.2.3 Inductive Sensors

There are two basic types of inductive sensors: one in which the inductance of the sensors changes in response to the input stimulus and second in which induction is used as a coupling mechanism to sense the stimulus [38,39]. Pressure and displacement sensors usually lie in the first category in which the spacing between the turns is changed in response to an external pressure that can result in change of inductance. The second type of inductive sensor is mostly used in electrical measurements and most commonly in current measurement. Inductors can pick up an electrical signal or a flowing current without even physical contact through a phenomenon known as electromagnetic induction. They are thus preferred as sensing elements in delicate and sensitive systems where the sensors should be physically isolated from the circuit loop.

2.3.2.4 Electrochemical Sensors

Electrochemical sensors work on the principle of change in electrical properties of the active area in response to a chemical reaction that started when the target analyte comes in contact with it [40–42]. Electrochemical sensors are mostly used in chemistry and biology to detect the amount of certain chemicals and bio-reagents in the sample. The surface of the active area is functionalized with a specific receptor that can only bind to the target analyte. The binding event results in the initiation of a reaction that releases or consumes ions and electrons (current carriers) depending on the nature of the reaction. This process results in change in magnitude of the flowing current through the circuit. The transducers used are simple conductive electrodes that collect and release those charge carriers and complete the circuit for the flow of current. Some gas sensors are also electrochemical based with polar gases resulting in variation of the mobility of the active region, thus affecting the amount of flowing current. That change in the amount of current can be translated to the amount of analyte using a calibration curve against a known parameter. Field effect transistors (FETs) and three electrode-based electrochemical cells are used as transducers while potentiostats are used as the interface circuit in this case.

2.3.2.5 Optical Sensors

Passive optical sensors are mostly based on change in color upon detection of the target body and are usually used for qualitative (absence or presence only) detection rather than quantitative detection (amount of analyte) [43,44]. They are mostly used in biology and chemistry to detect the presence or absence of certain chemicals and bio-reagents. No transducers are involved here, and the active region itself changes its color when it physically comes in contact with the target analyte. Different colored dyes are used for the active area fabrication that are sensitive to specific chemicals. Some common examples are pregnancy testers, influenza testers, humidity testing strips, etc.

Some passive optical sensors are also based on the change in refractive index of the active region upon the occurrence of the detection event [45,46]. A specifically coated optical fiber acting as the transducer is usually used for these types of sensors. When the physical target analyte comes in contact with the active region or the stimulus like temperature is applied, the refractive index of the materials changes, and that affects the behavior of light passing through the fiber. The change in the angle of refraction is then translated to the amount of analyte detected. These types of sensors are mostly used in optical systems and need a secondary transducer to convert that optical signal into an electrical signal for interfacing them with electrical systems.

2.3.3 Active Sensors

Active sensors do not need external power source to operate and produce their own energy upon the occurrence of the detection event. There is currently a very little collection of active sensors available in market, but a considerable portion of research is being carried out to make more and more self-powered sensors. The major advantage of these devices is their capability to even operate and provide readings in remote monitoring where providing a power source is not an easy task. Another advantage is their zero power consumption, thus saving power for the whole system and resulting in more environmental friendly devices. Most commonly known self-powered sensors are the solar cells that can measure the intensity of incident light and produce an output current whose magnitude changes with changing intensity [47,48]. Most of the current photodetectors work on the same principle but are tuned to respond to different wavelengths of light through the use of different active materials with selective bandgaps. Other prominent examples of active sensors include self-powered pressure, stress, and strain sensors. They work on the principles of triboelectric, piezoelectric, and pyroelectric nanogenerators that are composed of sheets of active materials that produce electrical charges when subjected to physical inputs of different forms [48–50]. The charges are collected by transducer conductive electrodes and result in current flow when the circuit is completed.

2.4 Actuators

An actuator is an active component responsible for motion and controlling a mechanism or a system using a defined signal. It connects the information processing part of a system with a technical or nontechnical process. The output of an actuator is always "energy" or "power". An actuator always consists of at least one or more energy controllers connected in series. An actuator is an energy conversion component, and the main challenge is its controllability. Thus, there are only three main components of actuator:

a. Input energy/input variable
b. Energy controller/converter (at least one or multiple connected in series)
c. Output energy/output variable.

This definition and explanation is not standard but is accepted and used by the scientific community around the globe. Based on the above three components, an actuator can be classified into multiple categories for various applications, e.g. in case of first component (input energy), it can be pressure, heat, light, water, electric potential, magnetic field, or pneumatic. The second component (controller) is used to control the min/max limit of the first component. The third component (energy) is the resultant output. In the start of this chapter, most common types of actuators are briefly described along with their applications, and in order to make this chapter more useful, more light is given on novel and new actuator technologies and their possible future applications. The most common actuators are

i. Hydraulic
ii. Pneumatic
iii. Electric
iv. Mechanical.

The novel actuators included in this chapter are

a. Smart soft
b. Electrochemical
c. Shape memory alloy (SMA)/shape memory polymer (SMP)
d. Piezoelectric
e. Magnetic.

2.4.1 Hydraulic Actuators

A hydraulic actuator consists of a piston that uses hydraulic force to enable mechanical operation. The mechanical motion produces an output in the form of linear, rotatory, or oscillatory motion. There is no compression in a hydraulic actuator because it is impossible or very difficult to compress a liquid. The piston in hydraulic actuator can slide freely inside a cylinder and can be moved in two directions, thus making them a single acting or double acting system. The direction of the motion is controlled by the respective side pressure. These actuators are used in applications where large strokes with large actuation power and speed are needed. The main limitation of hydraulic actuators is the sealing machinery. An excellent sealing machinery is desired in order to avoid driving fluid leaking.

2.4.2 Pneumatic Actuators

A pneumatic actuator changes energy of compressed air at high pressure into either linear or rotary mechanical motion. In pneumatic actuator, small pressure change produces significant large forces. Pneumatic actuators are quick to start and stop because their pneumatic energy only needs to be stored in the forward direction; therefore they are highly desirable in robotics. These actuators operate using compressed air which makes them safe compared to other conventional actuators. These actuators are also cheaper and relatively more precise and capable of handling high force operation.

2.4.3 Electric Actuators

An electric actuator is powered by a motor that converts electrical energy into mechanical torque. The electrical energy is used to actuate equipment such as multi-turn valves. Additionally, a brake is typically installed above the motor to prevent the media from opening a valve. If no brake is installed, the actuator will uncover the opened valve and rotate it back to its closed position. If this continues to happen, the motor and actuator will eventually become damaged. It is one of the cleanest and most readily available forms of actuator because it does not directly involve oil or other fossil fuels.

2.4.4 Mechanical Actuators

A mechanical actuator functions to execute movement by converting one kind of motion, such as rotary motion, into another kind, such as linear motion. An example is a rack and pinion. The operation of mechanical actuators is based on combinations of structural components, such as gears and rails or pulleys and chains.

2.4.5 Smart Soft Actuators

Smart soft actuators mainly consist of soft functional materials known as smart materials. The design of smart actuators can be inspired from nature, or it can be a precise engineering model. Smart actuators are a big part of devices that mimic the difficult and precise behavior which nature optimized in centuries. A large number of such devices which use smart actuators have been fabricated for various purposes such as soft robots, soft actuators, underwater soft structures, etc. [51]. Many animal inspired smart actuator designs are fabricated such as elephant trunks, octopus arms, squid tentacles, tri-legged spider, and snakes [52,53]. The fabrication of such actuators is the need of innovative soft functional devices with improved performance as compared to traditional design concepts, e.g. in terms of locomotion or grasping. The smart soft actuators are mainly composed of SMAs, SMPs, electroactive polymers, and magnetic responsive materials. There also exit smart materials which change their phase. Such phase changing materials are a good alternative to electromechanical actuators. The core functional concept of smart actuators composed of phase changing materials is the force generated by the fast expansion at phase transition temperature.

Soft actuators are difficult to control because of their infinite and complex degree of freedom. The main challenge of smart soft actuators is the lack of easily processed robust soft actuators with high strain density. Such actuators would be easy to produce and to mold, cut, and 3D print into a desired shape, yet would produce large macroscopic actuation at relatively low voltage and current. Today, soft actuation techniques are based on either electroactive polymers, SMAs and SMPs or compressed air and pressurized fluid actuators. Some examples of fabricated smart soft actuators are shown in Figure 2.4. Smart soft actuators are further discussed based on the fabrication technology.

2.4.5.1 Electrochemical Actuators

The core concept of the electrochemical actuator (ECA) is based on evolving gas during the application of a small voltage. The overall structure of such actuator is a closed system, the pressure builds up and is then transformed via appropriate design features into mechanical work. The generated pressure is stopped by short-circuiting or by pole change. By using different types of chemicals with distinct reactions, these actuators can be modified for various types of applications. An example of electrochemical actuator reaction is shown below.

$$Ag + H_2O \rightarrow AgO + H_2$$

In the reaction above, the electrochemical actuator is reset without insertion of new energy by short-circuiting the electrodes. Pressure builds up by the hydrogen produced on the counter electrode while the silver oxidizes. The main applications of these actuators are throttle control, regulation of gases and fluids, room heating systems, etc.

2.4.5.2 SMA/SMP Actuators

In 1950s, shape memory effect was discovered by chance in certain copper alloys. Later nickel titanium (NiTi) alloys exhibit the real effect that can be

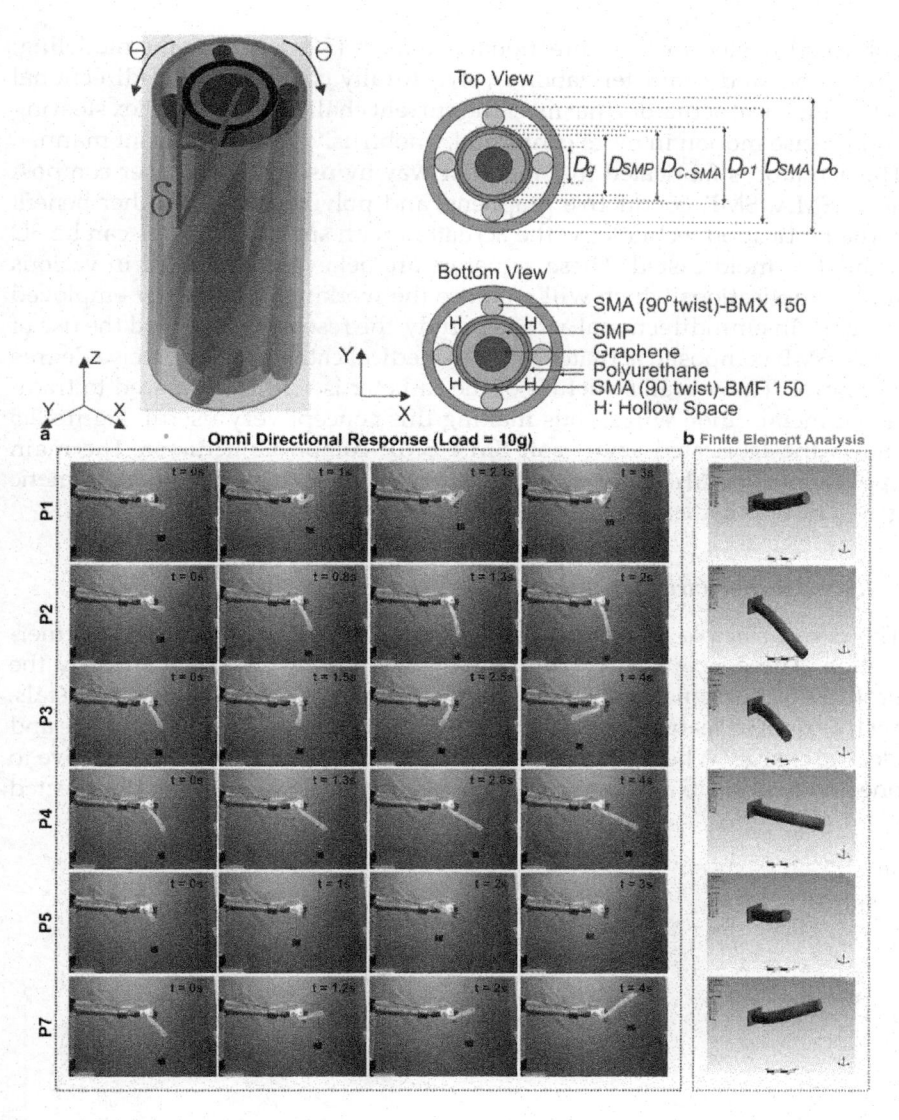

FIGURE 2.4
Soft actuator mechanical design concept. The fabricated soft actuator complex motion time-lapse with load.

used as an actuator for various applications. This unusual behavior is based on the reversible transformation from the martensite into the austenite phase of an SMA. Besides alloys, this effect is also observed in polymers, and many researchers enhanced the functionality of this effect by making the composite of SMA and SMP [53]. In order to produce complex motion with multiple degrees of freedom, this composite is widely used. For example, a work published related to omnidirectional actuators [54] describes the modeling, fabrication, and characterization of structurally controlled omnidirectional soft cylindrical actuators that meet the current challenges of complex steering and precise motion in more conformal, unobtrusive, and compliant manner. The actuator is fabricated in a modular way by using a multilayer composite of SMA, SMP, conductive graphene, and polyurethane. Another benefit is the fabrication technology; the actuators with such composites can be 3D printed or mold casted. These actuators are believed to be used in various future applications which will enhance the working of currently employed systems. In omnidirectional actuator study, the researchers proved the use of SMA–SMP composite actuator as a bio-medical catheter with precise degree of control. The overall structure of the catheter is soft as compared to traditional metal guide wires, thus making this concept very useful. Figure 2.5 shows the fabricated smart soft SMA–SMP composite actuator. The main applications of SMA–SMP actuators are biomimetic locomotion, biomimetic grippers, bio-medical devices, and miniature systems.

2.4.5.3 Piezoelectric Actuators

The core principle of piezoelectric actuator depends on electric polymerization. The concept of electric polymerization was first explained by the brothers Jacques and Pierre Curie in 1880. In piezoelectric actuators, crystals, such as quartz, feature a physical relationship between mechanical force and electric charge. When the crystal lattice ions are elastically shifted relative to one another due to an external force, an electric polarization can be detected

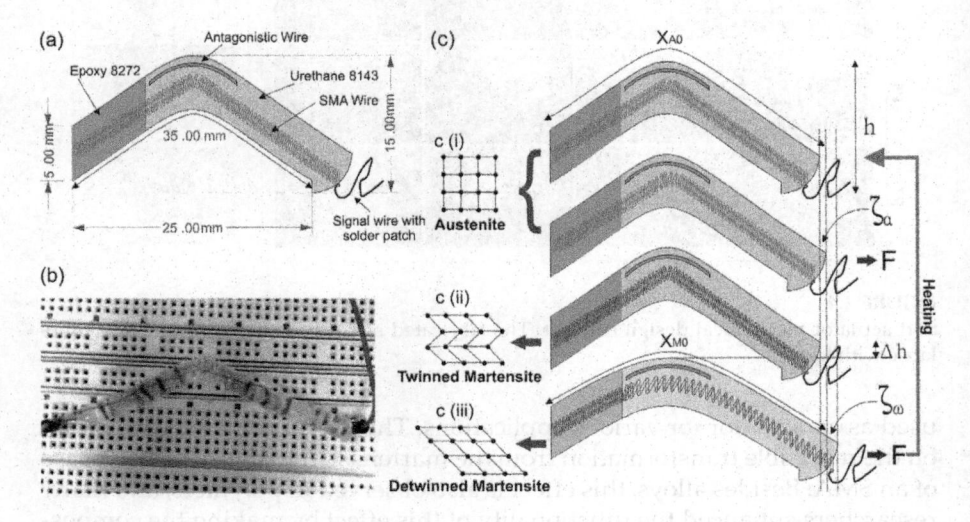

FIGURE 2.5
SMA–SMP smart soft actuator. The working principle of actuator is shown which when heated moves to austenite state and when relaxed moves to detwinned martensite state.

by means of metallic electrodes on the surface. This is called piezoelectric effect. The effect is reversible and is then called reciprocal or inverse piezoelectric effect. If, for instance, an electric voltage is applied to a disc-shaped piezo crystal, the thickness of the crystal changes due to the reciprocal piezoelectric effect. It is this property that is made use of in actuators. The main application of piezoelectric actuators is the position control free of hysteresis.

2.4.5.4 Magnetic Actuators

The magnetic actuator consists of ferromagnetic materials. When a ferromagnetic material is magnetized, its shape changes with increasing field strength. This phenomenon is labeled magnetostrictive effect. Joule effect is the key component of magnetostrictive effect. The Joule effect describes that the magnetic domains turn in the direction of magnetization and shift their borders. Strain in the range of 10–30 µm/m can be generated using conventional alloys such as iron or nickel. High strain in the range of 2,000 µm/m can also be achieved using big material samples of rare earth. Magnetic actuators are very useful for underwater operation.

2.5 Conclusions

Selection of sensors, transducers, and actuators solely depends upon the requirements of the target application. There are hundreds of options available for carrying out a single task at hand, but every one of them has its own pros and cons. There are always some trade-offs between certain performance parameters themselves and the prices of the devices that are to be considered carefully while selecting these devices for employment in various systems. In some systems, the top priority is reducing the system cost irrespective of the power requirements and the performance. Passive sensors and actuators with simple conductive electrode-based transducers can be used in this case. In some systems, accuracy and reliability are key issues where cost doesn't matter such as in high-end communication systems. More complex interfaces with highly reliable devices having excellent performance parameters are selected irrespective of the size, complexity, and power consumption. In other cases, miniaturization and low power consumption are the main issues. Here, MEMS-based active sensors and actuators with integrated self-powered transducers can be used with compromise on fabrication and cost. Also, in other cases, the sole purpose is to get the job done regardless of the size, power, and cost where multiple systems having different operation domains have to be interconnected in the form of an autonomous complex machine. A combination of active, passive, cheap, and expensive devices can be used for this purpose based on the available flexibilities in the given domain.

References

1. H. B. Kim, M. Sajid, K. T. Kim, K. H. Na, and K. H. Choi, "Linear humidity sensor fabrication using bi-layered active region of transition metal carbide and polymer thin films," *Sens. Actuators, B*, vol. 252, pp. 725–734, Nov. 2017.

2. M. Sajid, G. U. Siddiqui, S. W. Kim, K. H. Na, Y. S. Choi, and K. H. Choi, "Thermally modified amorphous polyethylene oxide thin films as highly sensitive linear humidity sensors," *Sens. Actuators A*, vol. 265, pp. 102–110, Oct. 2017.

3. M. Sajid, J. Z. Gul, S. W. Kim, H. B. Kim, K. H. Na, and K. H. Choi, "Development of 3D-printed embedded temperature sensor for both terrestrial and aquatic environmental monitoring robots," *3D Print. Addit. Manuf.*, vol. 5, no. 2, pp. 160–169, Jun. 2018.

4. M. Sajid, H. B. Kim, Y. J. Yang, J. Jo, and K. H. Choi, "Highly sensitive BEHP-co-MEH:PPV + poly(acrylic acid) partial sodium salt based relative humidity sensor," *Sen. Actuators B*, vol. 246, pp. 809–818, Jul. 2017.

5. M. Sajid, S. Aziz, G. B. Kim, S. W. Kim, J. Jo, and K. H. Choi, "Bio-compatible organic humidity sensor transferred to arbitrary surfaces fabricated using single-cell-thick onion membrane as both the substrate and sensing layer," *Sci. Rep.*, vol. 6, no. 1, p. 30065, Sep. 2016.

6. P. Boissy et al., "Carbon nanotubes (CNTs) based strain sensors for a wearable monitoring and biofeedback system for pressure ulcer prevention and rehabilitation," in *2011 Annual International Conference of the IEEE Engineering in Medicine and Biology Society*, Boston, MA, 2011, pp. 5824–5827.

7. Y.-M. Choi, E.-S. Lee, T.-M. Lee, and K.-Y. Kim, "Optimization of a reverse-offset printing process and its application to a metal mesh touch screen sensor," *Microelectron. Eng.*, vol. 134, pp. 1–6, Feb. 2015.

8. B. Su, S. Gong, Z. Ma, L. W. Yap, and W. Cheng, "Mimosa-inspired design of a flexible pressure sensor with touch sensitivity," *Small*, vol. 11, no. 16, pp. 1886–1891, Apr. 2015.

9. M. Sajid, H. W. Dang, K. H. Na, and K. H. Choi, "Highly stable flex sensors fabricated through mass production roll-to-roll micro-gravure printing system," *Sens. Actuators A*, vol. 236, pp. 73–81, 2015.

10. Y. Zhou, L. Wang, J. Wang, J. Pei, and Y. Cao, "Highly sensitive, air-stable photodetectors based on single organic sub-micrometer ribbons self-assembled through solution processing," *Adv. Mater.*, vol. 20, no. 19, pp. 3745–3749, 2008.

11. R. Usamentiaga, J. Molleda, and D. Garcia, "Structured-light sensor using two laser stripes for 3D reconstruction without vibrations," *Sensors*, vol. 14, no. 11, pp. 20041–20063, Oct. 2014.

12. G. U. Siddiqui, M. Sajid, J. Ali, S. W. Kim, Y. H. Doh, and K. H. Choi, "Wide range highly sensitive relative humidity sensor based on series combination of MoS_2 and PEDOT:PSS sensors array," *Sen. Actuators B*, vol. 266, pp. 354–363, Aug. 2018.

13. K. H. Choi, M. Sajid, S. Aziz, and B.-S. Yang, "Wide range high speed relative humidity sensor based on PEDOT:PSS–PVA composite on an IDT printed on piezoelectric substrate," *Sens. Actuators A*, vol. 228, pp. 40–49, Jun. 2015.

14. a. V. Mamishev, K. Sundara-Rajan, F. Yang, Y. Du, and M. Zahn, "Interdigital sensors and transducers," *Proc. IEEE*, vol. 92, no. 5, pp. 808–845, May 2004.

15. M. Sajid, M. Zubair, Y. H. Doh, K.-H. Na, and K. H. Choi, "Flexible large area organic light emitting diode fabricated by electrohydrodynamics atomization technique," *J. Mater. Sci. Mater. Electron.*, vol. 26, no. 9, pp. 7192–7199, Sep. 2015.
16. V. Reboud et al., "Enhanced light extraction in ITO-free OLEDs using double-sided printed electrodes," *Nanoscale*, vol. 4, no. 11, pp. 3495–3500, Jun. 2012.
17. K.-T. Lau et al., "A low-cost optical sensing device based on paired emitter–detector light emitting diodes," *Anal. Chim. Acta*, vol. 557, no. 1–2, pp. 111–116, Jan. 2006.
18. L.-W. Ji, Y.-J. Hsiao, S.-J. Young, W.-S. Shih, W. Water, and S.-M. Lin, "High-efficient ultraviolet photodetectors based on $TiO_2/Ag/TiO_2$ multilayer films," *IEEE Sens. J.*, vol. 15, no. 2, pp. 762–765, Feb. 2015.
19. T. Agostinelli, M. Campoy-Quiles, J. C. Blakesley, R. Speller, D. D. C. Bradley, and J. Nelson, "A polymer/fullerene based photodetector with extremely low dark current for X-ray medical imaging applications," *Appl. Phys. Lett.*, vol. 93, no. 20, pp. 2006–2009, 2008.
20. O. Kanoun et al., "Flexible carbon nanotube films for high performance strain sensors," *Sensors (Basel)*, vol. 14, no. 6, pp. 10042–10071, 2014.
21. V. Giurgiutiu, "Multi-mode damage detection methods with piezoelectric wafer active sensors," *J. Intell. Mater. Syst. Struct.*, vol. 20, no. 11, pp. 1329–1341, 2008.
22. S. C. Mukhopadhyay, *Next Generation Sensors and Systems*, vol. 16. Cham: Springer International Publishing, 2016.
23. H. Wang et al., "Label-free electrochemical immunosensor for prostate-specific antigen based on silver hybridized mesoporous silica nanoparticles," *Anal. Biochem.*, vol. 434, pp. 123–127, 2013.
24. N. Ronkainen and S. Okon, "Nanomaterial-based electrochemical immuno-sensors for clinically significant biomarkers," *Materials (Basel)*, vol. 7, no. 6, pp. 4669–4709, Jun. 2014.
25. M. Sajid, H. B. Kim, J. H. Lim, and K. H. Choi, "Liquid-assisted exfoliation of 2D hBN flakes and their dispersion in PEO to fabricate highly specific and stable linear humidity sensors," *J. Mater. Chem. C*, vol. 6, no. 6, pp. 1421–1432, 2018.
26. M. Sajid et al., "All-printed highly sensitive 2D MoS_2 based multi-reagent immunosensor for smartphone based point-of-care diagnosis," *Sci. Rep.*, vol. 7, no. 1, p. 5802, Dec. 2017.
27. Y. J. Yang, S. Aziz, S. M. Mehdi, M. Sajid, S. Jagadeesan, and K. H. Choi, "Highly sensitive flexible human motion sensor based on $ZnSnO_3/PVDF$ composite," *J. Electron. Mater.*, vol. 46, no. 7, pp. 4172–4179, Jul. 2017.
28. M. Sajid, H. B. Kim, G. U. Siddiqui, K. H. Na, and K.-H. Choi, "Linear bi-layer humidity sensor with tunable response using combinations of molybdenum carbide with polymers," *Sens. Actuators A*, vol. 262, pp. 68–77, Aug. 2017.
29. K. Shehzad et al., "Designing an efficient multimode environmental sensor based on graphene–silicon heterojunction," *Adv. Mater. Technol.*, vol. 2, no. 4, pp. 1–10, 2017.
30. H. Zhang, Y. Hong, B. Ge, T. Liang, and J. Xiong, "A novel readout system for wireless passive pressure sensors," *Photonic Sens.*, vol. 4, no. 1, pp. 70–76, 2014.
31. K. S. Karimov et al., "Effect of humidity on the NiPc based organic photo field effect," *Proc. Rom. Acad. Ser. A - Math. Phys. Tech. Sci. Inf. Sci.*, vol. 17, no. 1, pp. 84–89, 2016.

32. A. D. Smith et al., "Resistive graphene humidity sensors with rapid and direct electrical readout," *Nanoscale*, vol. 7, no. 45, pp. 19099–19109, 2015.

33. E. I. Ionete et al., "Graphene layers used as cryogenic temperature sensor," in *2014 International Conference and Exposition on Electrical and Power Engineering (EPE)*, Iasi, Romania, 2014, no. Epe, pp. 774–777.

34. C. Lee and M. Gong, "Resistive humidity sensor using phosphonium salt-containing polyelectrolytes based on the mutually cross-linkable copolymers," *Macromol. Res.*, vol. 11, no. 5, pp. 322–327, 2003.

35. M. Saari, B. Xia, B. Cox, P. S. Krueger, A. L. Cohen, and E. Richer, "Fabrication and analysis of a composite 3D printed capacitive force sensor," *3D Print. Addit. Manuf.*, vol. 3, no. 3, pp. 137–141, 2016.

36. Y. Cheng, R. Wang, H. Zhai, and J. Sun, "Stretchable electronic skin based on silver nanowire composite fiber electrodes for sensing pressure, proximity, and multidirectional strain," *Nanoscale*, vol. 9, no. 11, pp. 3834–3842, 2017.

37. W. Yao, X. Chen, and J. Zhang, "A capacitive humidity sensor based on gold-PVA core-shell nanocomposites," *Sensors Actuators, B Chem.*, vol. 145, no. 1, pp. 327–333, 2010.

38. A. J. DeRouin, B. D. Pereles, T. M. Sansom, P. Zang, and K. G. Ong, "A wireless inductive-capacitive resonant circuit sensor array for force monitoring," *J. Sens. Technol.*, vol. 3, no. 3, pp. 63–69, 2013.

39. P. Nicolay and M. Lenzhofer, "A wireless and passive low-pressure sensor," *Sensors*, vol. 14, no. 2, pp. 3065–3076, Feb. 2014.

40. S. Azzouzi et al., "Citrate-selective electrochemical μ-sensor for early stage detection of prostate cancer," *Sen. Actuators B*, vol. 228, pp. 335–346, Jun. 2016.

41. Z. Wang, J. Xu, Y. Yao, L. Zhang, and Y. Wen, "Facile preparation of highly water-stable and flexible PEDOT:PSS organic/inorganic composite materials and their application in electrochemical sensors," *Sens. Actuators B*, vol. 196, pp. 357–369, 2014.

42. H. Xu, G. Li, J. Wu, Y. Wang, and J. Liu, "A glucose oxidase sensor based on screen-printed carbon electrodes modified by polypyrrole," in *2005 IEEE Engineering in Medicine and Biology 27th Annual Conference*, Shanghai, China, 2005, vol. 2, pp. 1917–1920.

43. C. Y. Chao and L. J. Guo, "Biochemical sensors based on polymer microrings with sharp asymmetrical resonance," *Appl. Phys. Lett.*, vol. 83, no. 8, pp. 1527–1529, 2003.

44. J. Courbat, M. Linder, M. Dottori, D. Briand, J. Wollenstein, and N. F. de Rooij, "Inkjet printed colorimetric ammonia sensor on plastic foil for low-cost and low-power devices," in *2010 IEEE 23rd International Conference on Micro Electro Mechanical Systems (MEMS)*, Wanchai, Hong Kong, 2010, pp. 883–886.

45. P. Miluski, D. Dorosz, M. Kochanowicz, and J. Żmojda, "Optical fibre temperature sensor based on fluorescein and rhodamine codoped polymer layer," in *Photonics Applications in Astronomy, Communications, Industry, and High-Energy Physics Experiments 2013*, Wilga, Poland, 2013, vol. 8903, no. 2, p. 89030C.

46. F. Li, H. Murayama, K. Kageyama, and T. Shirai, "Guided wave and damage detection in composite laminates using different fiber optic sensors," *Sensors*, vol. 9, no. 5, pp. 4005–4021, 2009.

47. B. V. Misra et al., "Flexible technologies for self-powered wearable health and environmental sensing," *Proc. IEEE*, vol. 103, no. 4, pp. 665–681, 2015.

48. Y. Yang, Y. Zhou, J. M. Wu, and Z. L. Wang, "Single micro/nanowire pyroelectric nanogenerators as self-powered temperature sensors," *ACS Nano*, vol. 6, no. 9, pp. 8456–8461, Sep. 2012.
49. S.-H. Bae et al., "Graphene-P(VDF-TrFE) multilayer film for flexible applications," *ACS Nano*, vol. 7, no. 4, pp. 3130–3138, Apr. 2013.
50. J. Nunes-Pereira et al., "Energy harvesting performance of $BaTiO_3$/poly (vinylidene fluoride-trifluoroethylene) spin coated nanocomposites," *Compos. Part B Eng.*, vol. 72, pp. 130–136, Apr. 2015.
51. J. Z. Gul, K. Y. Su, and K. H. Choi, "Fully 3D printed multi-material soft bio-inspired whisker sensor for underwater-induced vortex detection," *Soft Robot*, vol. 5, no. 2, pp. 122–132, Apr. 2018.
52. J. Z. Gul et al., "3D printing for soft robotics – a review," *Sci. Technol. Adv. Mater.*, vol. 19, no. 1, pp. 243–262, Dec. 2018.
53. J. Z. Gul, B.-S. Yang, Y. J. Yang, D. E. Chang, and K. H. Choi, "In situ UV curable 3D printing of multi-material tri-legged soft bot with spider mimicked multi-step forward dynamic gait," *Smart Mater. Struct.*, vol. 25, no. 11, p. 115009, Nov. 2016.
54. J. Z. Gul, Y. J. Yang, K. Y. Su, and K. H. Choi, "Omni directional multimaterial soft cylindrical actuator and its application as a steerable catheter," *Soft Robot*, vol. 4, no. 3, pp. 224–240, Sep. 2017.

Part 2

Sensor Development

One of the major aims of the machine tool research group is the indigenous development of sensors, transducers, and actuators. Chapters 1–3 define state-of-the-art classification of this area. This part reports work carried by researchers at College of Material Science and Technology, Nanjing University of Aeronautics and Astronautics, China, pertaining to humidity sensor development for machine tool application. Work done by machine tool research group in the area of development of strain gauges, flexi sensors, and temperature and humidity sensors is reported elsewhere. The development of additive manufacturing systems used for the production of these sensors shall be reported in the next volume of the series.

3

Reverse Engineering of Titanium to Form TiO₂ for Humidity Sensor Applications

Azhar Ali Haidry and Linchao Sun

Nanjing University of Aeronautics and Astronautics

Ministry of Industry and Information Technology

CONTENTS

3.1 Introduction 55
3.2 Sensor System 56
3.3 Humidity Sensors Fundamentals 58
    3.3.1 Dynamic Response 59
    3.3.2 Sensitivity (S) 59
    3.3.3 Response Time (τ_{Res}) 59
    3.3.4 Recovery Time (τ_{Res}) 59
    3.3.5 Selectivity 59
    3.3.6 Stability or Durability 59
    3.3.7 Hysteresis 60
    3.3.8 Cost and Power Consumption 60
3.4 Experimental 60
3.5 Results and Discussion 60
    3.5.1 Sensing Mechanism 65
3.6 Summary and Conclusion 69
3.7 Acknowledgements 70
References 70

3.1 Introduction

In this modern world, the innovations in the machine tools seem infinite, since the technology has evolved steadily over the last few decades in this field. In order to keep pace with novel developments of the machine tools, professionals must ameliorate the subject knowledge. Meanwhile, several new topics in machine tools have been introduced for improved and safe performance. Among them, the sensor technology has recently been intensively applied in conventional and modern machine technology not only for precise and enhanced efficiency but also for the safety purposes [1–8].

Besides efficiency and safety, the sensor technology has also succoured in increasing business and production of many enterprises around the globe. In fact, the sensor technology is inspired by the natural sensory system of the living organisms. Human body, for instance, is an excellent illustration of complex bio-mechanical system that is sensitive to several parameters including temperature, humidity, motion, electric field, magnetic field, gravity, and many others [9–15].

3.2 Sensor System

In machine tools, two terms are commonly used to describe a sensor device: "sensor" and "transducer". But then the term "actuator" is sometimes confused with "sensor". Actuator can be broadly imagined as a component of a machine that controls a mechanical system. For better understanding, a situation can be considered in which a sensor device is employed to monitor the linear motion of an object; based on information received, actuator responds and converts linear motion to rotary motion [16]. These three terms are interlinked to each other.

The word "sensor" stems from Latin word "Sensus", which means to feel or perceive, while transducer is capable of converting and transforming received information into useful signal. In principle, a sensor is used to measure (both quantitative and qualitative) the physical, chemical, environmental, and medical parameters. Primarily, a sensor consists of two main functions: a receptor function and a transducer function. A sensor responds to information containing stimuli, in which the receptor recognizes it with the help of natural science laws, and then transducer converts it into an output signal. In signal processing, the output signal is refined to remove noise and disturbance. The typical schematic illustrating the working of a sensor is shown in Figure 3.1.

With the ever-increasing machine tool tasks, the application of intelligent and smart sensing devices has become a rapidly expanding research field. The progressive miniaturization technology has made it highly possible to combine the sensing element, transducing element, and signal processing

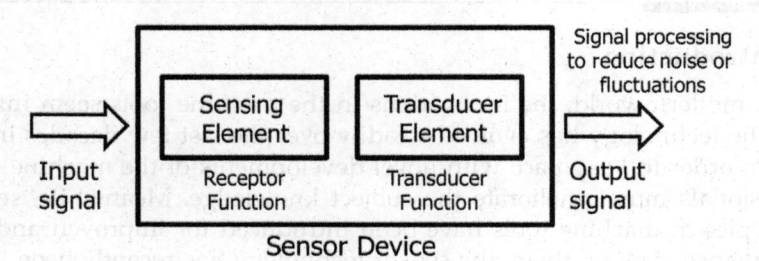

FIGURE 3.1
The illustration of the working principle of a sensor.

electronics to give rise to intelligent and smart sensor devices [17–19]. The objective of these smart sensor devices is not only to sense the stimuli but also make appropriate adjustment in their as well as machine functionality. To date, the precise definition of smart sensors is indefinite; however, a conventional sensor becomes intelligent if it is able to filter the output signal aided with the computer-controlled data processing to imitate multiple functions simultaneously adoptable with the given circumstance. The appropriate capabilities of these smart sensors include highly intelligent sensing and self-adopting mechanisms enabling error-free independent automation. Potential implementation of these smart sensors is extremely promising in computer numerical control (CNC) machines; various types of sensors used in machine tools are shown in Figure 3.2.

Among aforementioned kinds of sensors, humidity sensors play a significant role in manufacturing and machine tools [20]. For instance, the variation in temperature and humidity due to unstable seasons in many regions causes severe damage to the electronic components during the operation of CNC machines. If the relative humidity (% RH) level is high, the moisture in the air can accumulate and subsequently condense on electrical wiring leading to corrosion, but then the low humidity (below 30% RH) results in electrostatic discharge (ESD). In industrial manufacturing, ESD may result in oxidation, metal melting due to heat, and interface burning; the device in this case may be permanently damaged. Thus, to prevent any unwanted accident, the humidity level must be kept in the range 40%–60% RH. Such detriments force malfunctioning of several CNC machine components under

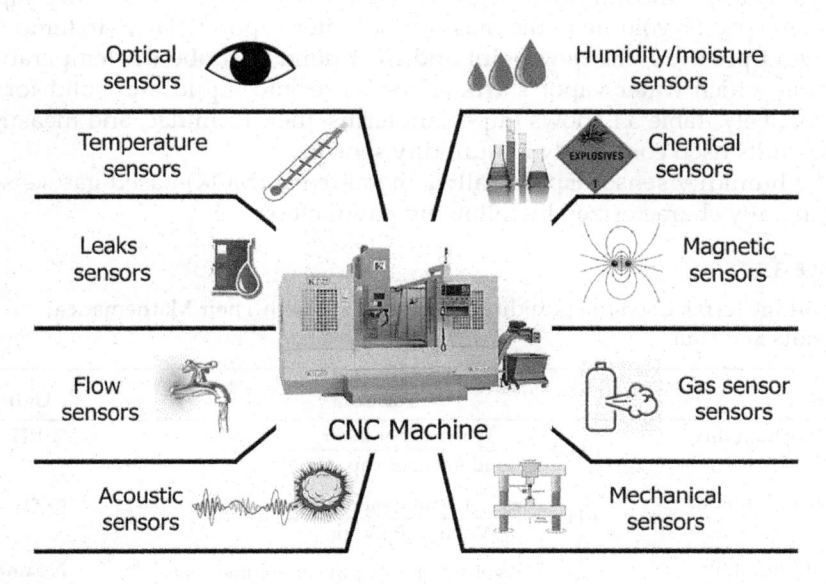

FIGURE 3.2
Various kinds of sensors used CNC machines.

these environments [21–25]. Many death casualties of workers have also been reported due to ESD. In order to avoid these casualties, a multifunctional smart sensor system embedded in wristwatch has been proposed to be worn by workers; these smart watches are equipped with typical humidity, pressure, and temperature sensors to monitor the working environment and for early warning of any forthcoming fatality [26–28].

3.3 Humidity Sensors Fundamentals

For better understanding and clarity, first the basic terms used in humidity and moisture sensing are defined as many of them are interwoven in this field. Basically, water is a molecule consisting of oxygen atom bonded with diatomic hydrogen atoms and is present in many different states on our planet. The amount of water in gaseous state present in air (or reference gases like oxygen, nitrogen, etc.) in the form of vapors is basically called humidity. On the other hand, the amount of water vapor diffused or accumulated on the surface of an object is called moisture. Humidity can be represented in terms of RH which is the ratio of the water vapor in air and saturation water content in air and is measured in % (i.e., % RH). Absolute humidity is defined as the mass of water vapor within specific volume of the air or gas. Humidity by weight or volume percent can be primarily expressed as parts per million (ppm). The specific humidity is given by the fraction of current water vapor mass in specific volume to the mass of the water vapor. Relative to temperature and pressure, the dew point and frost point describe the temperature level at which water vapor starts to condense into liquid and solid forms respectively. Table 3.1 shows important terms, their formulae, and measurement units used commonly in humidity sensors.

The humidity sensor (specifically a metal-oxide (MOX)-based gas sensor) is generally characterized by following parameters.

TABLE 3.1

Important Terms Used in Humidity Sensing Along with Their Mathematical Formula and Unit

Term	Formula	Unit
Relative humidity	$RH = \dfrac{\text{Water vapor in air}}{\text{Saturation water vapor in air}} = \dfrac{p_w}{p_{sw}}$	% RH
Absolute humidity	$AH = \dfrac{\text{Mass of water vapor in air}}{\text{Volume of the air}} = \dfrac{m_w}{V}$	g/m^3
Specific humidity	$SH = \dfrac{\text{Mass of water vapor in air volume}}{\text{Total mass of the air}} = \dfrac{m_w}{m_a}$	No unit

3.3.1 Dynamic Response

It is a typical graph that shows the real-time variation in the resistance (or conductance) of the MOX layer when exposed to humidity. The dynamic response shows us how sensitive the material is, what the response and recovery times are, and other characteristics of a sensor.

3.3.2 Sensitivity (*S*)

In our work, we define sensitivity as

$$s = \frac{\sigma_{RH} - \sigma_B}{\sigma_B} = \frac{R_B - R_{RH}}{R_{RH}}$$

Here σ_B and R_B are the baseline electrical conductivity and resistance of the sensing material in air while σ_{RH} and R_{RH} and are the saturated electrical conductivity and resistance of sensing material in specific humidity respectively.

3.3.3 Response Time (τ_{Res})

The time interval during which a sensor attains 90% change of its saturated value of electrical resistance signal after the sensor is exposed to humidity is response time.

3.3.4 Recovery Time (τ_{Res})

The time interval during which a sensor attains 90% recovery in its baseline resistance after turning off the gas is denoted by recovery time.

3.3.5 Selectivity

The selectivity, considered as a major problematic issue in MOX gas sensors, is defined as the ability of a sensor to respond for a particular gas in the presence of other gases.

3.3.6 Stability or Durability

It is very important for a sensor to have reproducible response; this is related with the stability of a sensor. Short-term stability is referred to the performance of a sensor under certain constant conditions for a short time period. It is very difficult to maintain the same experimental conditions for every measurement, so long-term stability is a time duration for a sensor to show same response for a significant number of measurements, which is generally estimated in years.

3.3.7 Hysteresis

For stability, the humidity hysteresis is also an important characteristic which is defined as the maximum difference between the adsorption (the process of incorporation of water molecules on sensor surface by means of physisorption or chemisorption) and desorption (the process referring to the release of water from sensor surface) curves.

3.3.8 Cost and Power Consumption

A sensor with low cost and power consumption is always preferred for sensing applications, and MOX gas sensors are considered to be cost-effective and consume less power.

3.4 Experimental

The fabrication method is by controlled thermal oxidation process of pure Ti discs to obtain rutile titanium dioxide (TiO_2). The raw material utilized for the preparation of rutile TiO_2 submicron particles was high-purity Ti-discs (99.9% purity), and the dimensions of the discs are 15 mm diameters and 1 mm thick. Firstly, ultrasonic bath in acetone and deionized water for 15 min was carried out to remove oil.

The abrasive metallographic paper was used to polish the cleaned surfaces until the surface was smooth. Subsequently, the discs were cleaned ultrasonically several times in alcohol and deionized water, respectively, in order to remove grease and dust. A porcelain combustion boat containing the cleaned Ti-discs was placed in tube furnace (Hefei Kejing, OTF-1200X). The oxidation process was carried out in technical air atmosphere ($O_2 = 20\%$ and $N_2 = 80\%$) with 200 sccm air flow rate. The temperature of oxidation process was maintained at 1,000°C for 1 hour to grow crystalline submicron particles at the rate of 10°C/min. The rutile TiO_2 submicron particles were obtained after cooling to room temperature. Finally, the TiO_2 submicron particles were separated from the the discs via ultrasonic vibration and dispersed in ethanol solution. The schematic of oxidation process is shown in Figure 3.3.

3.5 Results and Discussion

TiO_2 occurs naturally in three phases: anatase, rutile, and brookite. Rutile has the highest density and is also the more thermally stable form while anatase and brookite are thermally metastable phases [29–32]. Anatase (blue color)

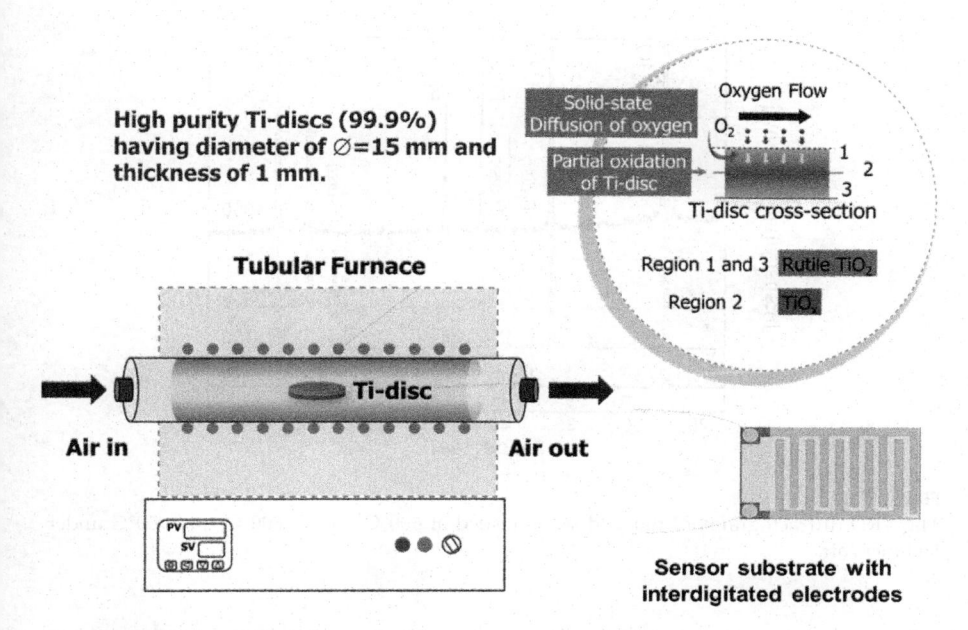

FIGURE 3.3
Thermal oxidation process of titanium to form highly stoichiometric TiO_2 for humdity sensor.

and rutile (reddish color) share many same chemical properties; however, they differ in crystal habitat and cleavage. Anatase has slightly higher energy gap (3.23 eV) than rutile (3.02 eV); the colors mentioned above for anatase and rutile originate due to impurities present naturally in parent ores. Depending on deposition parameters, the anatase and brookite transform to rutile when annealed at higher temperatures [33–35]. Generally, in temperature range of 500°C–800°C, anatase transforms completely to rutile phase [29–35]. It is noticeable from the X-ray diffraction (XRD) spectra presented in Figure 3.4 that no anatase phase has been observed. This indicates that the nanostructures grown by this method transform from metal Ti phase to rutile phase directly; the results are in agreement with previous reports [36–37]. The peaks match perfectly with the powder diffraction peaks of Joint Committee on Powder Diffraction Standards (JCPDS) card number 21-1276.

From Figure 3.5a, it is evident that there was formation of nanostructured TiO_2 even at 600°C, but the resistance of these discs was very low ($R_B \sim 248\,\Omega$) as the current passes through Ti-disc bulk which contains mainly the metallic titanium. These discs did not show any significant resistance variation to any humidity range due to low resistance.

The growth of nanostructured TiO_2 started at annealing temperature 1,000°C due to compressive stress created on the surface of Ti-discs as can be noticed from Figure 3.5b. The SEM micrographs of TiO_2 aggregates having submicron particles exhibit the grain size and surface morphology at 10.00 Kx magnifications. The average size of TiO_2 aggregates is nearly 200–600 nm.

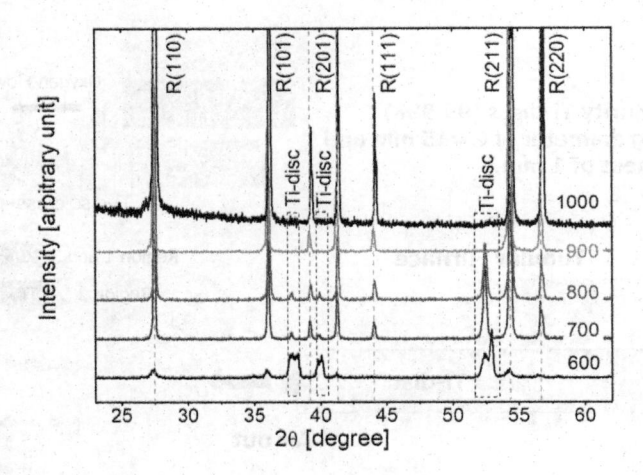

FIGURE 3.4
The XRD diffractograms of the Ti-discs annealed at 600°C, 700°C, 800°C, and 900°C under technical air.

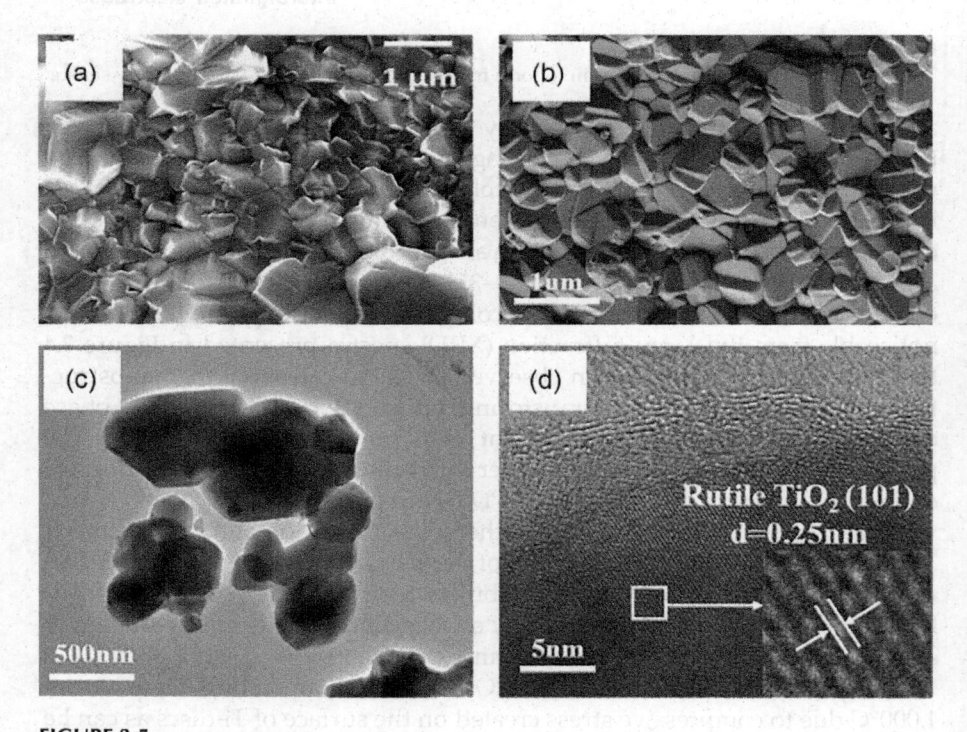

FIGURE 3.5
SEM micrographs of of Ti-discs annealed at 600°C (a), 1,000°C (b), low-magnification TEM image of Ti-discs annealed at 1,000°C (c), and high-magnification TEM image of Ti-discs annealed at 1,000°C (d).

The low-magnification Image result for transmission electron microscopy (TEM) images of TiO_2 are shown in Figure 3.5c, which are consistent with scanning electron microscopy (SEM) micrographs. The high-resolution transmission electron microscopy (HRTEM) image (Figure 3.5d) exhibits the lattice spacing $d = 0.25$ nm of TiO_2, which corresponds to the (101) planes of rutile phase.

The humidity sensing properties of the prepared samples were tested at room temperature, and the dynamic changes in the electrical resistance toward various humidities were measured by Keithley Dual-Channel Picoammeter/Voltage Source model 6482. The measurement of the electrical resistance involves applying external voltage on sensor and measuring the current. The resistance is caculated by Ohm's Law ($R = V/I$). As we all know, the stability of humidity sensors is determined by whether the sensing material surface gets contaminated and poisoned over several humidity cycles, caused by absorbed water and dust. Hence, well-designed measurement cycles were carried out to ascertain the stability of sensors in this work. The RH level is adjusted by the flow rate of the dry air and wet argon (99.99%) flowing through water. The total gas-flow rate is maintained at 400 sccm by regulating the mass flow controllers.

The resulting stable RH levels are determined by a commercial humidity sensor prior to each measurement. The detailed explanation of the experimental methods is reported elsewhere [38–41]. The dynamic resistance change curve toward different RHs ranging from 9% to 90% is shown in Figure 3.6a. The response of TiO_2-aggregate-based humdity sensor is extremely high ($S_{TiO_2} \sim 2.5 \times 10^6$) toward 90% RH. The dynamic response curve toward 75% RH is exhibited in Figure 3.6b. The initial resistance (R_B) remains stable in dry air; when the humid air enters, the value of resistance

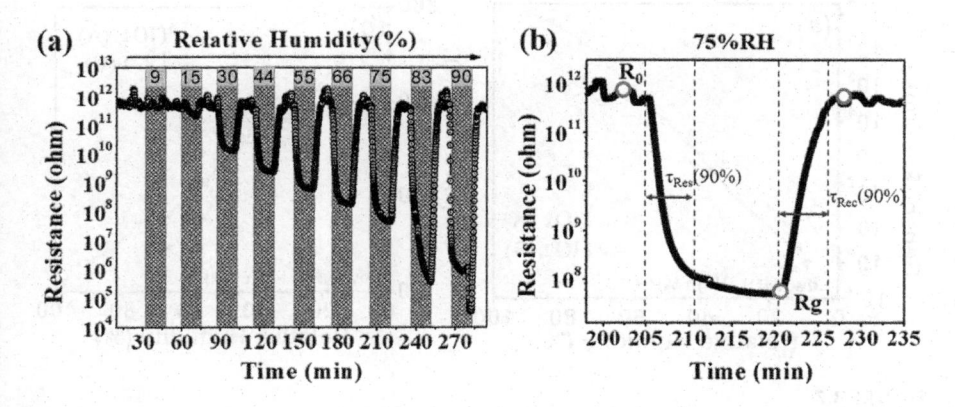

FIGURE 3.6
The dynamic responses of the of Ti-discs annealed at 1,000°C to large range of humidity from 9% to 90% RH (a) and zoomed view of one of the dynamic responses to 75% RH showing the response and recovery times.

drops sharply (R_{RH}); subsequently, the resistance increases to initial value when the dry air enters again. The sensor response is defined as $S = R_B/R_{RH}$. The response time (τ_{Res}) and recovery time (τ_{Rec}) are defined as 90% change in the baseline stable resistance.

Figure 3.6a shows the the dynamic responses of the of sensors based on TiO_2 aggregates toward large range of humidity from 9% to 90% RH, and it can be observed that humidity sensor based on TiO_2 aggregates exhibits extremely high response to high humidity level and a little response to low humidity level (below 30% RH). Ultimately, in order to check the long-term stability of the sensors, identical measurements were performed two months later as shown in Figure 3.7.

As it can be seen, the response curves maintain similar readings within a reasonable error range after two months, which indicates that the sensors possess good stability. Furthermore, the sensor response versus RH is an approximately linear correlation, indicating excellent reliability for detecting various RHs. Meanwhile, the response time is shown in Figure 3.7b, which decreases with the increase of RH level.

The response and recovery times in this case are longer than the other reports because of 3 m long gas pipeline prior to contacting with the tested sensors. It is worthy of mentioning that the response time of the sensors toward 75% RH is much shorter (as low as the 70 s) than that of commercial humidity sensor (around 250 s) when employing identical gas sensing system. In addition, the above results prove that the sensors based on TiO_2 aggregates show superior humidity sensing performance and directly demonstrate good reliability and long-term stability.

The hysteresis characteristics of humidity sensors based on typical step-wise humidity cycle measurement are in Figure 3.8. It can be observed that

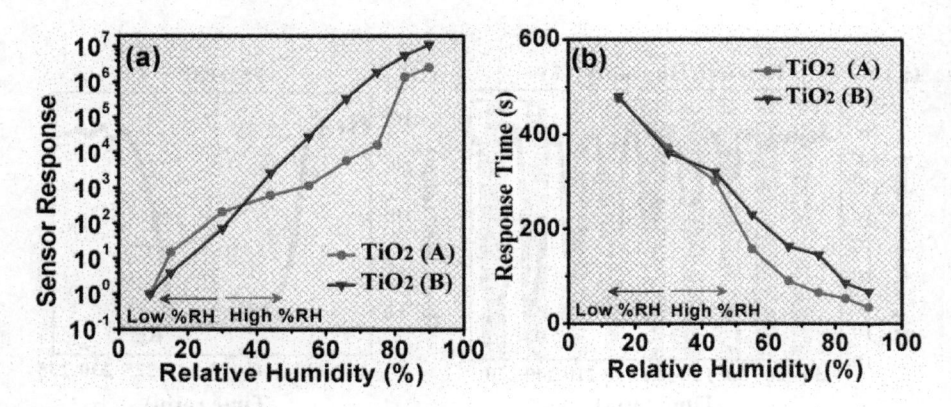

FIGURE 3.7
Sensor response (a) and response time (b) versus humidity level ranging from 9% to 90% RH. Here the x-axis scale (RH%) is same for both (a) and (b). Capital letters in parentheses (A) & (B) of the graph legends refer to initial humidity measurement and humidity measurement two months later, respectively.

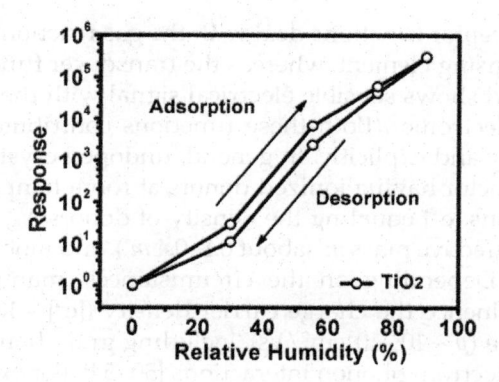

FIGURE 3.8
The hysteresis characteristics of humidity sensors based on TiO$_2$ aggregates.

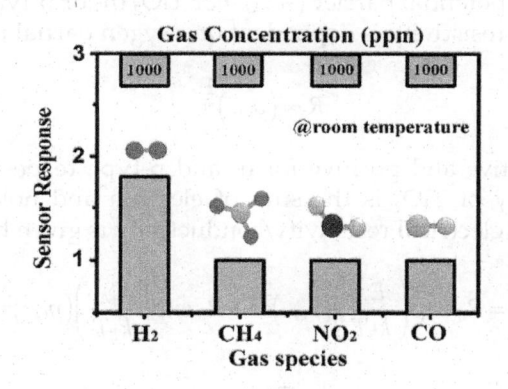

FIGURE 3.9
The selectivity of the gas sensor employing TiO$_2$.

the electrical resistance of sensors decreases with every increase of RH and vice versa. Furthermore, the deviation threshold for TiO$_2$-based sensors is small (~0.644) toward different humidity levels, which reveals the good reproducibility of TiO$_2$-based sensors.

Figure 3.9 shows the sensitivity of TiO$_2$-based sensors to 1,000 ppm H$_2$, CH$_4$, NO$_2$, CO. The results confirm that there is nearly no response to 1,000 ppm of reducing (such as H$_2$, CO, and CH4) as well as to strong oxidizing (NO$_2$) gases at room temperature, which meets practical application requirements of the sensors.

3.5.1 Sensing Mechanism

Metal oxide semiconductor materials have been widely investigated for humidity sensors, but the sensing mechanism remains debatable [42–49]. Basically, a gas sensor response is controlled by receptor and transducer functions. The receptor functions deal with the gas reaction and subsequent changes in the sensing element, whereas the transducer function recognizes these changes and shows sensible electrical signal with the help of metallic electrodes and electronics. Both these functions contribute to the sensing properties equally and explicitly. In general, undoped crystalline TiO_2 is an n-type semiconductor having ionized donors at room temperature with the density of electrons [e-] equaling the density of donors N_D ([e-] = N_D = 10^{17}–10^{20}cm^{-3}), small effective mass m^* (about 0.1–0.4 m_e), and mobility in the range $\mu = 1$–50cm^2/Vs. Depending on the circumstances, many factors can be considered to influence the charge carrier density ([e-] ~ 10^{16}–10^{18}cm^{-3}) and the mobility value (μ ~ 10^2–10^3cm^2/Vs), including grain boundaries, ionized impurities, and electron–phonon interactions [50–54]. For example, the grain boundaries in polycrystalline TiO_2 contain very high densities of surface states that result in depleted grains and capture (or scatter) free charge carriers and generate potential barrier (~eV_s). For TiO_2 (n- or p-type) semiconductor, the electrical resistivity is dependent on oxygen partial pressure,

$$R \propto \left(p_{O_2} \right)^{\frac{1}{m}},$$

where m is negative and positive for n- and p-type respectively. Since the total conductivity of TiO_2 is the sum of electron and hole conductivities ($\sigma = \sigma_e + \sigma_h$), the electrical resistivity/conductivity is given by

$$R = R_0 \exp\left(\frac{E_A}{k_B T} \right) \left(p_{O_2} \right)^{\frac{-1}{m}} + R_0 \exp\left(\frac{E_A}{k_B T} \right) \left(p_{O_2} \right)^{\frac{1}{m}},$$

$$\sigma = \sigma_n + \sigma_p + \sum_{pi} \sigma_{\text{ion}} \cong q\mu_n n + q\mu_p p,$$

$$\text{where } n = N_C \exp\left(\frac{-(E_C - E_F)}{kT} \right) \text{ and } p = N_V \exp\left(\frac{-(E_F - E_V)}{kT} \right),$$

$$\text{Fermi Energy} = E_F = \frac{E_C - E_F}{2} + \frac{kT}{2} \ln\left(\frac{N_V}{N_C} \right),$$

where μ_n is the mobility of electrons and μ_p is the mobility of holes, n is the electron density in the conduction band and p is the hole density in valence band, N_C is the effective density of state of the conduction band and N_V is the effective density of state the valence band, E_C is the energy of the bottom of the conduction band and E_V is the energy of the top of valence band, and q is the electric charge and k is the Boltzmann constant [55–61].

Since the existence of ideal structure is out of question, TiO_2 crystals may contain various structural imperfections and/or defects. These defects include point defects, line defects, and plane defects. The most important one is the point defect, which means empty sites (vacancies). These defects are mostly oxygen vacancies and probably caused by heating at higher temperatures. Vacancies are the places where constituent atoms are missing and/or these places are occupied by interstitial atoms. Line defects or dislocations are realized by the displacements in the periodic structure in certain directions. Plane defects comprise stacking faults, internal surfaces (grain boundary), and external surfaces [60,61].

The humidity sensing mechanism has previously been proposed on the basis of three parameters, i.e., (i) the interface between sensing layer and metallic electrodes, (ii) electron transport via grain boundaries, and (iii) depletion layer (or inversion) layer (as illustrated in Figure 3.10).

If all the mentioned factors are included, then the total resistance of the sensor is the sum of the contributions from the Pt/TiO_2 interface condition, electronic conduction, and ionic conduction; see the following equation:

$$R_{total} = 2R_{Pt/TiO_2} + R_0 \exp\left(\frac{E_A}{k_B T}\right)(p_{O_2})^{\frac{-1}{m}} + R_0 \exp\left(\frac{E_A}{k_B T}\right)(p_{O_2})^{\frac{1}{m}}$$

Here k_B is the Boltzmann constant, and T is the absolute temperature.

In addition to temperature, the contributions from R_{Pt/TiO_2} depend on the contact area and on Schottky barrier height. Meanwhile, the electronic contribution depends on electron mobility μ_e, number of free electrons [e$^-$], and intergranular surface potential $eV_s \sim E_A$. The ionic conduction part of the resistance plays an additional dominating role in high resistance change, which depends on a number of variations, such as the applied electric field E, ionic mobility μ_v, and number of oxygen vacancies [$V_{\ddot{O}}$] and ions [O_i].

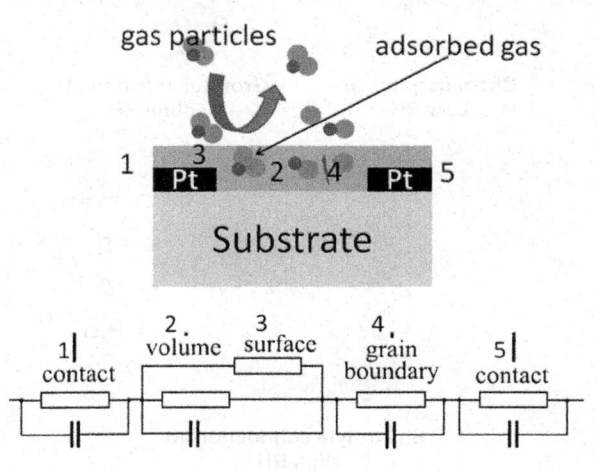

FIGURE 3.10
The schematics of gas reaction and total conductivity (or resistivity) contributions (equivalent circuit model) from various elements of the gas sensor device.

As it is known that the humidity sensitivity is related to the reaction of H_2O molecules at surface and interface, while the reaction process and conduction mechanism remain to be discussed further. At present, two recognized conduction mechanisms are proposed to explain the humidity sensing mechanism. One is electronic, and the other is ionic type. The electronic-type sensing mechanism depends on the chemisorption of water molecules acting as electron donors, which is determined by the conductivity type, the specific surface area, and the porosity of semiconductor materials. For the ionic mechanism, the free electrons on surface are captured by adsorbed oxygen, forming oxygen ions such as O_2^-, O^-, O^{2-}. When exposed to humid air, water molecules react with the previously ionized oxygen ions and then release the electrons e^- as carriers. The plausible sensing mechanism is discussed to explain the performance of the sensors, shown in Figure 3.11. It is noticeable that the TiO_2 aggregates show typical n-type semiconducting behavior; thus the charge carrier is dominated by electrons $[e^-]$. Initially, the free electrons are captured by adsorbed oxygen on the composites surface and ionized to form O_2^- (equation 3.1) at room temperature. Subsequently, in low humidity atmosphere, the water molecules react with the previously formed O_2^- and Ti^{3+} defect sites in nature (equations 3.2–3.4), and then form a chemisorbed layer. In addition, it is believed that the oxidized TiO_2 aggregates not only offer more active sites such as vacancies, defects, and oxygen

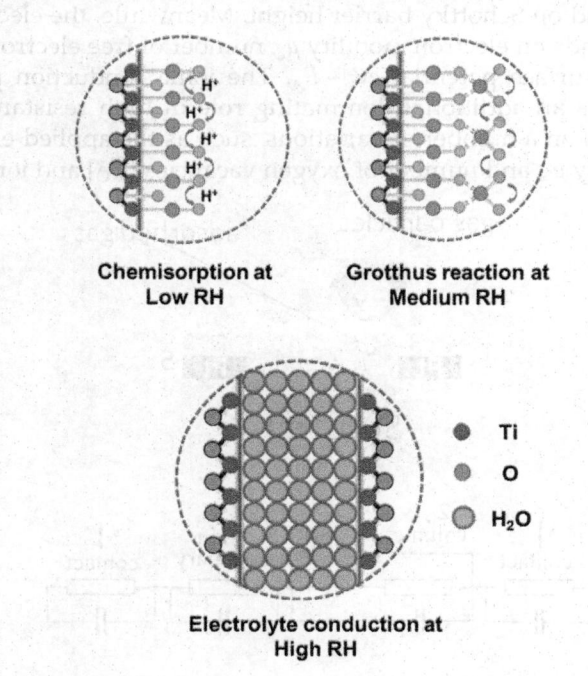

**Chemisorption at
Low RH**   **Grotthus reaction at
Medium RH**

● Ti
● O
● H_2O

**Electrolyte conduction at
High RH**

FIGURE 3.11
Water vapor adsorption on TiO_2 surface in low, medium, and high humidity ranges.

functional groups but also increase the high surface area and electrical conductivity [62–67]:

$$O_2 \text{ (gas)} + e^- \leftrightarrow O_2^- \text{ (surf)} \tag{3.1}$$

$$2H_2O \text{ (gas)} + O_2^- \text{ (ads)} \rightarrow 2H_2O_2 \text{ (gas)} + e^- \tag{3.2}$$

$$H_2O \text{ (gas)} + O_O + 2Ti \leftrightarrow 2\left(Ti^{\delta+} - \delta OH^-\right) + V_{\ddot{O}} + e^- \tag{3.3}$$

$$3H_2O \text{ (gas)} + 3V_{\ddot{O}} + 2Ti^{3+} \leftrightarrow 2\left(Ti^{3+} - 3OH^-\right) \tag{3.4}$$

With the increase of humidity level, owing to high electrostatic field of Ti^{4+} and OH^- species on the surface, H^+ ions can be ionized further. As the main charge carriers, the generated H^+ ions (equation 3.5) contribute to electrical conduction. The deviation between the adsorption and desorption curves (see Figure 3.8) also demonstrates the transition of the two sensing mechanisms.

$$4H_2O + Ti^{4+} \leftrightarrow \left(Ti^{4+} - 4OH^-\right) + 4H^+ \tag{3.5}$$

When exposed to higher RH level, water molecules will be physically adsorbed on the chemisorbed layer. This moment, the water molecules decompose into OH^- and H_3O^+ according to Grotthuss chain (equation 3.6 and 3.7) which will result in the increase of the electrical conductivity:

$$2H_2O \leftrightarrow H_3O^+ + OH^- \tag{3.6}$$

$$H_3O^+ \leftrightarrow H_2O + H^+ \tag{3.7}$$

The absorbed water vapor forms a network owing to hydrogen bonding between oxygen atoms and hydrogen atoms. Numerous protons will be transported along a series of hydrogen bonds in the form of proton hopping according to the Grotthuss chain [68–80]. The performance of our sensors is compared with previously reported literature, see Table 3.2.

3.6 Summary and Conclusion

In this work, microparticles based on TiO_2 were fabricated by facile thermal oxidation and thereafter used as humidity sensors. It is observed that the sensors have great potential to detect high range of humidity (30%–90% RH) even at room temperature with applied voltage 2 V. The sensor response is more than five orders of magnitude ($>10^5$) with excellent stability and selectivity against other interfering gases. The response and recovery times of

TABLE 3.2

The Comparison of Sensor Performance in this Work with Other Fabricated
Humidity Sensors Mentioned in Previous Studies

Materials Type	Humidity Range	Sensor Response (Impedance or Resistance Changes)
TiO_2–WO_3 composite [81]	12%–90% RH	Two orders of magnitude
TiO_2 nanotubes [82]	11%–95% RH	Two orders of magnitude
Porous TiO_2 [83]	11%–95% RH	Four orders of magnitude
TiO_2 and polystyrene sulfonic sodium composite film [44]	11%–95% RH	Four orders of magnitude
Ordered mesoporous Co-doped TiO_2 [40]	9%–90% RH	Five orders of magnitude
TiO_2 aggregates [this work]	9%–90% RH	$\geq 2.5 \times 10^6$

the sensor are below 20 s, which shows its potential to be used as commercial
humidity sensor. The hysteresis from adsorption and desorption kinetics
is very low ~3% RH. The SEM/TEM, XRD, and electrical characterization
enabled us to propose the relevant sensing mechanism based on results.

There are no conflicts to declare.

3.7 Acknowledgements

This work was supported by National Natural Science Foundation of China
(51850410506), Natural Science Foundation of Jiangsu Province (BK20170795),
Priority Academic Program Development of Jiangsu Higher Education
Institutions (PAPD) and Opening Project (56XCA17006-3) from Key
Laboratory of Materials Preparation and Protection for Harsh Environment
(Nanjing University of Aeronautics and Astronautics), and Ministry of
Industry and Information Technology. Dr. Azhar Ali Haidry thanks Ministry
of Education of Pakistan and Slovakia for providing PhD scholarship and
NUAA for providing start-up research funding.

References

1. A.C. Romain, J. Nicolas, Long term stability of metal oxide-based gas sensors
   for e-nose environmental applications: An overview, *Sens. Actuators B*, 146
   (2010), 502–506.
2. J.-M. Rivière, D. Luttenbacher, M. Robert, J.-P. Jouannet, Design of smart
   sensors: towards an integration of design tools, *Sens. Actuators A*, 47:1–3
   (March–April 1995), 509–515. doi:10.1016/0924-4247(94)00952-E.

3. K.J. Yoon, S.I. Lee, H. An, J. Kim, J.W. Son, J.H. Lee, H.J. Je, H.W. Lee, B.K. Kim, Gas transport in hydrogen electrode of solid oxide regenerative fuel cells for power generation and hydrogen production, *Int. J. Hydrogen Energy*, 39 (2014), 3868–3878.

4. Z. Zhang, L. Zhu, Z. Wen, Z. Ye, Controllable synthesis of Co_3O_4 crossed nanosheet arrays toward an acetone gas sensor, *Sens. Actuators B*, 238 (2017), 1052–1059.

5. C.Y. Lee, G.B. Lee, Humidity sensors: A review, *Sensor Lett.*, 3 (2005), 1–14.

6. X. Liu, J. Hu, B. Cheng, H. Qin, M. Jiang, Acetone gas sensing properties of SmFe1–xMgxO3 perovskite oxides. *Sens. Actuators B*, 134 (2008), 483–487.

7. Y. Yamamoto, S. Harada, D. Yamamoto, W. Honda, T. Arie, S. Akita, K. Takei, Printed multifunctional flexible device with an integrated motion sensor for health care monitoring, *Sci. Adv.*, 2 (2016), e1601473.

8. H.J. Kim, J.H. Lee, Highly sensitive and selective gas sensors using p-type oxide semiconductors: Overview. *Sens. Actuators B*, 192 (2014), 607–627.

9. Y. Kim, P. Rai, Y.T. Yu, Microwave assisted hydrothermal synthesis of Au@TiO_2 core–shell nanoparticles for high temperature CO sensing applications. *Sens. Actuators B*, 186 (2013), 633–639.

10. M.Z. Jacobson, Review of solutions to global warming, air pollution, and energy security, *Energy Environ. Sci.*, 2 (2009), 148–173.

11. D.V. Bavykin, A.A. Lapkin, P.K. Plucinski, J.M. Friedrich, F.C. Walsh, Reversible storage of molecular hydrogen by sorption into multilayered TiO_2 Nanotubes, *J. Phys. Chem. B*, 109 (2005), 19422–19427.

12. H. Zeng, Y. Zhao, Sensing movement: Microsensors for body motion measurement, *Sensors*, 11 (2011), 638–660. doi:10.3390/s110100638.

13. V. Dusastre, D.E. Williams, Selectivity and composition dependence of response of wolframite-based gas sensitive resistors $(MWO_4)_x([Sn–Ti]O_2)_{1-x}$ (0<x<1; M=Mn, Fe, Co, Ni, Cu, Zn), *J. Mater. Chem.*, 9 (1999), 965–971.

14. J. Bai, B. Zhou, Titanium dioxide nanomaterials for sensor applications, *Chem. Rev.*, 114 (2014), 10131–10176.

15. M.E. Franke, T.J. Koplin, U. Simon, Metal and metal oxide nanoparticles in chemiresistors: Does the nanoscale matter? *Small*, 2 (2006), 36–50.

16. Y. Koyama, M. Nishiyama, K. Watanabe, A motion monitor using hetero-core optical fiber sensors sewed in sportswear to trace trunk motion, *IEEE Trans. Instrum. Meas.*, 62:4, 828–836, doi:10.1109/TIM.2013.2241534.

17. J.A. Duro, J.A. Padget, C.R. Bowen, H. Alicia Kim, A. Nassehi, Multi-sensor data fusion framework for CNC machining monitoring, *Mech. Syst. Signal Proces.*, 66–67 (January 2016), 505–520. doi:10.1016/j.ymssp.2015.04.019.

18. U. Mönks, H. Trsek, L. Dürkop, V. Geneiß, V. Lohweg, Towards distributed intelligent sensor and information fusion, *Mechatronics*, 34 (March 2016), 63–71. doi:10.1016/j.mechatronics.2015.05.005.

19. S.S. Yao, A. Myers, A. Malhotra, F.Y. Lin, A. Bozkurt, J.F. Muth, Y. Zhu, A wearable hydration sensor with conformal nanowire electrodes, *Adv. Health. Mater.*, 6 (2017), 1601159.

20. E. Bracken, Combating humidity - The hidden enemy in manufacturing, *Sensor Rev.*, 17:4 (1997), 291–298. doi:10.1108/02602289710185108.

21. P. Tamminen, T. Viheriäkoski, L. Sydänheimo, L. Ukkonen, ESD qualification data used as the basis for building electrostatic discharge protected areas, *J. Electrost.*, 77 (2015), 174–181. doi:10.1016/j.elstat.2015.08.009.

22. S.Y. Park, Y.H. Kim, S.Y. Lee, W. Sohn, et al., Highly selective and sensitive chemoresistive humidity sensors based on rGO/MoS$_2$ van der Waals composites, *J. Mater. Chem. A*, 6 (2018), 5016–5024.

23. J.J. Zhang, L. Sun, C. Chen, M. Liu, W. Dong, W.B. Guo, S.P. Ruan, High performance humidity sensor based on metal organic framework MIL-101(Cr) nanoparticles, *J. Alloys Compd.*, 695 (2017), 520–525.

24. Y. Zhang, B. Jiang, M.J. Yuan, P.W. Li, X.J. Zheng, Humidity sensing and dielectric properties of mesoporous Bi$_{3.25}$La$_{0.75}$Ti$_3$O$_{12}$ nanorods, *Sens. Actuators B*, 237 (2016), 41–48.

25. E. Traversa, Ceramic sensors for humidity detection: The state-of-the-art and future developments, *Sens. Actuators B*, 23 (1995), 135–156.

26. J.-S. Kim, K.-Y. Chun, C.-S. Han, Ion channel-based flexible temperature sensor with humidity insensitivity, *Sens. Actuators A*, 271 (2018), 139–145. doi:10.1016/j.sna.2018.01.025.

27. D. Zhang, Y. Sun, P. Li, Y. Zhang, Facile fabrication of MoS$_2$-Modified SnO$_2$ hybrid nanocomposite for ultrasensitive humidity sensing, *ACS Appl. Mater. Interfaces*, 22 (2016), 14142–14149.

28. M.H. Mamat, M.Z. Sahdan, Z. Khusaimi, A.Z. Ahmed, S. Abdullah, M. Rusop, Influence of doping concentrations on the aluminum doped zinc oxide thin films properties for ultraviolet photoconductive sensor applications, *Opt. Mater.*, 32 (2010), 696–699.

29. T. Roch, E. Dobročka, M. Mikula, A. Pidík, P. Ďurina, A.A. Haidry, T. Plecenik, M. Truchlý, B. Grančič, A. Plecenik, P. Kúš, Strong biaxial texture and polymorph nature in TiO$_2$ thin film formed by ex-situ annealing on c-plane Al$_2$O$_3$ surface, *J. Cryst. Growth*, 338:1 (2012), 118–124. doi:10.1016/j.jcrysgro.2011.10.053.

30. A.A. Haidry et al., Hydrogen gas sensors based on nanocrystalline TiO$_2$ thin films, *Cent. Eur. J. Phys.*, 9 (2011), 1351–1356. doi:10.2478/s11534-011-0042-3.

31. A.A. Haidry, J. Puškelová, T. Plecenik, P. Ďurina, J. Greguš, M. Truchlý, T. Roch, M. Zahoran, M. Vargová, P. Kúš, A. Plecenik, G. Plesch, Characterization and hydrogen gas sensing properties of TiO$_2$ thin films prepared by sol gel method, *Appl. Surf. Sci.*, 259 (2012), 270–275. doi:10.1016/j.apsusc.2012.07.030.

32. N. Savage, B. Chwieroth, A. Ginwalla, B.R. Patton, S.A. Akbar, P.K. Duttta, Composite n–p semiconducting titanium oxides as gas sensors, *Sens. Actuators B*, 79 (2001) 17–27.

33. A.A. Haidry, P. Ďurina, M. Tomášek, J. Greguš, P. Schlosser, M. Mikula, M. Truchlý, T. Roch, T. Plecenik, A. Pidík, M. Zahoran, P. Kúš, A. Plecenik, Effect of post-deposition annealing treatment on the structural, optical and gas sensing properties of TiO$_2$ thin films, *Key Eng. Mater.*, 510–511 (2012), 467–474. www.scientific.net/KEM.510-511.467.

34. A.A. Haidry, C. Cetin, K. Kelm, B. Saruhan, Sensing mechanism of low temperature NO$_2$ sensing with top-bottom electrode (TBE) geometry, *Sens. Actuators B*, 03 (2016). doi:10.1016/j.snb.2016.03.016.

35. A.A. Haidry, A. Ebach-Stahl, B. Saruhan, Effect of Pt/TiO$_2$ interface on room temperature hydrogen sensing performance of memristor type Pt/TiO$_2$/Pt structure, *Sens. Actuators B*, 253 (December 2017), 1043–1054. doi:10.1016/j.snb.2017.06.159.

36. L. Sun, A.A. Haidry, Q. Fatima, Z. Li, Z.J. Yao, Improving the humidity sensing below 30% with TiO$_2$/GO composite structure, *Mater. Res. Bull.*, 99, 124–131. doi:10.1016/j.materresbull.2017.11.001.

37. A.A. Haidry, L. Sun, B. Saruhan, A. Plecenik, T. Plecenik, Z.J. Yao, Cost-effective fabrication of polycrystalline TiO_2 with tunable n/p conductivity for selective hydrogen monitoring. *Sens. Actuators B*, 274 (20 November 2018), 10–21. doi:10.1016/j.snb.2018.07.082.

38. Z. Li, A.A. Haidry, L. Sun, Q. Fatima, Z.J. Yao, Facile synthesis of N-doped ordered mesoporous TiO_2 structure with improved humidity sensing, *J. Alloys Compd.*, 742, 814–821. doi:10.1016/j.jallcom.2018.01.361.

39. L. Sun, A.A. Haidry, Q. Fatima, Z. Li, Z. Yao, Improving the humidity sensing below 30% RH of TiO_2 with GO modification, *Mater. Res. Bull.*, 99 (2018), 124–131.

40. Z. Li, A.A. Haidry, B. Gao, T. Wang, Z.J. Yao, The effect of Co-doping on the humidity sensing properties of ordered mesoporous TiO_2, *Appl. Surf. Sci.*, 412 (2017), 638–647. doi:10.1016/j.apsusc.2017.03.156.

41. Q. Fatima, A.A. Haidry, Z. Yao, Y. He, Z. Li, L. Sun, L. Xie, On the critical role of hydroxyl groups on water vapour sensing of graphene oxide, *Nanoscale Adv.*, 1 (2018), 1319–1330.

42. A.S. Ismail, M.H. Mamat, N.D.M. Sin, M.F. Malek, A.S. Zoolfakar, A.B. Suriani, A. Mohamed, M.K. Ahmad, M. Rusop, Fabrication of hierarchical Sn-doped ZnO nanorod arrays through sonicated sol-gel immersion for room temperature, resistive-type humidity sensor applications, *Ceram. Int.*, 42 (2016), 9785–9795. doi:10.1016/j.ceramint.2016.03.071.

43. X. Wang, Z. Jian, Z. Zhu, J. Zhu, Humidity sensing properties of Pd^{2+}-doped ZnO nanotetrapods, *Appl. Surf. Sci.*, 253 (2007), 3168–3173. doi:10.1016/j.apsusc.2006.07.033.

44. X. Wang, J. Li, Y. Li, L. Liu, W. Guan, Emulsion-templated fully three-dimensional interconnected porous titania ceramics with excellent humidity sensing properties, *Sens. Actuators B*, 237 (2016) 894–898. doi:10.1016/j.snb.2016.07.014.

45. B. Karunagaran et al., TiO_2 thin film gas sensor for monitoring ammonia, *Mater. Charact.*, 58 (2007), 680–684. doi:10.1016/j.matchar.2006.11.007.

46. U. Diebold, The surface science of titanium dioxide, *Surf. Sci. Rep.*, 48 (2003), 53–229. doi:10.1016/S0167-5729(02)00100-0.

47. A.A. Mosquera, J.M. Albella, V. Navarro, D. Bhattacharyya, J.L. Endrino, Effect of silver on the phase transition and wettability of titanium oxide films, *Sci. Rep.*, 6 (2016), 32171. doi:10.1038/srep32171.

48. Z. Zhang, J.T. Yates, Band bending in semiconductors: Chemical and physical consequences at surfaces and interfaces, *Chem. Rev.*, 112 (2012), 5520–5551.

49. J. Bai, B. Zhou, Titanium dioxide nanomaterials for sensor applications, *Chem. Rev.*, 114 (2014), 10131–10176.

50. W. Guo, Q.Q. Feng, Y.F. Tao, L.J. Zheng, Z.Y. Han, J.M. Ma, Systematic investigation on the gas-sensing performance of TiO_2 nanoplate sensors for enhanced detection on toxic gases, *Mater. Res. Bull.*, 73 (2016), 302–307. doi:10.1016/j.materresbull.2015.09.012.

51. A. Gil, M. Fernández, I. Mendizábal, S.A. Korili, J. Soto-Armañanzas, A. Crespo-Durante, C. Gómez-Polo, Fabrication of TiO_2 coated metallic wires by the sol-gel technique as a humidity sensor, *Ceram. Int.*, 42 (2016) 9292–9298, doi:10.1016/j.ceramint.2016.02.074

52. Z. Wang, L. Shi, F. Wu, S. Yuan, Y. Zhao, M. Zhang, The sol-gel template synthesis of porous TiO_2 for a high performance humidity sensor, *Nanotechnology*, 22 (2011), 275502. doi:10.1088/0957-4484/22/27/275502.

53. T. Plecenik, M. Mosko, A.A. Haidry, P. Durina, M. Truchly, B. Grancic, M. Gregor, T. Roch, L. Satrapinskyy, A. Moskova, M. Mikula, P. Kus, A. Plecenik, Fast highly-sensitive room-temperature semiconductor gas sensor based on the nanoscale Pt–TiO$_2$–Pt sandwich, *Sens. Actuators B*, 207 (2015), 351–361. doi:10.1016/j.snb.2014.10.003.

54. S. Kozhukharov, Z. Nenova, T. Nenov, N. Nedev, M. Machkova, Humidity sensing elements based on cerium doped titania-silica thin films prepared via a sol–gel method, *Sens. Actuators B*, 210 (2015) 676–684. doi:10.1016/j.snb.2014.12.119.

55. K. Zakrzewska, *Titanium Dioxide Thin Films for Gas Sensors and Photonic Applications*, AGH Ucelniane Wydawnictwa Naukowo-Dydadaktyczne, Kraków, 2003.

56. K. Zakrzewska, Nonstoichiometry in TiO$_{2-y}$ studied by ion beam methods and photoelectron spectroscopy, *Adv. Mater. Sci. Eng.*, 2012, Article ID 826873, 13 pages. doi:10.1155/2012/826873.

57. T. Bak, J. Nowotny, M. Rekas, C.C. Sorrell, Defect chemistry and semiconducting properties of titanium dioxide: II. Defect diagrams, *J. Phys. Chem. Solids*, 64 (2003), 1057–1067. doi:10.1016/S0022-3697(02)00480-8.

58. T. Bak, J. Nowotny, M.K. Nowotny, Defect disorder of titanium dioxide, *J. Phys. Chem. B*, 110:43 (2006), 21560–21567. doi:10.1021/jp063700k.

59. K. Hashimoto et al., TiO$_2$ photocatalysis: A historical overview and future prospects, *Jpn. J. Appl. Phys.*, 44:12 (2005). doi:10.1143/JJAP.44.8269.

60. J. Nowotny, T. Bak, E.C. Dickey, W. Sigmund, M.A. Alim, Electrical conductivity, thermoelectric power, and equilibration kinetics of Nb-doped TiO$_2$, *J. Phys. Chem. A*, 120:34 (2016), 6822–6837. doi:10.1021/acs.jpca.6b04104.

61. M.A. Alim, T. Bak, A. Atanacio, J. Du Plessis, M. Zhou, J. Davis, J. Nowotny, Electrical conductivity and defect disorder of tantalum-doped TiO$_2$, *J. Am. Chem. Soc.*, 79:2–4, 47–154; 100:9 (September 2017), 4088–4100. doi:10.1111/jace.14959.

62. Z. Chen, C. Lu, Humidity sensors: A review of materials and mechanisms, *Sens. Lett.*, 3 (2005) 274–295. doi:10.1166/sl.2005.045.

63. H. Farahani, R. Wagiran, M.N. Hamidon, Humidity sensors principle, mechanism and fabrication technologies: A comprehensive review, *Sensors*, 14 (2014), 7881–7939. doi:10.3390/s140507881.

64. Q. Wang, Y. Pan, S. Huang, S. Ren, P. Li, J. Li, Resistive and capacitive response of nitrogen-doped TiO$_2$ nanotubes film humidity sensor, *Nanotechnology*, 22 (2011), 025501. doi:10.1088/0957-4484/22/2/025501.

65. D. Li, J. Zhang, L. Shen, W. Dong, C. Feng, C. Liu, S. Ruan, Humidity sensing properties of SrTiO$_3$ nanospheres with high sensitivity and rapid response, *RSC Adv.*, 5 (2015), 22879–22883. doi:10.1039/C5RA00451A.

66. B.C. Yadav, N. Verma, S. Singh, Nanocrystalline SnO$_2$–TiO$_2$ thin film deposited on base of equilateral prism as an optoelectronic humidity sensor, *Opt. Laser Technol.*, 44 (2012), 1681–1688. doi:10.1016/j.optlastec.2011.12.041.

67. C. Lai, X. Wang, Y. Zhao, H. Fong, Z. Zhu, Effects of humidity on the ultraviolet nanosensors of aligned electrospun ZnO nanofibers, *RSC Adv.*, 3 (2013), 6640–6645.

68. J.H. Anderson, G.A. Parks, The electrical conductivity of silica gel in the presence of adsorbed water, *J. Chem. Phys.*, 72 (1968) 3666–3668. doi:10.1021/j100856a051.

69. S. Karthick, H. Lee, S. Kwon, R. Natarajan, V. Saraswathy, Standardization, calibration, and evaluation of tantalum-nano rGO-SnO$_2$ composite as a possible candidate material in humidity sensors, *Sensors*, 16 (2016), 2079. doi:10.3390/s16122079.

70. J. Lee, K. Akash, J. Kim, S.S. Kim, Growth of networked TiO_2 nanowires for gas-sensing applications, *J. Nanosci. Nanotechnol.*, 16 (2016), 11580–11585.
71. M. Urbiztondo, I. Pellejero, A. Rodriguez, M.P. Pina, J. Santamaria, Zeolite-coated interdigital capacitors for humidity sensing, *Sens. Actuators B*, 157 (2011), 450–459. doi:10.1016/j.snb.2011.04.089.
72. I.C. Cosentino, E.N.S. Muccillo, R. Muccillo, Development of zirconia-titania porous ceramics for humidity sensors, *Sens. Actuators B*, 96 (2003), 677–683. doi:10.1016/j.snb.2003.07.013.
73. E. Traversa, Ceramic sensors for humidity detection: The state-of-the-art and future developments, *Sens. Actuators B*, 23 (1995), 135–156.
74. Q. Kuang, C. Lao, Z. Wang, Z. Xie, L. Zheng, High-sensitivity humidity sensor based on a single SnO_2 nanowire, *J. Am. Chem. Soc.*, 129 (2007), 6070–6071. doi:10.1021/ja070788m.
75. N. Yamazoe, Y. Shimizu. Humidity sensors: Principles and applications. *Sens. Actuators*, 10 (1986), 379–398. doi:10.1016/0250-6874(86)80055-5.
76. L. Sun, A.A. Haidry, Z. Li, L. Xie, Z. Wang, Q. Fatima, Z. Yao, Effective use of biomass ash as an ultra-high humidity sensor, *J. Mater. Sci.: Mater. Electron.*, 29 (2018), 18502–18510.
77. B.H. Lee, W.H. Khoh, A.K. Sarker, C.H. Lee, J.D. Hong. A high performance moisture sensor based on ultra large graphene oxide, *Nanoscale*, 7 (2015), 17850–17811. doi:10.1039/C5NR05726D.
78. W.M. Sears, The effect of oxygen stoichiometry on the humidity sensing characteristics of bismuth iron molybdite, *Sens. Actuators B*, 67 (2000), 161–172.
79. Z.Y. Wang, L.Y. Shi, F.Q. Wu, S. Yuan, Y. Zhao, M.H. Zhang, The sol-gel template synthesis of porous TiO_2 for a high-performance humidity sensor, *Nanotechnology*, 22 (2011), 275502.
80. M.A. Henderson, The interaction of water with solid surfaces: Fundamental aspects revisited, *Surf. Sci. Rep.*, 46 (2002), 1–308.
81. W. Lin, D. Lai, M. Chen, R. Wu, F. Chen, Evaluate humidity sensing properties of novel TiO_2-WO_3 composite material, *Mater. Res. Bull.*, 48 (2013), 3822–3828.
82. Y. Zhang, W. Fu, H. Yang, et al., Synthesis and characterization of TiO_2 nanotubes for humidity sensing, *Appl. Surf. Sci.*, 254 (2008), 5545–5547.
83. A. Sun, L. Huang, Y. Li, Study on humidity sensing property based on TiO_2 porous film and polystyrene sulfonic sodium, *Sens. Actuators B*, 139 (2009), 543–547.

Part 3

Re-Manufacturing

This part provides a chapter on metal deposition using additive manufacturing for the repair of machine tool parts. This chapter provides the background, presents the technique in general, and describes the application in detail.

4

Manufacturing, Remanufacturing, and Surface Repairing of Various Machine Tool Components Using Laser-Assisted Directed Energy Deposition

Muhammad Iqbal, Asif Iqbal, Malik M. Nauman, Quentin Cheok, and Emeroylariffion Abas

Universiti Brunei Darussalam

CONTENTS

4.1 Introduction 79
4.2 Baseline Technology 80
    4.2.1 Feeding Ways 82
    4.2.2 Process Parameters 82
4.3 Application Materials 83
4.4 Industrial Applicability 83
4.5 Sustainability 84
4.6 Experimental Work 85
References 86

4.1 Introduction

Under intense friction and humidity in the environment, aluminum, its alloys, and steel surfaces fail and suffer wear and corrosion [1]. To get rid of corrosion and fatigue damage and to prolong the service life of metal parts and machine tools, various surface treatment technologies have been introduced for improved performance [2]. Laser cladding among others is an advantageous thermal technique due to the non-equilibrium fast cooling characteristics over conventionally used coating methods to repair, protect, or remanufacture surfaces by depositing metallurgical sound and dense layer [3]. Melting of both the substrate and depositing layer due to high temperature and intense power laser beam results in strong metallurgical bond [4]. Surface modification and treatment using laser beam includes laser surface remelting [5], laser surface alloying, [6] and laser cladding [7]. The performance of the cladding layer depends on the material, reinforced element, and the process parameters used in laser cladding [8]. Amorphous alloys are particularly important because of their regular atomic structure and compositional homogeneity, but they have small critical dimensions and hence restricted applications [9]. To broaden the industrial applications of these alloys, different amorphous alloys such as Ni, Cu, Zr, Fe, and Co-based alloys are extensively used as powder materials in laser cladding coating technology. Due to the low price, wettability, wear resistance, self-melting, lubrication effect and higher amorphous formation, Ni-based and Co-based alloys are increasingly used as powder materials [10]. Formation of pores and cracks during and after laser cladding is considered a barrier to broaden their applicability which occurs when contraction stress becomes higher than the interfacial tension of the film [11]. Occurrence of cracks is observed mostly between substrate and coating, between hard cladding particle and matrix, or within the interface between successive cladding tracks [12]. High stress and low material strength also result in crack formation which can be minimized by reducing stress and using specialized materials for laser cladding with good ductility [13]. Laser technology is increasingly used for material cutting [14], manufacturing [15], remanufacturing [16], and surface treatment in various materials including copper [17], steel [18], magnesium [19], and titanium [20], because of focused beam, excellent resolution, accurate directionality, high energy density, and coherence. Using optimized parameters and materials, researchers produced best coatings on aluminum and its alloys with more hardness and improved wear and corrosion properties [21]. This chapter is organized into different sections: in Section 4.2, baseline technology is discussed with various process parameters and feeding methods. Application materials are explained in Section 4.3, while industrial applicability, sustainability, and experimental work are presented in Sections 4.4–4.6 respectively.

4.2 Baseline Technology

The general architecture of experimental setup for laser cladding process is shown in Figure 4.1. In the setup, a signal modulator usually converts the laser beam pulse to continuous sinusoidal wave, which after passing through the laser beam tube falls on a focusing lens to be reflected on the substrate. Laser cladding is a technique which combines laser technology, control system, and computer aided manufacturing (CAM) [22]. The laser beam is used as heat source for deposition of preplaced clad material on a moving substrate, and the moving stages are controlled by an operating system. The material properties and composition of the coating and substrate remain unchanged during mixing by laser surface cladding. The specimen prepared as a result of laser cladding can be divided into four distinct parts as the substrate, cladding zone, heat affected zone, and interfacial zone [23] (Figure 4.2).

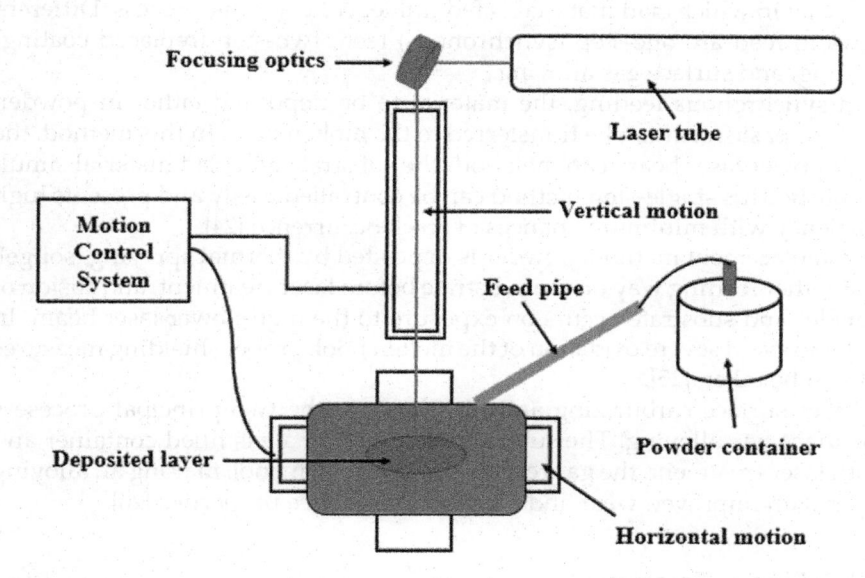

FIGURE 4.1
Schematic illustration of laser cladding setup.

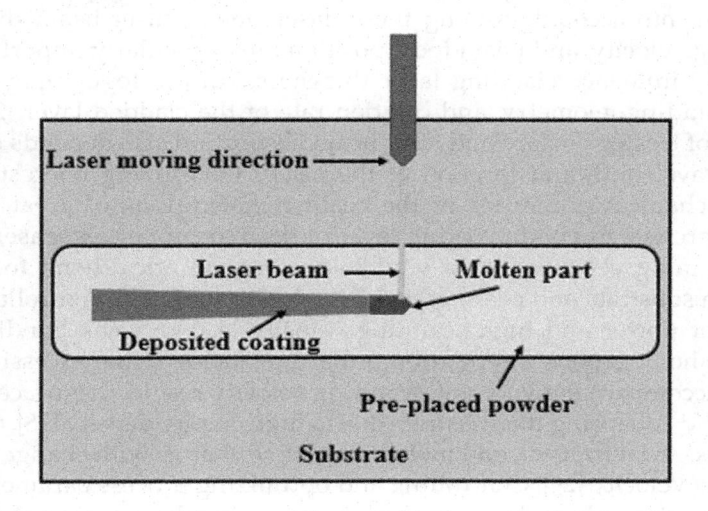

FIGURE 4.2
Schematic diagram of laser cladding process.

The key factors affecting laser cladding process include powder feeding ways, feeding rate, prepalced coating thickness, process parameters, and cladding material.

4.2.1 Feeding Ways

The way in which clad material is fed influences cladding process. Different ways to feed are one-step (synchronous) feed, two-step (replaced coating) feeding, and surface gas alloying.

In synchronous feeding, the material to be deposited either in powder, wire, or paste form can be transferred to the molten pool. In this method, the high-power laser beam can melt both the substrate and clad material simultanously. This single-step method can be controlled easily and presents high efficiency with minimum chances of crack occurrence [24].

In replacement method, powder is deposited by thermal spraying, sol-gel, or another feeding way on the substrate before laser treatment, and fusion of powder and substrate occurs on exposure to the high-power laser beam. In order to avoid severe oxidation of the molten pool, proper shielding measures should be taken [25].

Laser surface carburizing and nitriding are the two principal procesess in suface gas alloying. The substrate is placed in a gas filled container and with laser treatment; the gas reacts with the molten pool, making an alloying layer with improved wear and corrosion resistance properties [26].

4.2.2 Process Parameters

Some parameters that greatly affect the laser cladding process should be taken into account. Among these, laser power, laser beam diameter, scanning velocity, and beam focal position are particularly important and evidently influence cladding layer thickness, surface topography, aspect ratio, cladding geometry, and dilution rate of the cladded layer [27]. The quality of treated surface and laser beam absorption also depends on laser beam wavelength and location of the beam. Optimizing microstructure and mechanical properties of the coating materials is of great interest for researchers to obtain coating layer of desired properties. Laser power and scanning velocity play a vital role in metallurgical bond formation between substrate and coating, and the process should be controlled. Very low laser power and high scanning velocity lead to weak bonding and hence inhomogenous distribution of hard particles, while excessive laser power accompanying very low scanning velocity results in surface evaporation and collapsing the substrate due to high energy density [28]. Cooling rate, mass–heat transfer, and molten-pool time change with change in laser scanning velocity [29]. Controlling and optimizing process parameters can guarentee desired results for laser cladding in metals including aluminum and its alloys.

4.3 Application Materials

Various metals, such as nickel (Ni), cobalt (Co), iron (Fe), stainless steel (AISI), carbides (WC), carbon (C), boron (B), chromium (Cr), manganese (Mn), silicon (Si), molybdenum (Mo), and their alloys are commonly used in laser cladding, but aluminum (Al) and magnesium (Mg) alloys among them are particularly important because of their special mechanical and chemical properties. Increased demand of lightweight materials in aerospace and automotive industries such as Mg, Al, and their alloys encouraged manufacturers and researchers due to their low densities and high specific strengths as compared to other alloys. Mg has the lowest density, but its highest chemical reactivity with water and air restricts its use in many applications, and that is why Al-alloys are still considered ideal for use in automotive and aerospace applications, but considerable research is carried out to enhance chemical as well as mechanical properties of Mg using coating technology. However Al-alloys on the other hand have good strength-to-weight ratio and present better wear, corrosion, and creep resistances. Titanium is also increasingly used due its low density which is about half that of steel, and when it reacts with water, it forms a protective corrosion-resistant layer. Four different classes of stainless steel namely austinitic, ferritic, duplex, and martensitic are extensively used because of corrosion resistance and low maintenance cost.

4.4 Industrial Applicability

Laser cladding was increasingly applied in the last two decades in manufacturing industries for repair, remanufacturing, and even new tool design due to the ease with which cladding could be performed with laser hardening machine by simply installing laser feeding device. Wear resistance of important machine tools and modification of dies and molds by depositing a resistant layer are important applications of laser cladding technology. Critical machine tool areas like pinch edges, contact surfaces of hemming nests, cutting tool edges, injection molds endings, and drawing tool radius can easily be repaired and modified. High-power diode laser and solid-state lasers are used as state-of-the-art technology in improving wear resistance of selected tool areas due to their cost effectiveness and efficient output [30]. As discussed earlier, a modern laser head can be used for laser cladding and laser surface hardening by simply changing the focusing mirror and tieing up or removing powder feed nozzle. During laser hardening and cladding, temperature control is very much important to avoid pores, cracks, and melting of the specimen to be treated at very high temperature and collapsing

TABLE 4.1

Summary of the the Process Parameters in Laser Cladding and Laser Hardening Process

Parameter	Unit	Laser Cladding	Laser Hardening	Ref
Laser power	kW	2–6	2–6	[32]
Laser intensity	W/cm^2	10^4–10^5	10^3–10^4	[33]
Wavelength	nm	1,027	2,090	[34]
Process temperature	°C	<1,800	<1,000	[35]
Scanning velocity	m/s	0.008–0.041	0.005	[36]
Clad material	None	Powder or wire	N/A	[37]
Clad layer geometry	N/A	Round	Rectangle	[38]
Layer width	mm	0.5–5	5–60	[39]
Gas flow rate	l/min	Ar or He	Not necessary	[40]

of the same at very low operating temperature. Therefore, temperature measurement meter (pyrometer) is now an essential part in moderen laser heads to ensure excellent desired results. These pyrometers are resistant to the high-power laser emissions and present excellent accuracy in measurement. The operation temperature for laser cladding (1,800°C–1,900°C) is more than laser hardening (1,000°C–1,300°C) and can be measured with pyrometer (Table 4.1) [31].

Processing of heavy machine tools like cutting tool, injection molding tool, or multi-layer drawing tool can be done easily with the help of a robot machine as a guide on the working table. Most of the automobile industry tools are prepared using laser technology nowadays because of low cost, compatibility, and better durability. Machining is an essential process after laser cladding for achieving smooth and dimensionally accurate finished part.

4.5 Sustainability

Sustainable manufacturing is unquestionably of great importance toward achieving economic and environmental targets and to renew a product life cycle. Repair and remanufacturing of damaged machine tools using laser cladding technology can save considerable amount of material, energy, time, and it costs only a fraction of newly manufacturing a tool [41]. Remanufacturing a unit reduces the cost more than 50% that of new one in terms of labor work, raw material, and energy dissipation. Sustainable engineering design and manufacturing ensures efficient resource consumption in the form of recycling and reuse with minimal adverse environmental impact. Machine tool's life cycle has been changed from one cycle to multiple service life cycles with the advent of increasing awareness about environmental load through material recovery. By remanufacturing, used products and assemblies can be brought into their new working state with minimum waste production, expenditure, and loss of material. It is basically a full new life cycle for a product to be remanufactured with more or less extension in its service life. During the process, parts are cleaned, examined, and any damaged or missing part is replaced, reconditioned, or rectified by finishing and machining process. Various machine tools can be reused by cleaning, replacing, and dismantling products if not worn out and repaired in timely manner contributing to material recovery [42]. In automobile parts, such as engine block, head, gears, shafts, connecting rods, bearing, and piston, remanufacturing is very common and accounts for almost two-third of the total remanufacturing process. Consumers are well aware of the fact that by remanufacturing most of the automobile parts like rods, engines, alternators, starters, and CV joints can be replaced.

4.6 Experimental Work

The ever-growing demand for laser cladding is an increasingly developed area, and due to its vast applications and viability in various important prducts, tools, units, and parts of high-value-added engineering components, it is applied in many manufacturing industries. Shibang et al. [2] used SiC and Ti powder with Ni60A as bonding phase on grade 45 steel substrate to study the effects of changing process parameters on clad layer using Ni-based reinforcement on steel. Powder was mixed completely to ensure homogeneity and for strong bonding; 5% polyvinyl alcohol solution was used. After the powder dried, its layer thickness was measured as 1 mm and cladded using diode laser. The scanning velocity was considered constant as 125 mm/min because of no significant change with changing speed, and power of laser was increased slowly from 400 to 1,200 W which affected the coating layer width, and the hardness was increased with increased power; however, an optimal laser power was determined as 800–1,000 W for desired results.

Mg was used as a substrate for single-step coating (1.5 mm thick) of an amorphous alloy $Zr_{65}A_{17.5}Ni_{10}Cu_{17.5}$ in argon gas atmosphere [24]. Cladding was performed using CO_2 continuous wave laser of power 3.7 W with scanning speed as 5 mm/s in controlled environment observing microstructure using high-performance X-ray spectroscopy and XRD system. The crystallization and thermal stability was examined with scanning calorimeter. A thick amorphous alloy layer of 1.5 mm with up to 1.1 mm pure amorphous structure was cladded on Mg substrate having good corrosion resistance and no cracks.

References

1. Y. Chi, G. Gu, H. Yu, and C. Chen, "Laser surface alloying on aluminum and its alloys: A review," *Opt. Lasers Eng.*, vol. 100, no. May 2017, pp. 23–37, 2018.
2. D. Z. Ma Shibang1, X. Zhenwei, X. Yang, L. Chao1, and and S. Houjie, "Parameter optimization and microstructure evolution of in-situ TiC particle reinforced Ni-based composite coating by laser cladding," *J. Eng. Sci. Technol. Rev.*, vol. 11, no. 2, pp. 88–95, 2018.
3. L. Zhang, C. Wang, L. Han, and C. Dong, "Influence of laser power on microstructure and properties of laser clad Co-based amorphous composite coatings," *Surf. Interfaces*, vol. 6, pp. 18–23, 2017.
4. Z. Liu, P. H. Chong, P. Skeldon, P. A. Hilton, J. T. Spencer, and B. Quayle, "Fundamental understanding of the corrosion performance of laser-melted metallic alloys," *Surf. Coat. Technol.*, vol. 200, pp. 5514–5525, 2006.
5. N. Pagano, V. Angelini, L. Ceschini, and G. Campana, "Laser remelting for enhancing tribological performances of a ductile iron," *Procedia CIRP*, vol. 41, pp. 987–991, 2016.
6. G. F. Sun et al., "Microstructure and corrosion characteristics of 304 stainless steel laser-alloyed with Cr–CrB_2," *Appl. Surf. Sci.*, vol. 295, pp. 94–107, 2014.
7. S. Iwatani, Y. Ogata, K. Uenishi, K. F. Kobayashi, and A. Tsuboi, "Laser cladding of Fe–Cr–C alloys on A5052 aluminum alloy using diode laser," *Mater. Trans.*, vol. 46, no. 6, pp. 1341–1347, 2005.
8. J. Liu, H. Yu, C. Chen, F. Weng, and J. Dai, "Research and development status of laser cladding on magnesium alloys: A review," *Opt. Lasers Eng.*, vol. 93, no. February, pp. 195–210, 2017.
9. H. Zhang, Y. He, and Y. Pan, "Enhanced hardness and fracture toughness of the laser-solidified FeCoNiCrCuTiMoAlSiB$_{0.5}$ high-entropy alloy by martensite strengthening," *Scr. Mater.*, pp. 1–4, 2013.
10. R. Li, Z. Li, J. Huang, P. Zhang, and Y. Zhu, "Effect of Ni-to-Fe ratio on structure and properties of Ni–Fe–B–Si–Nb coatings fabricated by laser processing," *Appl. Surf. Sci.*, vol. 257, no. 8, pp. 3554–3557, 2011.
11. H. Hu, A. Iqbal, X. Wang, and H. Zhang, "Experimental investigation of residual stress relief using pre- and post-clad heating in the laser cladding process," *Lasers Eng.*, vol. 31, pp. 11–27, 2015.
12. S. Zhou, X. Zeng, Q. Hu, and Y. Huang, "Analysis of crack behavior for Ni-based WC composite coatings by laser cladding and crack-free realization," *Appl. Surf. Sci.*, vol. 255, pp. 1646–1653, 2008.
13. F. Shu, B. Yang, S. Dong, H. Zhao, B. Xu, and F. Xu, "Effects of Fe-to-Co ratio on microstructure and mechanical properties of laser cladded FeCoCrBNiSi high-entropy alloy coatings," *Appl. Surf. Sci.*, vol. 450, no. August, pp. 538–544, 2018.
14. V. Veiko et al., "Development of complete color palette based on spectrophotometric measurements of steel oxidation results for enhancement of color laser marking technology," *Mater. Des.*, vol. 89, pp. 684–688, 2016.
15. A. S. Hamada, A. Järvenpää, E. Ahmed, P. Sahu, and A. I. Z. Farahat, "Enhancement in grain-structure and mechanical properties of laser reversion treated metastable austenitic stainless steel," *Mater. Des.*, vol. 94, pp. 345–352, 2016.

16. J. P. Oliveira, F. M. B. Fernandes, R. M. Miranda, N. Schell, and J. L. Ocaña, "Residual stress analysis in laser welded NiTi sheets using synchrotron X-ray diffraction," *Mater. Des.*, vol. 100, pp. 180–187, 2016.

17. H. Yan, P. Zhang, Z. Yu, Q. Lu, S. Yang, and C. Li, "Microstructure and tribological properties of laser-clad Ni–Cr/TiB$_2$ composite coatings on copper with the addition of CaF$_2$," *Surf. Coat. Technol.*, vol. 206, no. 19–20, pp. 4046–4053, 2012.

18. C. T. Kwok, F. T. Cheng, and H. C. Man, "Laser surface modification of UNS S31603 stainless steel. Part I : microstructures and corrosion characteristics," *Mater. Sci. Eng. A*, vol. 290, pp. 55–73, 2000.

19. T. M. Yue, H. Xie, X. Lin, H. O. Yang, and G. H. Meng, "Solidification behaviour in laser cladding of AlCoCrCuFeNi high-entropy alloy on magnesium substrates," *J. Alloys Compd.*, vol. 587, pp. 588–593, 2014.

20. I. Watanabe, M. Mcbride, P. Newton, and K. S. Kurtz, "Laser surface treatment to improve mechanical properties of cast titanium," *Dent. Mater.*, vol. 5, pp. 629–633, 2009.

21. V. Y. Zadorozhnyy et al., "Synthesis of the Ni-Al coatings on different metallic substrates by mechanical alloying and subsequent laser treatment," *J. Alloys Compd.*, vol. 707, pp. 351–357, 2017.

22. S. F. C. Ehsan Toyserkani, A. Khajepour, *Laser Cladding*, 1st Edition. Boca Raton: CRC Press, 2005.

23. K. Liu, Y. Li, J. Wang, and Q. Ma, "Effect of high dilution on the in situ synthesis of Ni–Zr/Zr–Si (B, C) reinforced composite coating on zirconium alloy substrate by laser cladding," *Mater. Des.*, vol. 87, pp. 66–74, 2015.

24. T. M. Yue, Y. P. Su, and H. O. Yang, "Laser cladding of Zr$_{65}$Al$_{7.5}$Ni$_{10}$Cu$_{17.5}$ amorphous alloy on magnesium," *Mater. Lett.*, vol. 61, pp. 209–212, 2007.

25. S. R. Paital et al., "Improved corrosion and wear resistance of Mg alloys via laser surface modification of Al on AZ31B," *Surf. Coat. Technol.*, vol. 206, no. 8–9, pp. 2308–2315, 2012.

26. C. Meneau and P. Andreazza, "Laser surface modification: Structural and tribological studies of AlN coatings," *Surf. Coat. Technol.*, vol. 101, no. 97, pp. 12–16, 1998.

27. Y. Mao, Z. Li, K. Feng, X. Guo, Z. Zhou, and Y. Wu, "Corrosion behavior of carbon film coated magnesium alloy with electroless plating nickel interlayer," *J. Mater. Process. Tech.*, vol. 219, pp. 42–47, 2015.

28. G. Casalino, "[INVITED] Computational intelligence for smart laser materials processing," *Opt. Laser Technol.*, vol. 100, pp. 165–175, 2018.

29. Y. Jun, G. P. Sun, H. Wang, S. Q. Jia, and S. S. Jia, "Laser (Nd:YAG) cladding of AZ91D magnesium alloys with Al+Si+Al$_2$O$_3$," *J. Alloys Compd.*, vol. 407, pp. 201–207, 2006.

30. E. Kennedy, G. Byrne, and D. N. Collins, "A review of the use of high power diode lasers in surface hardening," *J. Mater. Process. Technol.*, vol. 156, pp. 1855–1860, 2004.

31. P. Hoffmann and R. Dierken, "Laser technology for tool production application," *Laser Tech. J.*, vol. 15, no. 3, pp. 32–35, 2018.

32. C. G. Luo Fang, Y. Jian-Hua, H. Xia-Xia, "Effect of laser power on the cladding temperature field and the heat affected zone," *J. Iron Steel Res. Int.*, vol. 18, no. 1, pp. 73–78, 2011.

33. J. Lin, "Laser attenuation of the focused powder streams in coaxial laser cladding," *J. Laser Appl.*, vol. 28, no. pp. 28–33, May 2014, 2006.

34. J. Nilsson et al., "High-power cladding-pumped Tm-doped silica fiber laser with wavelength tuning from 1860 to 2090 nm," *Opt. Fiber Technol.*, vol. 10, no. 22, pp. 5–30, 2004.

35. L. Song and J. Mazumder, "Feedback control of melt pool temperature during laser cladding process," *IEEE Trans. Control Syst. Technol.*, vol. 19, no. 6, pp. 1349–1356, 2011.

36. R. M. Mahamood, E. T. Akinlabi, M. Shukla, and S. Pityana, "Scanning velocity influence on microstructure, microhardness and wear resistance performance of laser deposited Ti6Al4V/TiC composite," *Mater. Des.*, vol. 50, pp. 656–666, 2013.

37. S. N. Li et al., "Crossmark," *Ceram. Int. J.*, vol. 43, no. July 2016, pp. 961–967, 2017.

38. G. Lian, M. Yao, Y. Zhang, and C. Chen, "Analysis and prediction on geometric characteristics of multi-track overlapping laser cladding," *Int. J. Adv. Manuf. Technol.*, vol. 97, no. 5–8, pp. 2397–2407, 2018.

39. G. Zhu, S. Shi, G. Fu, J. Shi, S. Yang, and W. Meng, "The influence of the substrate-inclined angle on the section size of laser cladding layers based on robot with the inside-beam powder feeding," *Int. J. Adv. Manuf. Technol.*, vol. 88, pp. 2163–2168, 2016.

40. P. Cirp, T. Lane, P. Stief, J. Dantan, A. Etienne, and A. Siadat, "A new methodology to analyze the functional and physical architecture of existing products for an assembly oriented product family identification," *Procedia CIRP*, vol. 74, pp. 719–723, 2018.

41. P. Zhang and Z. Liu, "On sustainable manufacturing of Cr-Ni alloy coatings by laser cladding and high-efficiency turning process chain and consequent corrosion resistance," *J. Clean. Prod.*, vol. 161, pp. 676–687, 2017.

42. S. Statham, "Remanufacturing – Towards a more sustainable future," *Electronics-Enabled Products Knowledge-Transfer Network*, Loughboroug, UK, 2006.

Part 4

Metrology

Tool wear is an important factor that affects the component geometry leading to decision on component acceptance at quality check. It is a micro-level decision parameter that may influence the entire supply chain if the remedy is not incorporated at the machine tool where the tool causing adverse effect is mounted. Authors of the chapter "Dynamic Prediction of tool wear" have proposed a solution for tool wear detection using machine vision and support vector machine. The second chapter in this part proposes a concept of real-time in-process measurement also using machine vision, thus merging manufacturing and quality control together implementing automated Shigeo Shingo method. Both techniques incorporate a high level of automation and strive for savings in cost, lead time, energy, and environment.

5

Investigation on Dynamical Prediction of Tool Wear Based on Machine Vision and Support Vector Machine

Chen Zhang and Jilin Zhang

Nanjing University of Aeronautics and Astronautics

CONTENTS

5.1 Introduction 91
5.2 Dynamical Prediction Method of Tool Wear 94
    5.2.1 Basic Idea of Dynamical Prediction 94
    5.2.2 Flow Process of Dynamical Prediction 94
    5.2.3 Online Tool Wear Measurement Based on Machine Vision 95
    5.2.4 The Establishment of Tool Wear Model Based on LS-SVM 96
5.3 Experimental Works and Discussion 96
    5.3.1 Experimental Setup 96
    5.3.2 Design of Experiments 97
    5.3.3 Analysis and Discussion of Experimental Results 98
        5.3.3.1 Experimental Results 98
        5.3.3.2 The Result Analysis Considering Dynamical
                Prediction 98
5.4 Conclusions 103
References 103

5.1 Introduction

Tool wear investigations in machining have been amongst the important research areas during the last several decades as the quality of the machined product largely depends on the tool wear. Tool wear is defined as the change in the tool shape from its original shape during cutting in machining process. If the machining process continues with a worn tool, the dimensional accuracy, surface quality of finished component, and even process stability will be deteriorated. Therefore, the prediction of tool wear in machining process is very meaningful in order to improve machining quality and increase productivity. For actual machining system, the estimated tool wear can be used to optimize tool replacement and compensate machining error induced by tool wear. Thus, the tool wear estimation becomes crucial in machining processes.

Recent progress in digital image processing technology has led to attention for tool wear estimation using machine vision. Tool wear can be measured directly, achieving higher levels of precision and reliability by using machine vision. These computer vision approaches are mainly based on the wear shape contour, the wear shape properties, the texture of the wear region, or combinations of some of them [1–3].

Jurkovic et al. [4] used modern image processing techniques and machine vision system to acquire tool wear directly, and the presented system is characterized by its measurement flexibility, high spatial resolution, and good accuracy. Castejon et al. [5] proposed a new method based on a computer vision and statistical learning system to estimate wear level in cutting inserts, and every wear flank region has been described by means of nine geometrical descriptors. Tugrul et al. [6] developed a multiple linear regression model and a neural network model for predicting surface roughness and tool flank wear. A reliable machine vision system to automatically detect inserts and determine if they are broken is presented [7]. Unlike the machining operations studied in the literature, they deal with edge milling head tools for aggressive machining of thick plates (up to 12 cm) in a single pass. Antic et al. [8] apply a texture filter bank over the image formed by the Short-Term Discrete Fourier Transform (STDFT) spectra of vibration sensors to get descriptors for tool wear monitoring.

When obtaining the tool wear, the prediction and modeling technology is necessary to deal with these measured wear data. In this aspect, some promising methods are used such as artificial neural network, fuzzy logic, support vector machine (SVM), etc. Jacob and Joseph [9] proposed an artificial-neural-network-based in-process tool wear prediction system and used artificial neural networks to process the relation between tool wear and sensory signals. Wen-Tung and Chung-Shay [10] used the back-propagation neural network (BPN) to construct the tool wear predictive model, and the genetic algorithm was used in the optimization model. In these methods, these signals come from the machining points or fields, such as vibration, sound, acoustic emission (AE), cutting force, temperature, etc. The detected signal with tool wear will be compared with those of the unworn tool to access tool wear states. The relations between tool wear and the characteristic parameters must be prepared preliminarily in different cutting conditions through hardware monitoring equipment. Srinivas and Kotaiah [11] developed a neural network model to predict tool wear and cutting force in turning operations for cutting parameters such as cutting speed, feed, and depth of cut. Kuo et al. [12] proposed an online estimation system applied in the area of tool wear monitoring through integration of two promising technologies: artificial neural network and fuzzy logic. The proposed system is able to accurately predict the amount of tool wear. The results showed that the proposed system can significantly increase the accuracy of the predicted profile when compared to the conventional approaches.

Choa et al. [13] developed and implemented a tool breakage detection system using SVR in a milling process. To fulfill the diverse customer needs, multiple sensors have been used to establish the model. The breakage detection rate has been compared to a model developed using the traditional multiple variable regression approach. It is observed that the proposed SVR model performs well with a tight threshold value for tool breakage determination. Salgado et al. [14] investigated surface roughness estimation based on least squares support version vector machines (LS-SVMs) and proposed an in-process surface roughness estimation procedure based on LS-SVMs for turning processes. Brezak et al. [15] proposed a new type of continuous hybrid tool wear estimator, which is structured in the form of two modules for classification and estimation. Recently, tool wear studies in frequency and time–frequency domains gained some attention, and machine learning techniques have been widely used by researchers to predict tool wear. Support Vector Method, Hidden Markov Model (HMM), and Self-Organizing Map (SOM) have been used for predicting or categorizing tool wear in various machining processes such as drilling and milling. Kalman filter is another method that shows good noise suppression characteristic, and it has been used by some researchers in machining process control. Our studies have conducted the tool wear modeling and monitoring based on machine vision and showed that LS-SVM and Kalman filter can be used in machining process monitoring [16,17].

The objective of this study is to improve the tool wear estimation with low cost in machining hard to machine alloys by using dynamical prediction idea. Machine vision is used to measure tool wear online, and tool wear model is established considering the joint effect of machining conditions for predicting tool wear, namely models by using LS-SVMs. Then a dynamical prediction method is designed to adjust the established tool wear model by using the obtained tool wear online. Experimental works, which include experiments to train LS-SVM-based tool wear model and to verify dynamical prediction results are performed in MIKRON UCP710 five-axis milling center using stainless steel 0Cr17Ni12Mo2 in order to validate the proposed dynamical prediction method of tool wear. The measured tool wears in experiments are used to train the established LS-SVM-based tool wear model, and the inter-connection relationship between input and output parameters is determined after training. The tool wears obtained online by using machine vision are used to adjust the LS-SVM-based tool wear model on real-time basis.

5.2 Dynamical Prediction Method of Tool Wear

5.2.1 Basic Idea of Dynamical Prediction

Typical cutters exhibit two main wear zones known as flank wear and crater wear. A tool can be considered to have reached the end of its useful life when flank wear reaches a specified dimension. The extent of flank wear (V_b) can be measured as the distance between the top of the cutting edge and the bottom of the area where flank wear occurs. In order to obtain the tool wear more precisely, a dynamical prediction method of tool wear is proposed, which takes the measured value of the previous time as the input of the next time in the prediction of tool wear. The detailed idea can be explained as follows.

First, the tool wear is obtained by the established online tool wear measurement system using machine vision. Then, the tool wear model is established off-line by using LS-SVM method and considering the measured tool wear. The tool wear model based on LS-SVM can be used to predict tool wear. On the one hand, tool wear model can be used for off-line prediction, and the relationship between different machining parameters and tool wear can be easily obtained. Therefore, reasonable machining parameters can be selected and machining process can be conducted according to machining requirements. On the other hand, tool wear model can be used for online tool wear prediction.

When the tool wear is predicted online by using the established tool wear model, the error of predicted tool wear will continue cumulatively with the progress of machining process because of the randomness of the actual machining situation. If the predicted error cannot be fed back without delay, the estimated tool wear will significantly deviate from the actual tool wear. Therefore, in order to make the online prediction more suitable to the actual machining process, the effective feedback of prediction error should be fully considered. According to the difference between the predicted value and the measured value, the model can be adjusted appropriately to improve the output precision of the model. Thus in this paper, the initial tool wear is considered as the input when establishing the tool wear model, and the actual tool wear from online measurement system is also included in the established tool wear model to obtain the current tool wear. The error feedback is input into the tool wear model, and the dynamical prediction of tool wear is realized.

5.2.2 Flow Process of Dynamical Prediction

The dynamical prediction process of tool wear is established as shown in Figure 5.1. First, the tool wear model is used to predict the tool wear. The inputs of the model are cutting parameters and the initial tool wear.

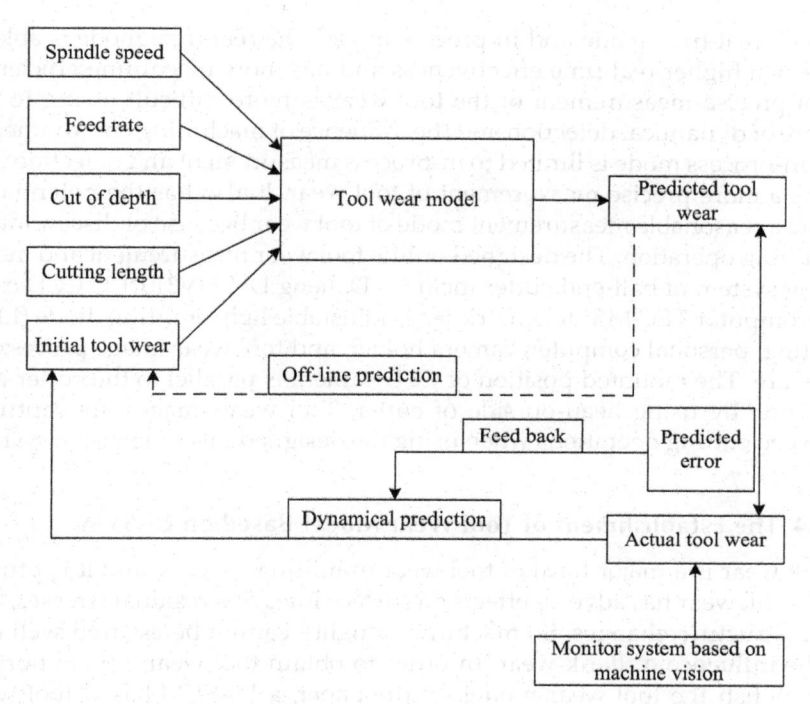

FIGURE 5.1
Dynamical prediction process of tool wear.

The difference is that the initial tool wear is used as the input of tool wear prediction. The cutting parameters include spindle speed, feed speed, cutting depth, and cutting length. The cutting parameters are determined by cutting requirements. The initial tool wear is the amount of tool wear before cutting. In fact, there must be some errors between the predicted tool wear and the measured value. When the machining parameters are selected to preliminarily predict tool wear before machining, the off-line prediction method can be used. In order to ensure the prediction accuracy, the dynamical prediction method is adopted, and the tool wear measured online by using machine vision is used as the input of the model, so that the prediction error can be effectively fed back. The prediction error of the previous stage will not be accumulated into the next stage. Thus, more accurate prediction of tool wear can be obtained based on the above dynamical prediction ideas.

5.2.3 Online Tool Wear Measurement Based on Machine Vision

In order to measure and monitor tool wear, an online tool wear measurement and monitoring system based on machine vision is designed and developed according to the characteristics of ball-end cutter [17]. There are two possible working principles for the design of tool wear measurement and monitoring system: real-time mode and in-process mode. The real-time mode is able to achieve a higher real-time effectiveness and has more machining efficiency, but a precise measurement of the tool wear is more difficult owing to the nature of dynamical detection and the influence of machining environments. The in-process mode is limited to in-process measurement and detection and offers a more precise measurement of tool wear. It also has the potential to create a reasonable measurement mode of tool wear because of discontinuity of milling operation. The designed online tool wear measurement and monitoring system of ball-end cutter includes Daheng DH-HV3103UC CCD camera, computar TEC-M55 telemetric lens, adjustable light emitting diode (LED) lighting, personal computer, camera holder, and tool wear image processing software. The mounted position of CCD camera is parallel to the cutter-axis direction by using head-on side of cutter. Tool wear images are captured between cutting operations when using the designed machine vision system.

5.2.4 The Establishment of Tool Wear Model Based on LS-SVM

Flank wear is a major form of tool wear in milling process, and it is proved that flank wear has adverse effects on surface integrity, residual stresses, and microstructure changes. So machining quality cannot be assured well due to the influence of flank wear. In order to obtain tool wear, it is important to establish the tool wear model. In the paper, a LS-SVM-based tool wear model [16] is used to conduct the tool wear prediction. The LS-SVM-based tool wear model structure includes three parts: input vector part, model part, and output vector part. The cutting conditions and the tool geometric information are taken as the input vector x_i in order to apply LS-SVM method in tool wear estimation. The number of output vectors is taken to be one, so as to indicate the value of tool wear at cutting edge position. The middle layer is LS-SVM-based tool wear model, which is used to describe the relationship between the input vector and tool wear V_b (the output of the model y). The input vector and output vector constitute the input–output pairs to train LS-SVM-based tool wear model and determine the relationship between the input vector and output vector.

5.3 Experimental Works and Discussion

5.3.1 Experimental Setup

In order to verify the validity of the proposed online tool wear dynamical prediction method, a series of milling wear experiments were conducted on MIKRON UCP710 five-axis milling machining center. Stainless steel 0Cr17Ni12Mo2, which is a difficult-to-cut material, was selected as work material of tool wear machining experiments. 0Cr17Ni12Mo2 stainless steel workpiece material was milled with coated carbide ball-end cutter in dry cutting condition. After cutting full length job, the tools were imaged as 1024×768-pixel gray-level digital images using a CCD camera. The corresponding average flank wear (V_b average) of the cutter was also measured using JSZ6S microscope with 8–50× magnification equipped with image acquisition software.

Experiments were performed to verify the validity of the proposed method. The stainless steel is machined by two-flute ball-end tools of cemented carbides with the diameter 16 mm. To perform the tool wear milling experiments, a workpiece in the form of a 50 mm thick box was prepared from 0Cr17Ni12Mo2 tool steel. The workpiece was fixed on standard fixture of MIKRON UCP710 five-axis milling machining center.

5.3.2 Design of Experiments

In this study, the orthogonal experimental method was used to design the milling experiment, and then the representative tool wear data were obtained, which were used to establish the tool wear model and verify the dynamical prediction method. Two groups of experiments were conducted as follows:

1. Experiment 1: In order to establish the tool wear model, the factors affecting tool wear should be considered comprehensively, including spindle speed, feed speed, cutting depth, and cutting length. The orthogonal experiment method was adopted to select three levels for each factor. The complete design is shown in Table 5.1.

2. Experiment 2: The purpose was to validate the established wear prediction model. The machining conditions employed are shown in Table 5.2.

TABLE 5.1

Orthogonal Factor Table for Ball-End Milling Cutter ($L = 75$ mm)

No.	Depth of Cut a_p (mm)	Spindle Speed n (r/min)	Feed Speed V (mm/min)	Cutting Length L (mm)
1	0.6	800	60	L
2	0.6	1,000	80	$3L$
3	0.6	1,200	100	$5L$
4	0.8	800	80	$5L$
5	0.8	1,000	100	L
6	0.8	1,200	60	$3L$
7	1.0	800	100	$3L$
8	1.0	1,000	60	$5L$
9	1.0	1,200	80	L

TABLE 5.2

Experimental Conditions for Group 2

Depth of Cut a_p (mm)	Spindle Speed n (r/min)	Feed Speed V (mm/min)	Cutting Length L (mm)
0.8	1,000	80	225

5.3.3 Analysis and Discussion of Experimental Results

5.3.3.1 Experimental Results

The developed software was used to process the tool wear image and measure the tool wear amount of each experiment. The maximum and average wear of the cutter were extracted by the tool wear extraction method discussed above.

For experiment 1, nine groups of experiments were carried out twice; each time the wear of left and right cutting edges was measured, and four groups of experimental data were obtained, as shown in Tables 5.3 and 5.4. The experimental number corresponds to the experimental number in the table. Table 5.3 is the maximum wear volume measurement, and Table 5.4 is the average wear volume measurement result. The wear values before machining and wear values after machining are given for each group. The wear measurement data in these two tables will be used for training tool wear models.

Then machining experiments are conducted using a tool based on the machining parameters in Table 5.2. After cutting a certain length such as L, the wear image of the tool is collected, and the wear trend of the tool with time is observed. The results are compared with those predicted by the wear model. The results of the measured wear data for the tool are shown in Table 5.5.

5.3.3.2 The Result Analysis Considering Dynamical Prediction

For the tool wear model based on LS-SVM after training by using the wear data in Tables 5.3 and 5.4, the optimal parameters of the maximum tool wear model are $C = 3.9702$, $\sigma = 17.697$, while the optimal parameters of the average tool wear model are $C = 5.2936$, $\sigma = 29.4951$. Based on the wear data in Table 5.2, the predicted average error is 4.48% and the standard deviation is 1.23% for the maximum tool wear model while the predicted average error is 2.65% and the standard deviation is 0.74% when using wear data in Table 5.3 for the average tool wear model. The above experimental results show that the tool wear model based on LS-SVM has better prediction accuracy. Thus, the established tool wear model can be used to predict tool wear off-line under certain machining scope. But the actual machining process is complicated and will change dynamically with the progress of machining process. In order to predict tool wear more accurately, the dynamical machining process should be considered in the established tool wear model. In this study,

TABLE 5.3

Results of Maximum Tool Wear (Unit: mm)

No	Data 1		Data 2		Data 3		Data 4	
	Before Machining	After Machining	Before Machining	After Machining	Before Machining	After Machining	Before Machining	After Machining
1	0.0475	0.051	0.0492	0.0534	0.051	0.0546	0.0534	0.0572
2	0.0546	0.0611	0.0572	0.065	0.0643	0.0709	0.0702	0.077
3	0.0611	0.0717	0.065	0.0765	0.0713	0.0826	0.0838	0.0964
4	0.0831	0.0948	0.0984	0.1105	0.0525	0.0633	0.0556	0.0666
5	0.0719	0.081	0.0765	0.0862	0.0701	0.079	0.075	0.0841
6	0.075	0.0831	0.0901	0.0984	0.0466	0.052	0.0494	0.0556
7	0.0708	0.085	0.0749	0.0902	0.0503	0.0637	0.0438	0.0569
8	0.0501	0.0639	0.0513	0.0647	0.0645	0.0784	0.0633	0.0769
9	0.0633	0.0708	0.0666	0.0749	0.0406	0.0458	0.0413	0.0476

TABLE 5.4

Results of Average Tool Wear (Unit: mm)

No.	Data 1		Data 2		Data 3		Data 4	
	Before Machining	After Machining	Before Machining	After Machining	Before Machining	After Machining	Before Machining	After Machining
1	0.0295	0.0323	0.037	0.0402	0.0323	0.0366	0.0398	0.0435
2	0.0366	0.0411	0.0435	0.0493	0.0482	0.0528	0.0458	0.0492
3	0.0411	0.0486	0.0493	0.0586	0.0514	0.0593	0.0646	0.0747
4	0.0676	0.0732	0.082	0.0901	0.0367	0.041	0.0412	0.0483
5	0.0556	0.059	0.0586	0.0612	0.0488	0.0517	0.0492	0.0516
6	0.059	0.0651	0.0777	0.082	0.0339	0.0377	0.0354	0.0412
7	0.0454	0.0562	0.0542	0.067	0.0301	0.0359	0.0266	0.036
8	0.0266	0.0373	0.0285	0.0346	0.0442	0.0513	0.0386	0.0459
9	0.041	0.0454	0.0483	0.0542	0.0278	0.0304	0.0287	0.0305

TABLE 5.5

Experimental Data

No.		Initial Value	Machining L	Machining 2L	Machining 3L	Machining 4L	Machining 5L	Machining 6L	Machining 7L
1	Maximum	0.0381	0.04175	0.04625	0.04938	0.05375	0.05958	0.06617	0.08175
	Average	0.0194	0.025	0.02816	0.0322	0.03523	0.03804	0.04192	0.04497
2	Maximum	0.0388	0.04417	0.0475	0.05283	0.05612	0.06013	0.06699	0.08325
	Average	0.0256	0.02808	0.03127	0.03496	0.03704	0.03817	0.04294	0.04767

a dynamical prediction method was designed to adjust the established tool wear model by using the tool wear obtained on online basis.

The dynamical prediction method takes each measured tool wear as the input of the tool wear model for the next wear prediction. Based on Group 2 experiments, the dynamical prediction effectiveness was validated. The comparison between the predicted and measured results for maximum wear and average wear is shown in Figures 5.2 and 5.3, where the maximum wear

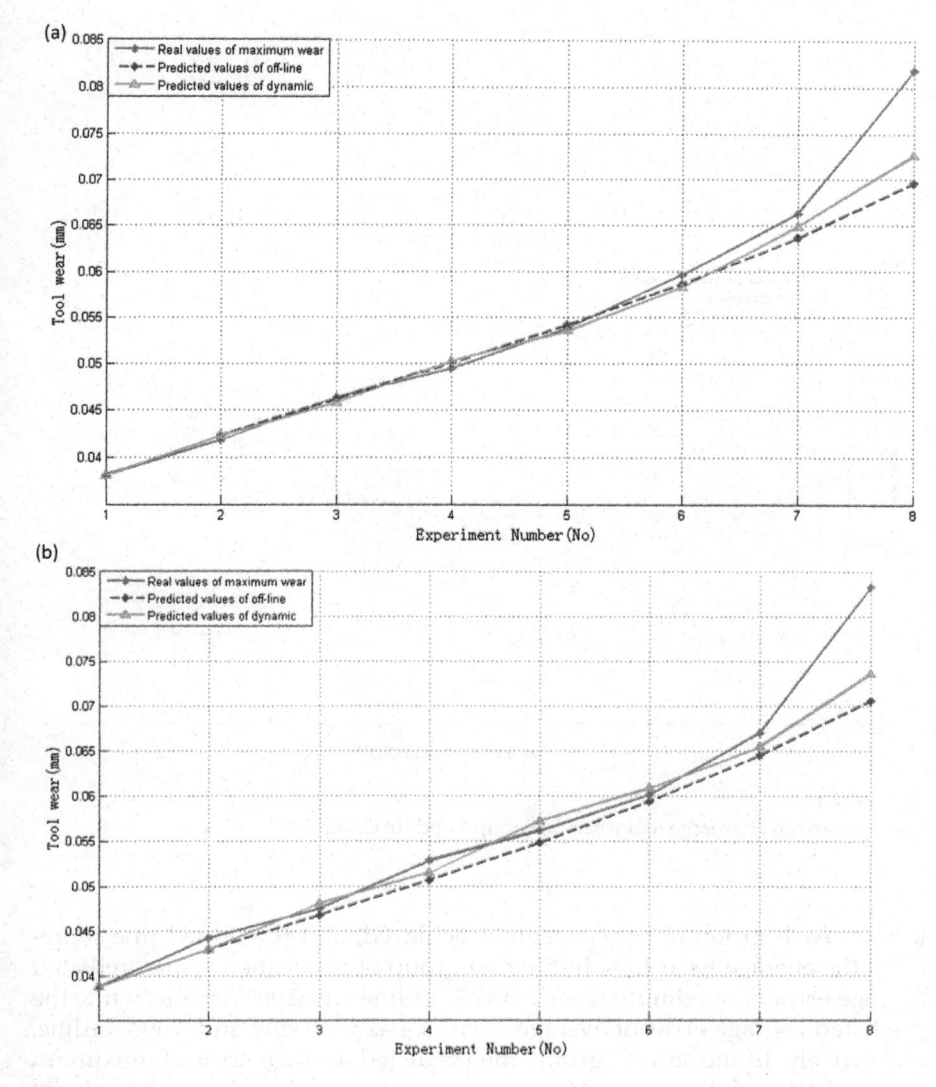

FIGURE 5.2
The comparison of the maximum tool wear: (a) Group 1 and (b) Group 2.

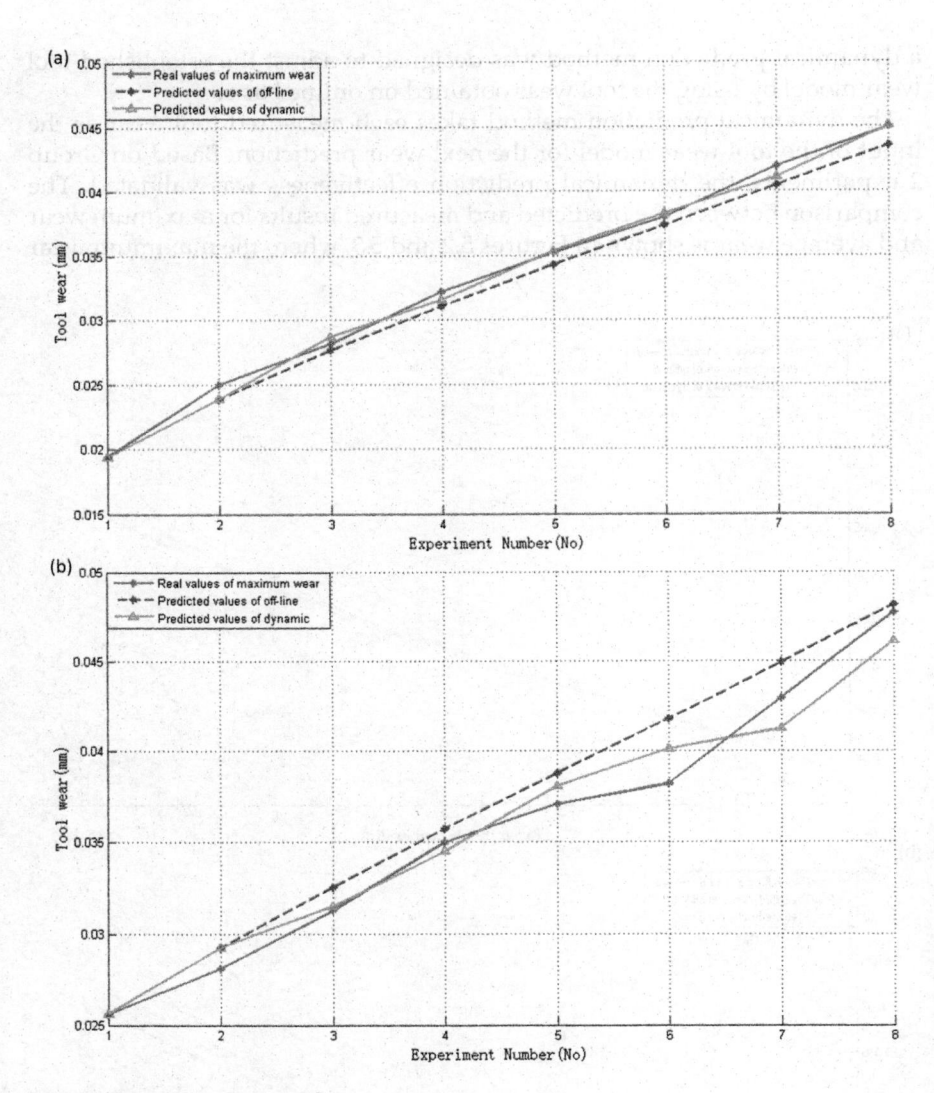

FIGURE 5.3
The comparison of average tool wear: (a) Group 1 and (b) Group 2.

loss for each group of experiments is achieved, and the dotted line represents the average wear loss. In the first group of experiments, the predicted average error of maximum wear is 3.41% off-line and 2.90% online, while the predicted average error of average wear is 4.42% off-line and 3.39% online, respectively. In the second group, the predicted average error of maximum wear is 2.94% off-line and 1.66% online, and the predicted average error of average wear is 4.27% off-line and 3.03% online.

The experimental results show that the dynamical prediction value is closer to the actual measurement results. In conclusion, the dynamical prediction method feeds back the online measured tool wear to the wear prediction model working on LS-SVM and further modifies the predicted tool wear value by considering the in-process tool wear in machining process.

5.4 Conclusions

In this chapter, a new dynamical prediction method of tool wear is proposed based on machining vision and SVM. The following is a summary of findings from this work:

- Machine vision is used to measure the tool wear on online basis. The tool wear model is established considering the joint effect of machining conditions, namely model by using LS-SVMs.
- In order to predict tool wear more accurately, a dynamical prediction method is designed to adjust the established tool wear model by using the online measured tool wear.
- The proposed model is validated through extensive experimentation. The tool wears measured in experiments are used to train the established LS-SVM-based wear model, and the inter-connection relationship between input and output parameters is determined after training. The online tool wear measured by using machine vision is used to adjust the LS-SVM-based real-time wear model.
- The experimental comparisons show that the proposed online dynamical prediction method of tool wear is satisfactory and provides a basis for industrial application for online tool wear monitoring.

References

1. S. Soleimani, J. Sukumaran, A. Kumcu, P.D. Baets, W. Philips, (2014). Quantifying abrasion and micro-pits in polymer wear using image processing techniques, *Wear*, 319(1), 123–137.
2. M. Castejón, E. Alegre, J. Barreiro, L. Hernández, (2007). On-line tool wear monitoring using geometric descriptors from digital images, *International Journal of Machine Tools and Manufacture*, 47(12), 1847–1853.

3. S. Dutta, S.K. Pal, R. Sen, (2016). Progressive tool flank wear monitoring by applying discrete wavelet transform on turned surface images, *Measurement*, 77, 388–401.

4. J. Jurkovic, M. Korosec, J. Kopac, (2005). New approach in tool wear measuring technique using CCD vision system, *International Journal of Machine Tools & Manufacture*, 45(9), 1023–1030.

5. M. Castejon, E. Alegre, J. Barreiro, L.K. Hernandez, (2007). On-line tool wear monitoring using geometric descriptors from digital images, *International Journal of Machine Tools and Manufacture*, 47(12–13), 1847–1853.

6. T. Ozel, Y. Karpat, L. Figueira, (2007). Modeling of surface finish tool flank wear in turning of AISI D2 steel with ceramic wiper inserts, *Journal of Materials Processing Technology*, 189(1–3), 192–198.

7. L. Fernández-Robles, G. Azzopardi, E. Alegre, N. Petkov, (2017). Machine-vision-based identification of broken inserts in edge profile milling heads, *Robotics and Computer-Integrated Manufacturing*, 44, 276–283.

8. A. Antic, B. Popovic, L. Krstanovic, R. Obradovic, M. Milosevic, (2018). Novel texture-based descriptors for tool wear condition monitoring, *Mechanical Systems and Signal Processing*, 98, 1–15.

9. J.C. Chen, J.C. Chen, (2005). An artificial-neural-networks-based in-process tool wear prediction system in milling operations, *The International Journal of Advanced Manufacturing Technology*, 25(5–6), 427–434.

10. W.-T. Chien, C.-S. Tsai, (2003). The investigation on the prediction of tool wear and the determination of optimum cutting conditions in machining 17-4PH stainless steel, *Journal of Material Processing Technology*, 140(1–3), 340–345.

11. J. Srinivas, K. Rama Kotaiah, (2005). Tool wear monitoring with indirect methods, *Manufacturing Technology Today India*, 4, 7–9.

12. R.J. Kuo, (2000). Multi-sensor integration for on-time monitoring tool wear estimation through artificial neural networks and fuzzy neural network, *Engineering Application Artificial Intelligent*, 13, 249–261.

13. S. Choa, S. Asfoura, A. Onarb, N.N. Kaundinyaa, (2005). Tool breakage detection using support vector machine learning a milling process, *International Journal Machine Tools and Manufacturing*, 45, 241–249.

14. D.R. Salgado, F.J. Alonso, I. Cambero, A. Marcelo, (2009). In-process surface roughness prediction system using cutting vibrations in turning, *The International Journal of Advanced Manufacturing Technology*, 43(1–2), 40–51.

15. D. Brezak, D. Majetic, T. Udiljak, J. Kasac, (2012). Tool wear estimation using an analytic fuzzy classifier and support vector machines, *Journal of Intelligent Manufacturing*, 23(3), 797–809.

16. C. Zhang, H. Zhang, (2016). Modelling and prediction of tool wear using LS-SVM in milling operation, *International Journal of Computer Integrated manufacturing*, 29(1), 76–91.

17. C. Zhang, J. Zhang, (2013). On-line tool wear measurement for ball-end milling cutter based on machine vision, *Computer in Industry*, 64(6), 708–719.

6

In-Process Measurement in Manufacturing Processes

Ahmad Junaid, Muftooh Ur Rehman Siddiqi, and Riaz Mohammad
CECOS University

Muhammad Usman Abbasi
University of Oxford

CONTENTS

6.1 Introduction ... 106
6.2 Metrology ... 107
    6.2.1 Contact Measurement ... 109
    6.2.2 Non-contact Measurement ... 109
6.3 System Error and Variation ... 109
6.4 Closed-Loop Manufacturing ... 111
    6.4.1 Remote Measurement ... 112
    6.4.2 *In-situ* Measurement ... 112
    6.4.3 In-process Measurement ... 113
        6.4.3.1 Process Improvement Loop ... 113
        6.4.3.2 Process Control Loop ... 114
        6.4.3.3 Design-for-Manufacturability Loop ... 114
6.5 Machine Vision System (MVS) ... 114
    6.5.1 One-dimensional (1-D) Machine Vision ... 116
    6.5.2 Two-dimensional (2-D) Machine Vision ... 116
    6.5.3 Three-dimensional (3-D) Machine Vision ... 116
    6.5.4 Machine-Vision-Related Terms ... 116
        6.5.4.1 Field of View (FOV) ... 116
        6.5.4.2 Working Distance (WD) and Resolution ... 116
        6.5.4.3 Depth of Field ... 116
6.6 Research Gap ... 118
6.7 Methodology ... 120
    6.7.1 Intrinsic Parameters ... 121
    6.7.2 Extrinsic Parameters ... 121
    6.7.3 Application of Camera Calibration ... 122
    6.7.4 Calibration Process ... 122
        6.7.4.1 Adding Calibrated Images ... 122

      6.7.4.2 Mapping World Units 122
      6.7.4.3 Adjusting Radial Distortion Coefficients 122
      6.7.4.4 Tangential Distortion 123
      6.7.4.5 Projection Error 124
      6.7.4.6 Visualizing Extrinsic Parameters 124
      6.7.4.7 Removing Lens Distortion 125
6.8   Digital Image Processing 125
     6.8.1   Image Representation and Modeling 125
     6.8.2   Types of Image Processing 126
     6.8.3   Low-Level Processing 126
     6.8.4   Mid-Level Processing 126
     6.8.5   High-Level Processing 126
6.9   Image Acquisition 126
     6.9.1   Segmentation Techniques 127
     6.9.2   Analyzing the Histogram of an Image 127
     6.9.3   Selecting Threshold 127
     6.9.4   Labeling of Object 127
           6.9.4.1 2-D Connectivity 128
           6.9.4.2 3-D Connectivity 128
6.10 Blob Analysis 128
     6.10.1   Centroid 129
     6.10.2   Area 129
     6.10.3   Bounding Box 129
     6.10.4   Label 129
     6.10.5   Major Axis Length 130
     6.10.6   Minor Axis Length 130
     6.10.7   Perimeter 130
     6.10.8   Equivalent Diameter Squared 130
     6.10.9   Eccentricity 130
     6.10.10 Blob Count 130
6.11 Integrating Intelligence 131
6.12 Conclusion 131
References 131

6.1 Introduction

Advancement in technology leads to bring change in the manufacturing techniques and processes in order to achieve the highest quality products. To meet this challenging situation, installing new technologies becomes more essential for industrial sectors. Manufacturing errors in high-value components are totally intolerable in sectors such as automotive, aerospace, nuclear power, shipbuilding, and large science facilities. Integrated touchpads with fast interface capabilities are the new working technologies that are available in machine tools. Without producing any interruption in the manufacturing process, they can inspect the measurement of a workpiece, during and after the manufacturing process. However, for a control process, uncertainties in the measurement of data are still not endured. For any quality production, the goal must be to keep the cost to the minimum level. Consequently, the production stage which is directly related to technical requirements is also assessed in terms of cost. The qualitative production success is closely subject to the quality control of production. Moreover, efficiency, cost, and quality of the manufacturing process are affected when extreme level of changes occurs in the manufacturing process as different uncertainties are related to machining operation process, i.e., temperature, calibration repeatability, uncertainty ranges move toward the higher values because a small change that occurs in process parameters affects the production quality of the workpiece. The performance characteristics of machining operation are randomly distributed over the given normal range value that regulates the tolerance in the given set of operations. Moreover, a heavy workload that causes an increase in operating temperature, machine tool component wear, and coolant degradation remains to continue over time till adjustment is made because of their non-random pattern [1]. The perfect solution to the above-mentioned problem is to build a system that it is resistant to any external internal system disturbance in the manufacturing process, but it's not realistically practical. The more real-world approach is to design a manufacturing model that can respond to any parameter change in the operating process. This leads to the observation of the whole operating system and tells what exactly is happening in the operating process. This approach applies well because of the deterministic nature of the manufacturing process, as cause and effect relationship exists between the output and the process parameters. Different events take place that cause a specific change in the process; even the observer remains unaware of the driving force behind the particular action. But maintaining the steady-state level of important characteristics of the process will uphold the output relatively in the constant state. Conversely, when the process parameter changes drastically, it produces a significant impact on the output. Product quality and lower production cost can be achieved by measuring and adjusting the key process parameters in real time [2]. For this purpose, in manufacturing industry, closed-loop manufacturing (CLM) methods are widely applied.

6.2 Metrology

Measurement processes are critical to industries such as automobile, aerospace, etc., where strict regulations regarding customer requirements are followed. Similarly, precise measurement of uncertainty is a serious concern for both accreditation bodies and clients. An uncertainty evaluation must be completed for all procedures that need to add an uncertainty statement to the estimation results. With a specific end goal to keep up characterized quality gauges, aerospace and automobile industries need to recognize all measurement disciplines that benefit from the expression of uncertainty level and characterization of strategies to ascertain it for the complex measurement processes. Qualitative measurement of a certain product or procedure depends upon requirements for its evaluation. With every measurement in real-world scenario, the physical condition alters, thus generating inaccurate results. Therefore, measurement accuracy is taken into consideration that expresses the degree to which the measured values have deviated from the standard value. Based on the characteristics of measurement, accuracy is divided into trueness and precision. The term trueness is the closeness of agreement between the average values obtained from a large series of test results and an accepted value [3], whereas precision is the closeness of agreement between the independent results and precise results that shows the standard deviation from the trueness. Precision is based upon repeatability and reproducibility. Repeatability in turn is further categorized, i.e., repeatability of measurement and repeatability of measuring equipment. The repeatability of measurement is the deviation in the measured values under the same operating condition such as equipment, calibration, environment, and a constant time span between each measurement, while the repeatability of equipment is the ability of the tool to produce the same results for the same test for every single measurement. On the other side, reproducibility belongs to the precision when analogous techniques are utilized for the measurement of same system using altering operators and measurement tools. For on-machine measurements, which is the focus of this research, there are different error sources which contribute to measuring errors, but not all sources contribute significantly to measuring error in all cases.

In the manufacturing process, measurement uncertainties have a profound impact on high-quality productions. For this, different standards are available for assuring quality production. According to ISO 14253-1 [4], the decision whether a manufactured part lies within the specified tolerance is based on expanding the measurement uncertainty of the entire measurement process. A measured feature can be accepted within the specified tolerance if the measurement result is above the lower specification limit plus the measurement uncertainty or if it is below the upper specification limit minus the measurement uncertainty. Likewise, it is outside the specification if the measurement result is less than the lower specification limit minus the measurement uncertainty or if it exceeds the upper specification limit by more than the measurement uncertainty. The measurement results fall into the range of the specification limits plus or minus the measurement uncertainty, the so-called uncertainty regions. The final decision whether the feature meets the desired specification cannot be based upon the performed measurement. Therefore, the region of conformity decreases with the increase in measurement uncertainty.

In industries, a number of forming processes are in practice. The geometry of the formed part is precisely measured once it is manufactured. There are various contact and non-contact-based measuring techniques which can be employed for inspecting the geometry of the formed part, e.g., contact measurement machine (CMM), optical scanner, etc. All these metrology technologies are used for post-process inspection. The main focus of this research is to perform the in-process inspection during the manufacturing process using the non-contact measurement-based metrology technique.

6.2.1 Contact Measurement

In contact measurement, the sensing part and the workpiece both are in contact. The most commonly used contact technique is a mechanical method in which the measuring tool is directly in contact with the workpiece. Calliper and friction roller type instruments are some of its examples [5,6].

6.2.2 Non-contact Measurement

In the non-contact measurement, the workpiece is not in contact with the sensing part. There are various methods for doing non-contact measures such as electromagnetic induction, laser imaging, optical method, electrical method, ultrasonic method, etc. Using any of these non-contact measurement methods, different features of workpiece can be measured. The main advantage in using non-contact measurement is that it measures the workpiece without producing any deformation in it, which is caused by the contact forces using contact sensors when measurements are highly subtle to sub-micrometre accuracy resulting in lessening the disturbance in the manufacturing process.

6.3 System Error and Variation

The manufacturing operation cost and quality are affected by the process variability since the manufacturing systems are unable to perform the exact similar operations under all circumstances. The process control is the main characteristic of machining operations, but still variability exists in operations such as coolant conditions, part's diameter, temperature control, etc. The variation in the data can be collected by inspection and monitoring of the parts and process parameters in different operating conditions. Information collected through process parameters can be used to check the state of manufacturing operations in addition to the input control signal to a feedback algorithm. The deviation in the key process parameters causes an error. The error can be generated from two different sources: machine accuracy and tooling accuracy. Machine accuracy is related to the tooling accuracy path. This characteristic generally depends on two types of errors: dynamic errors and quasi-static errors. Dynamic errors are associated with machine's servo system such as axis motion error, vibration error, spindle error, etc. Quasi-static errors are the slow varying disturbances which are present for a long time. Improper tool positioning with the workpiece is caused by such errors. Tooling accuracy is divided into two types: cutting tool errors and workpiece fixture errors. Cutting tool errors occur due to the alignment gap between machine axis, cutter, and error in the tool. Workpiece fixture error is caused by applying inappropriate clamping force that results in deformation of the workpiece.

The important process parameters can be detected using a correct manufacturing system error model. These parameters give an idea of how to adjust and control the manufacturing process. This can be done by having a constant check on key parameters, in response to a more accurate and a stable manufacturing system, e.g., errors in machine geometry; the tool offsets built on the condition of the cutting tool can increase the system performance. Still, if historical data are used in terms of adjustments rather than doing the real-time monitoring, unwanted changes in the factors such as machine characteristics and tool performance that are produced in manufacturing system cause a delay between inspection and production. Therefore, in-process measurement provides the best alternative solution to the above-mentioned problem as data is collected in real time with the assurance in maintaining the quality of the workpiece within the desired limit. Besides measuring the dimension of the workpiece, in-process measurement can use the data for the analysis of process consistency. In comparison with the other two CLM measurement techniques, in-process measurements have many advantages such as continuous, direct, and real-time measurements of the geometric parts within the desired tolerance range. There are two different techniques that should be utilized for in-process measurement, i.e., process qualification and process control. By performing statistical analysis on the collected data, the performance of manufacturing processes can be improved. This research is focused on improvement in quality of the processes.

On-Machine Measurement (OMM) in combination with the in-process inspection is the major part of the closed-loop manufacturing system. In-process inspection with OMM operation is implemented where the workpiece is being made. This is a part of the production facility and therefore impacts the overall cost estimation. Other than cost of production tools, there are some factors such as maintenance, instrument calibration, training, process time, production environment, and labor associated with the inspection system which should be included in the overall cost [7]. Precise manufacturing of the workpiece is possible; hence, in-process inspection can be found in many industrial applications such as

- Measuring geometric parts with laser machining [8]
- Integration of textile production machines with a vision system for inspection of fabric by detecting various defects [9]
- Vision-system-based automatic deburring of various components of aero engine driven by cutting tool [10]
- Integrated laser-based roundness measurement using in-process inspection in a turning process [11].

6.4 Closed-Loop Manufacturing

Manufacturing process efficiency is optimized by closed-loop manufacturing method. It incorporates metrology techniques (measurement technology), such as CMMs and non-contact measurement devices, to collect the data of machine tool characteristics and most importantly actual component dimensions and consists of various elements, e.g., efficient analysis, automatic data acquisition, continuous improvement, calibrated instruments, and reliable machines. When each element of CLM is combined with other different methods, it gives a complete closed-loop solution. The CLM cycle is not complete without various important manufacturing elements such as process judgment, data acquisition, geometry measurement, and data analysis. It is not enough to have only controlled measurements, but the loop is closed when the measurement data is utilized to produce efficiently manufactured components with low uncertainties. The data which is acquired during the measurement phase has a great importance in optimizing a different kind of loop targeting a different aspect of the manufacturing process.

There are several benefits of using CLM in the manufacturing process:

a. The reliability of the machine improves.

b. It provides a more controlled environment.

c. It reduces the human error while modifying the offsets.

d. It provides an acceptable accuracy level of the machinery assurance.

e. It provides automated tool and workpiece setup.

CLM can be verified by the geometric measurement method, but still, there are some issues such as where to measure, what to measure, and when to measure. This is the main focus of this research. Three different types of measurement can be done in the manufacturing process, and all of them can be used in the CLM process to measure the data:

- Remote Measurement
- *In-situ* Measurement
- In-process Measurement.

6.4.1 Remote Measurement

Remote measurement also known as post-processing measurement is the standard inspection method used in the manufacturing process. In industries, process validation is carried out by performing an inspection method on the formed part at the end of the manufacturing process cycle. Geometric measurements of the formed part are compared with those of the nominal specification part in inspection method. For this purpose, two different coordinate-based measurement instruments are adopted, i.e., contact and non-contact measurement instruments. Most of the industries such as aerospace, automotive, etc. are employing the post-process measurement method to evaluate the validation of the process and the formed component, e.g., coordinate measuring machine. Furthermore, for precise measurement of the component, a controlled environment with conditions, e.g., humidity, pressure, and specific temperature, is needed during the inspection. It is a relatively slower process in comparison with the other two measurement processes because of the shifting of a workpiece from one machine to the inspection area part. Though post processing accomplishes the precise measurement, still it is difficult to avoid the faulty metal parts caused by various factors such as machine tool, production environment, and tool and workpiece deflection which are produced during the machining process. *In-situ* process measurement could lessen this factor as the measurement of the workpiece is done during the manufacturing process.

6.4.2 *In-situ* Measurement

Advancement in technology has produced a huge impact on the industrial manufacturing process to raise the bar of product quality. To meet the criteria for forming components of highest level of quality, it becomes necessary to adopt the inspection techniques which manufacture the product under a definite tolerance level and minimize the uncertainties to be closer to given tolerance range to achieve the desired product quality. Industry 4.0 is the new chauffeur that controls not just efficiency but the heads to provide those technical solutions where product quality is the top priority. By knowing the measurement uncertainties of the part geometry, a precise product can be manufactured within the right specified tolerance range. *In-situ* measurement brings us closer to this functionality. Evolving of digitization in modern technology and its impact on most of the important manufacturing processes make it possible to utilize advanced technological methods such as *in-situ* measurements, which improve product quality standards. *In-situ* measurement process performs the measurement operation during the formation of the workpiece. It measures the geometry of the part while the workpiece is being formed on part forming machine. Machining process stops when the workpiece is displaced from its original position to the machine tool for measurement. As the machine tool is closer to the workpiece, the process takes less time for completing the measurement task. That is why this is also called on-machine or process intermittent measurement [12]. Besides having many advantages, *in-situ* measurement has some disadvantages: for example, in forming process, it produces a pause state while measuring the part. This leads us toward the most important and dynamic approach to in-process measurement. In-process measurement methods inspect or measure the workpiece without any pauses during the manufacturing process.

6.4.3 In-process Measurement

In the manufacturing process, the inspections, including dimensional inspection, are commonly performed at the end of the process rather than during the process. In many cases, several machining steps are completed before the actual part is measured. Therefore, if the geometric measurement of a part deviates from the given tolerance, it is either rejected or reworked. In both cases, the overall manufacturing cost increases. In-process measurements provide the best way solution to the aforementioned problem. In manufacturing, the measurement operation (metrology) is performed when the inspection system becomes the essential part of a manufacturing system. In-process measurement is the combined set of software, hardware, activities, and procedures which are incorporated in the manufacturing system to output the measurement of dimensional characteristics of the manufactured products and tools during the manufacturing process. During the manufacturing process, real-time inspection is carried out; that is, the inspected data is collected and measured simultaneously during the manufacturing process. Errors and deviation during the manufacturing process are measured by digital sensors which help in controlling the process before it exceeds the tolerance range. Advancement in the digital computer, sensor technology, and control systems are allowing a drastic increase in the applications of in-process measurement.

The main advantages of using in-process measurement are that it performs measurements in real time, takes measurements continuously, and manufactures the part in the defined tolerance range.

CLM systems are of different types; therefore, to collect data, they use different measurement operations.

In a manufacturing system, three types of CLM loops are used.

6.4.3.1 Process Improvement Loop

This is not an instant and automated improvement loop. Processes improvement loop provides the understanding of process repeatability and its accuracy. *In-situ* measurements are normally recommended for processes improvement loop, but in-process measurement data can also be used for evaluation purpose.

6.4.3.2 Process Control Loop

Process control loop subsists between the measuring system and machine tool. It is also called intermediate closed loop. Metrology instruments take measurements and then evaluate the results with the nominal tolerance limits to further compensate the geometry deviation from the nominal shape. In-process measurement is normally used for this type of closed loop.

6.4.3.3 Design-for-Manufacturability Loop

Data collected by this loop gives the idea of the degree of difficulty in manufacturing and designing features of various parts. Remote measurement is used for this type of closed loop. However, for analyzing the data, *in-situ* measurement and in-process measurement can also be employed for this closed loop.

The most important CLM system among all the three types stated above is the process control loop because the manufacturing process can only produce high-quality parts if the manufacturing system is more flexible and adaptive to the rapid changes in the manufacturing industry. The important expect of using process control loop is to minimize the factors of process variability and bias in a manufacturing system. Preceding one is the difference between the design value and average parameter value. Bias errors are the constant deviations from the actual target value; they can be managed by using calibration processes. The process variable is a continuously varying phenomenon caused by the changing of any one of the parameters of the manufacturing process. It is hard to be fixed because of its unpredictable nature. However, these unwanted errors can be handled by knowing the inside information of manufacturing system parameters, and this can be achieved through real-time processing in CLM. The concept of CLM is not complicated, but the essential process of data collection is a challenging task. Knowing the knowledge of process variation and error resources helps to decide when and where to apply in-process measurement.

Therefore, the focus of our research is based on employing in-process measurement using non-contact measurement techniques which initiate the concept of machine vision system.

6.5 Machine Vision System (MVS)

In-process measurement is about the design and methods of manufacturing processes, which allow real-time in-process measurement to control and monitor the manufacturing process to get real-time quality assurance.

Implementing adaptive control system into the production lines for real-time optimization is one of the techniques used for monitoring and controlling of the manufacturing process. Several schemes were proposed to develop a decision-making system based on sensor signals that would derive a conclusion on machining process conditions. The most frequently employed are the neural networks and fuzzy logic for monitoring the machining process. In the past few decades, the progress in machine vision was exceptional. Most of the industrial activities benefit from a vision-inspection-based system such as quality textile production, circuit manufacturing, printing, etc. Previously, machine vision technology was not extensively applied at industry level due to high-cost equipment and limitations of image quality resolution. However, with continuous improvement in technology, the power of the monitoring systems and the resolution of the cameras were improved. This makes machine vision system technology to be widely used in many applications. In vision systems, the image quality and reliability of the vision algorithm depend on the placement and selection of the light sources and camera. Machine vision is a superior replacement for human vision in applications such as high-speed verification and testing. Machine vision uses cameras, computers, and algorithms to replace human vision in performing inspection tasks that require precise, repetitive, and high-speed verification and testing. Machine vision is the technology that provides an imaging-based automatic inspection and analysis for applications such as process control, automatic inspection, robotic guidance, etc. Although automotive industry, semiconductor industry, food packaging industry, and electronic/computer industry are some of the industrial applications of machine vision, in this research, machine vision technology is applied in the manufacturing process for in-process measurement. Machine vision system has many benefits and strategic goals which include automation of quality monitoring, higher quality products, increased productivity, production flexibility, contactless measurement, measurement of dimensions and tolerances, less machine downtime, less setup time, lower production costs, and scrap rate reduction. Machine vision systems dramatically improved the performance, ease-of-use, intelligence, and cost in past few years. With the dramatic increase in processor performance, many more applications can now be economically solved with machine vision.

The success of the MVS depends upon how smartly the key components of MVSs are managed. The main components of MVSs are lighting, camera, sensors, and processing. Features of the object are illuminated by the lighting which can easily be seen by the camera. After sensing the data by the camera sensor, the information is provided to the system for further processing.

Selecting the right vision system plays the central part to meet the need of various vision applications. There are different types of MVSs.

6.5.1 One-dimensional (1-D) Machine Vision

One-dimensional (1-D) vision system uses line scan cameras for continuous inspection and classification of defects on materials which are manufactured in a continuous process such as plastics, non-woven sheets, and roll goods.

6.5.2 Two-dimensional (2-D) Machine Vision

Most of the two-dimensional (2-D) capturing performed by the camera uses area scan inspection. Line scan is another type of 2-D machine vision in which the 2-D image is built line by line. The encoder is required to track the movement in a short exposure time.

6.5.3 Three-dimensional (3-D) Machine Vision

Three-dimensional (3-D) machine vision consists of more than one camera or sensor. In order to scan the entire part, the camera or object must be moved which is similar to line scanning. Most of its applications are found in robotics.

6.5.4 Machine-Vision-Related Terms

Some important terms related to MVSs are given below.

6.5.4.1 Field of View (FOV)

The part which is seen by the vision system at a given instant is defined as field of view (FOV). The FOV depends on the lens of the camera and the working distance between the camera and object.

6.5.4.2 Working Distance (WD) and Resolution

Working distance (WD) is the distance between the target and the lens front, whereas the resolution is the minimum feature size that the camera can detect and has a direct relation with FOV.

6.5.4.3 Depth of Field

It is the distance in front of the object which appears to be in focus.

The performance of MVSs can be improved if all the above-mentioned factors are considered while selecting the right machine vision optics. Light structure is one of the important aspects used in the MVS. There are different ways to structure the light. Direction and angles of the light produce a high impact on camera sensor while inspecting the object and determine how the object is seen by the camera.

Machine vision helps to inspect a variety of items such as printed circuit boards, electronic components, integrated circuits, and other machine tools. In the past few decades, machine vision was widely used in manufacturing to inspect the quality of printed circuit boards [13,14], electric components [15], chip alignment, integrated circuits' wire bonding [16,17], machine tools [18,19], metal parts [20–24], and semiconductor wafers [25,26]. For the inspection of these applications, a number of techniques have been used and reviewed [27–30]. An automated vision inspection system was developed to efficiently inspect the microdrill blades [31]. The artificial neural network was developed to identify the defective dies of LEDs using an automated inspection system [32]. In [33], an automated system that detects damages on ceramic machined surfaces using pattern recognition, image processing, and segmentation method of machine vision techniques was adopted to detect the defects using Gabor filter using fast and supervised feature-based segmentation of the algorithm fuzzy C-Mean to identify organic light emitting diodes (OLEDs) faults [34]. Inline inspection of the surface was done based on texture analysis of statistical image processing methods for a cold form of microparts [35]. Different types of defects such as deformation, scratches, and rust on the bearing cover were detected using an MVS [36]. A vision-based method was proposed to detect various brightness-level defects in liquid crystal display (LCD) panels, in addition to inconsistencies in the shape and size [37]. An automated Bayesian inspection method was introduced for printed circuit board assemblies, and it deals simultaneously with different circuit elements on various scales [38]. In [39], the authors presented vision-based detection of a cylindrical object that measures the diameter of a small cylindrical object using only one camera with high precision. In [40], the authors presented an approach for detecting the object and incomplete object poses and eliminating those poses using the noisy RGB-D sensor.

Similarly, in [41], the authors presented the embedded vision system for underwater object detection design. Both [42] and [43] represent a system which finds the diameter and center of cylindrical objects. A method was developed, which finds the coordinates of tube-shaped objects using a mobile robot armed with a laser range finder; besides measuring the orientation and diameter successfully, these techniques have limitations while using them in actual or in some specific condition especially in case of interferences. A non-contact method using machine vision technology was implemented to measure the roundness of a camshaft. The system consisted of a camera, image processing software, lighting devices, and holding tools, but still, the measurement-based concept needs improvement to evaluate roundness error more accurately [44].

The integrated vision system is one of the types of MVS used for an application that requires defect detection, alignment, measurements, etc. The human vision system can be easily replaced by an MVS in scenarios where the geometry part complexity increases and cannot be found easily.

Integrating MVS with in-process measurement can make fast, less expensive, and high-quality products which are the main focus of this research.

6.6 Research Gap

Conventional quality control procedures are performed after the workpiece is manufactured. The disadvantage is that if the workpiece dimensions lie outside the given tolerance range in quality control inspection, the product gets rejected or recycled. This not only affects the cost, lead time, energy, and machine and labor utilized to develop the workpiece but also has long-lasting environmental effects.

Real-time intelligent monitoring system (RIMS) will investigate, design, and develop a real-time intelligent measurement manufacturing system. This system will use machine vision technology using state-of-the-art image processing algorithms to identify and measure the real-time dimensions of the workpiece during the manufacturing process.

The scope of the research lies at the zenith of manufacturing processes, industrial internet of things (IIOT), MVS, metrology, and quality control. Figure 6.1 defines the scope of the research.

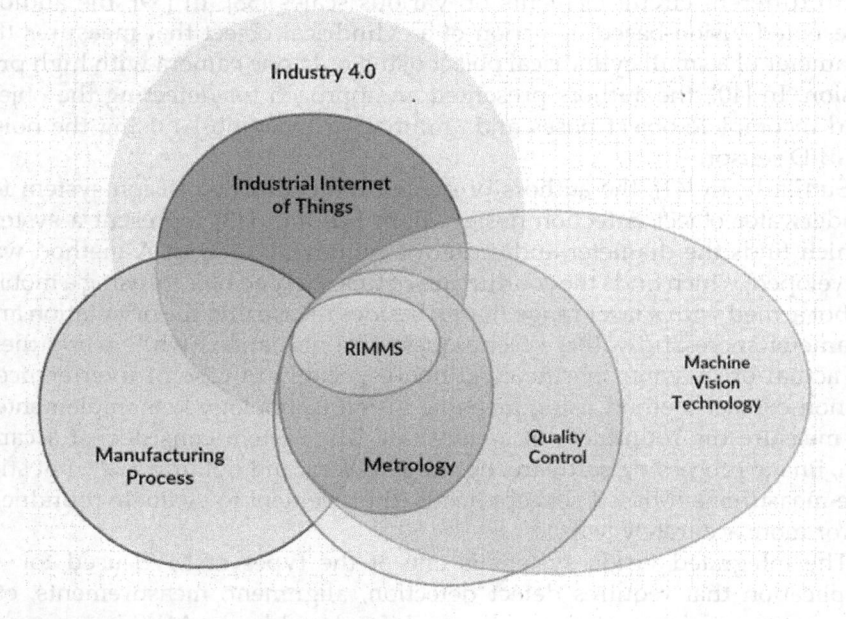

FIGURE 6.1
Scope of the research.

In the present scenario, determining the quality of the product, i.e., the geometric measurement, is done through post-processing measurement techniques. These methods are not only slow but also have a high impact on the final product cost. Using vision technology, the quality control and manufacturing process can be integrated into the same segment. This research project aims to implement in-process measurement for developing a real-time inspection system in which the quality control operation is performed during the manufacturing process.

The proposed in-process measurement flow diagram is shown in Figure 6.2. In the initial step, computer aided design (CAD) designed data will be input to the machine controller. In the manufacturing phase, the information is fetched from the machine control, and manufacturing process will be performed on the given raw product based on the CAD data information. At the same time during the manufacturing process, the metrology operation will be functional by integrating the quality control in the manufacturing phase. By merging the quality control and the manufacturing process, real-time metrology operation can be performed as both the processes will work simultaneously, and the result will be forwarded for comparison. In the comparison section, the comparison operation is performed between the manufactured data dimension and the defined CAD data dimensions. After performing the comparison, if the dimensions of the workpiece lie within the defined tolerance range of the input CAD model, the feedback control

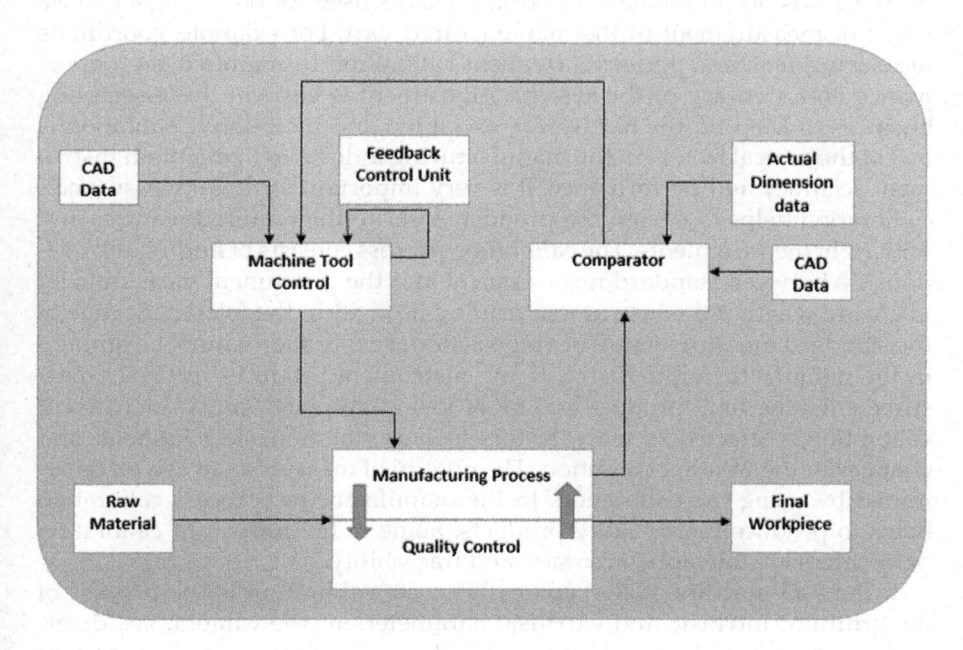

FIGURE 6.2
Proposed technique in-process measurement.

inputs the machine control to stop the process. Otherwise, the process will continue until the desired tolerance level is attained. By performing a real-time comparison of the manufacturing product, any kind of tolerance range can be achieved. The proposed aforementioned in-process measurement technique can be implemented using different hardware and software components, e.g., workpiece, machine tools, camera, CAD software, and image processing software.

6.7 Methodology

This section is divided into two parts. The first part is related to the calibration of its types and the workflow. The second part is about digital image processing with different tools and techniques related to image processing.

As this research focuses on real-time measurements of the workpiece during the manufacturing process, using a different measurement technology gives more precise and accurate results without disturbing the manufacturing process. MVS is one of the non-contact-based measurement techniques that can be used for metrology operation. Different image processing tools can be applied using Matlab for analysis. In manufacturing industries, calibration is one of the most important factors used for obtaining a precise result of measurement of the manufactured part. For example, coordinate measuring machine performs frequent calibration to maintain the performance and accuracy of the system. Adjustment is basic in the assembling business to keep up the hardware's execution and precision. Calibration is one of the critical factors in the manufacturing industries to maintain instrument accuracy and performance. It is very important for quality assurance. Calibration helps to design the product closer to the required features and with right measurements. The calibration process consists of finding the relationship between standard measurement and the instrument measurement to ensure whether the instrument reading lies within the tolerance range of the standard measurement. The importance of calibration cannot be ignored in the manufacturing industry. If any instrument produces incorrect measurements, the final product will be of low quality and faulty. Instrument calibration is affected by many factors such as vibration, electric shock, and changes in the weather condition. The quality of measurement can be determined by using the calibration. In the manufacturing, process calibration helps to produce high-quality products. Some of the important calibration parameters are tolerance, accuracy, and traceability.

In the 3-D machine vision context, camera calibration is the process of determining intrinsic and extrinsic parameters of the camera, i.e., determining the optical characteristics of the camera and internal camera geometric parameters for intrinsic parameter, whereas, the extrinsic parameter is the orientation 3-D position of the camera frame relative to a definite world coordinate system [8]. The overall machine vision system performance in many cases depends on the accuracy of the camera calibration. Physical parameters of the camera are divided into extrinsic and intrinsic parameters. Extrinsic parameters transform the object coordinates to a camera-centered coordinate frame. Moreover, the relationship between different cameras in a multi-camera system is also described by the extrinsic parameters.

Camera calibration is the process of finding the characteristics of the camera and its location in space with respect to the fixed object. This is critical to correct the lens distortion of the camera or measure the size of the object in world units. Intrinsic and extrinsic parameters are two important parameters related to the camera.

6.7.1 Intrinsic Parameters

The internal characteristics of the camera are known as intrinsic parameters. These include the focal length of the lens, lens distortion, optical center, distortion coefficient, and a scaling factor of pixels. Knowledge of these parameters can enhance the image quality, correct the lens distortion, and map the real-world distance to the pixels.

In the matrix containing the intrinsic camera parameters, $\begin{bmatrix} cx_1 & cy_1 \end{bmatrix}$ coordinates represent the principal point in pixels. The skew parameter is equal to zero when both the axes (x, y) are perpendicular to each other.

$$M = \begin{bmatrix} fx_1 & 0 & 0 \\ s & fy_1 & 0 \\ cx_1 & cy_1 & 1 \end{bmatrix},$$

where
$fx_1 = F*sx_1$
$fy_1 = F*sy_1$
F = focal length in world units, expressed in millimetres
$[sx_1, sy_1]$ = number of pixels per world unit in the x-direction and y-direction, respectively
fx_1 and fy_1 are expressed in pixels.

6.7.2 Extrinsic Parameters

Extrinsic parameters of the camera tell the camera location in the space in relation to the object. Camera calibration is useful in applications such as machine vision, where the actual size of the object is measured using images.

The following equation provides the transformation that relates a world coordinate in the checkerboard's frame $[X \quad Y \quad Z]$ and the corresponding image point $[x \quad y]$:

$$s\begin{bmatrix} x & y & 1 \end{bmatrix} = \begin{bmatrix} X & Y & Z & 1 \end{bmatrix} \begin{bmatrix} R \\ t \end{bmatrix} \cdot K,$$

where

 R = 3-D rotation matrix
 t = translation vector
 K = intrinsic matrix.

6.7.3 Application of Camera Calibration

In robotics, calibrated cameras are used in navigation systems and 3-D scenery construction, where different images taken with the calibrated camera are used to improve the 3-D structure of the scene.

6.7.4 Calibration Process

The computer vision system toolbox provides a function and app for camera calibration. It helps to complete the camera calibration workflow.

6.7.4.1 Adding Calibrated Images

The first step is to add the calibrated images of the checkerboard calibration pattern. Thirty to forty images are used for accurate calibration results. A checkerboard is used because of its regular pattern and to make it easy to detect automatically.

6.7.4.2 Mapping World Units

Next step is to enter the size of checkerboard square in world units. It could be in millimetres, iches, and centimeters. This is necessary to find the mapping between the image pixels and world unit. The checkerboard detection is then detected.

6.7.4.3 Adjusting Radial Distortion Coefficients

Radial distortion coefficient can also be specified before doing the calibration. Radial distortion occurs when a light ray bends at a greater amount near the edges of the lens than it does at its optical centre. There are two types of distortion: negative radial distortion and positive radial distortion. Typically, two distortion coefficients are sufficient, but for severe distortion, as in wide angle lens case, three distortion coefficients might be necessary.

The camera parameters of the object calculate the radially distorted location of a point.

Distorted points as (x_distorted, y_distorted) follow as

$$x\_distorted = x\left(1 + n_1 * r^2 + n_2 * r^4 + n_3 * r^6\right),$$

$$y\_distorted = y\left(1 + n_1 * r^2 + n_2 * r^4 + n_3 * r^6\right),$$

x, y = undistorted pixel locations
n_1, n_2, and n_3 = radial distortion coefficients of the lens

$$r^2 = x^2 + y^2$$

Typically, two coefficients are sufficient. For severe distortion, n_3 can be included. The undistorted pixel locations appear in normalized image coordinates with the origin at the optical center. The coordinates are expressed in world units.

6.7.4.4 Tangential Distortion

Tangential distortion is another type of distortion that can be computed. Tangential distortion occurs when the lens and camera center are not parallel. After computing the entire distortion, next step is to apply the calibration to solve the camera parameters.

Tangentially distorted location is calculated by camera parameters of the object as follows:

$$Distorted\ points = \left(x\_distorted, y\_distorted\right),$$

$$x\_distorted = x + \left[2 * m_1 * x * y + m_2 * \left(r^2 + 2 * x^2\right)\right],$$

$$x\_distorted = y + \left[m_1 * \left(r^2 + 2 * y^2\right) + 2 * m_2 * x * y\right],$$

x, y = undistorted pixel locations
m_1 and m_2 = tangential distortion coefficients of the lens

$$r^2 = x^2 + y^2$$

The undistorted pixel locations appear in normalized image coordinates with the origin at the optical center. The coordinates are expressed in world units.

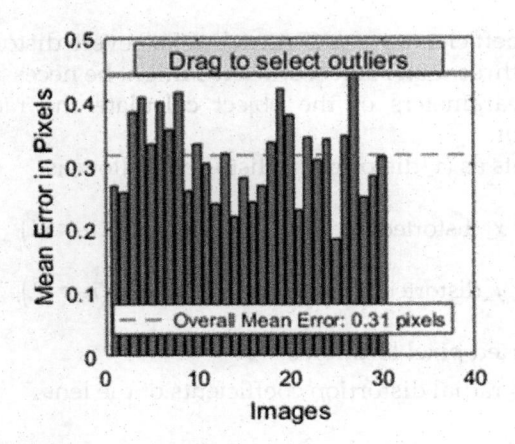

FIGURE 6.3
Reprojection error.

6.7.4.5 Projection Error

Once calibration is done, it can be evaluated by visualizing the projection errors. Projection errors are the global measures of calibration error, and it's the difference between points reprojected back on to the image and points detected in the image using the camera parameters that was calculated. It helps to identify noisy images, which can be removed and recalibrated for better results shown in Figure 6.3.

6.7.4.6 Visualizing Extrinsic Parameters

Extrinsic parameters can also be visualized to see which angles calibration images are taken from. This is sometimes useful to find out when the calibration images aren't taken from enough angles and more images might be needed to improve calibration results, as shown in Figure 6.4.

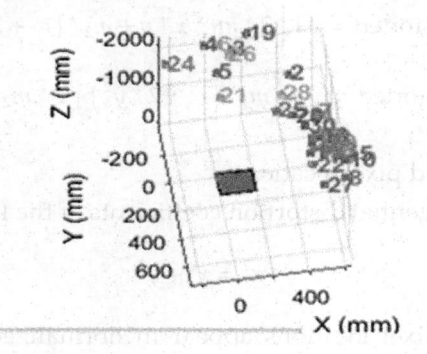

FIGURE 6.4
Visualizing extrinsic parameters.

(a) (b)

FIGURE 6.5
Lens distortion.

6.7.4.7 Removing Lens Distortion

Lens distortion is the common problem; it causes straight lines to appear as curves. By knowing the camera's intrinsic parameters, an undistortion function is applied to remove the lens distortion so that the edges in the images which appear as curves are straightened out, i.e., lens distortion has been removed. Correcting lens distortion is very useful in computer vision as shown in Figure 6.5.

Now as the camera is calibrated, and all the parameters of the camera are solved, the next step is to acquire the images for real-time processing.

6.8 Digital Image Processing

The rapid growth of digital technology and the development of algorithms have made digital image processing one of the most important components of digital applications. There is no technical area in which digital image processing applications are not required, e.g., in data communication, remote sensing, computer vision such as robotics, assembly line, autonomous systems, etc.

6.8.1 Image Representation and Modeling

An image can be represented in the transform or spatial domain. An important consideration in image representation is measuring the image quality such as contrast, the sharpness of the edge information, and color. Images represented in the spatial domain demonstrate luminance of an object. In the spatial domain, for modeling the image, linear statistical model can be used to develop different algorithms. Alternatively, image transformation techniques can be applied to digital images to extract characteristics such as spectral content, bandwidth, or other features for many applications including filtering, compression, and object recognition.

6.8.2 Types of Image Processing

Image processing is done at three levels:

- Low-level processing
- Mid-level processing
- High-level processing.

6.8.3 Low-Level Processing

Different operations can be performed in low-level processing such as noise reduction, sharpening of image, etc. No optical imaging system will give images of perfect quality. The aim of image enhancement is to improve the quality of the image. Examples of image enhancement (low-level processing) are contrast and grayscale improvement, pseudo coloring, edge sharpening, image restoration, and noise removal.

6.8.4 Mid-Level Processing

Mid-level processing involves partitioning an image into objects or regions. Image transformation includes the mapping of digital images to the transform domain. In various applications, image transformation is required as different useful characteristics of the images can be detected.

6.8.5 High-Level Processing

Both image analysis and computer vision are performed in high-level processing. It includes feature extraction, segmentation, and classification. Segmentation techniques are performed to separate the target object in the image from background so that the features of the desired object can be measured easily and accurately. The most important features are then extracted from the targets (segmented objects). Quantitative estimation of these features allows the classification and description of the object.

6.9 Image Acquisition

Image acquisition for real-time processing is the most important step. For this, Matlab program functions are used to capture images of the target object frequently until the required condition of the target object is satisfied.

This can be done by iterating over a variable several times to get continuous images using a webcam.

6.9.1 Segmentation Techniques

Image segmentation techniques are useful in order to identify the target object or to get desired information in the digital image and to separate the image into various parts. Segmentation subdivides an image into its constituent objects. The level to which the segmentation is carried depends on the problem being solved. Segmentation uses different techniques like morphing and point, line, and edge detections. Problems with illumination like shadows on the background and the difference in the intensities of the objects in the image may arise in segmentation.

Image segmentation can be performed in many ways that include

1. Otsu's method
2. K-means clustering
3. Watershed segmentation
4. Texture filters.

6.9.2 Analyzing the Histogram of an Image

The histogram shows the spread of intensity values across the various pixels of the image. For a grayscale image, intensity values range from 0 to 255. In any given image, the background and foreground of the image can be separated using the specified threshold value which can be selected by analyzing the histogram of the image.

6.9.3 Selecting Threshold

Otsu's method chooses the threshold to minimize the intra-class variance of the black and white pixels. Otsu's segmentation technique is used to separate the objects in the image using the histogram to separate the different objects based on their intensity values.

Segmentation of the image can be then done using the appropriate threshold value. This segmentation technique generates the threshold value used to convert the image into the binary image.

6.9.4 Labeling of Object

After converting the grayscale image into the binary image using Otsu's segmentation technique, the next step is to find the target object in the image. For this, the objects were labeled using the connected component tool. There are different ways to connect the objects using different connectivity.

A connected group formed in a binary image by the set of pixels gives rise to the connected component. The identification of the connected components and assigning each object a unique label is done by using the process of connected component labeling.

Based on dimensions, connectivity is divided into two types:

    a. 2-D connectivity

    b. 3-D connectivity.

6.9.4.1 2-D Connectivity

2-D connectivity includes 4- and 8-connected neighborhood pixels for connectivity of connected components.

6.9.4.2 3-D Connectivity

3-D connectivity includes 6-, 18-, and 12-connected neighborhood pixels for connectivity of connected components.

Currently, the 8-connected neighborhood is used. This labeling of an object helps to identify the target object.

6.10 Blob Analysis

Once the target object is identified, the diameter of the object can be calculated using blob analysis by using the computer vision system tool in Matlab. Blob analysis is a fundamental technique of machine vision based on analysis of consistent image regions. Main advantages of this technique include high flexibility and excellent performance. It also computes the statistics of the labeled region in the binary image.

The blob analysis solution consists of the following steps:

    a. Extraction: By applying image threshold techniques, pass the target object through inspection.

    b. Refinement: Enhancement of extracted region.

    c. Analysis: In the final step, the refined region is subject to measurements, and the final results are computed.

The subset of image pixel is called region. A blob analysis algorithm is used to find and count objects and to make a basic measurement of their characteristics. The purpose of the analysis is to determine whether the results obtained from an operation are accurate, logical, and true. The block returns quantities such as the centroid, area, bounding box, label, major axis, minor axis, perimeter, eccentricity, and blob count.

6.10.1 Centroid

This part of blob analysis block gives the centroid coordinates of a matrix. The center of mass for a binary image is the average x- and y-positions of the binary object. It is defined as a point, whose x-value is calculated by summing the x-coordinates of all pixels in the blob and then dividing by the total number of pixels. Similarly, for the y value, y-coordinates are summed up and then divided by the total number of pixels. In mathematical terms, the centroid (x_c, y_c) is calculated as

$$x_c = \frac{1}{N} \sum_{i=1}^{N} x_i$$

$$y_c = \frac{1}{N} \sum_{i=1}^{N} y_i$$

N = number of pixels in the blob

x_i = x-coordinates of N pixels

y_i = y-coordinates of the N pixels.

6.10.2 Area

Specifies the actual number of pixels in the object or region.

$$A = \sum_{(x,y) \in R} 1$$

R = set of pixels in a region.

6.10.3 Bounding Box

Returns the smallest rectangle containing the region, specified as a 1-by-Q*2 vector.

where

Q = number of image dimensions.

6.10.4 Label

This port gives the label matrix.

6.10.5 Major Axis Length

Specifies the length (in pixels) of the major axis of the ellipse in pixels.

6.10.6 Minor Axis Length

Specifies the length of the minor axis of the ellipse in pixels.

6.10.7 Perimeter

For each blob, this port contains the perimeter length in pixels.

6.10.8 Equivalent Diameter Squared

This port of blob analysis block represents equivalent diameters squared in a vector matrix. For circular objects, it is computed as

$$\text{sqrt}(4*\text{Area}/\text{pi}).$$

6.10.9 Eccentricity

Specifies the eccentricity of the ellipse. The eccentricity is the ratio of the distance between the foci of the ellipse and its major axis length. The value is between 0 and 1. An ellipse with 0 eccentricity is a circle, while an ellipse whose eccentricity is 1 is a line segment.

6.10.10 Blob Count

In an image, blob count gives a scalar matrix value which shows the definite number of labeled regions. Some additional techniques were also combined with the aforementioned tools for accurately finding the target object dimensions at the pixel level.

For performing the real-time metrology operation, the first in using the machine technology is to adjust the calibration of the instrument (camera), i.e., adjusting the intrinsic and extrinsic parameters of the camera. After adjusting the instrument calibration, the next step is to capture the image while the workpiece is in the manufacturing phase in order to perform the real-time metrology operation. This can be done by acquiring different features, information (dimensions of the workpiece), analyzing those images using the aforementioned techniques, and comparing the computed result with the actual value, i.e., the acquired data (dimensions) will be compared with the CAD generated data during the manufacturing process. After calculating the measurements of the targeted object in real-time processing, the program terminates if the given set of the conditions is satisfied. Else, the whole process continues until the desired output is achieved.

6.11 Integrating Intelligence

The focus of the research is to highlight the integration of manufacturing process, machines, and control inspection system to make the processes and system intelligent enough so that real-time computation can be done. After implementing the proposed in-process measurement technique, the system can be made intelligent as big data sets will be gathered during the process. Storing these data sets will help in learning the machine, and more intelligence can be inducted into the overall system using technologies such as artificial intelligence and deep learning. Different neural network codes can be used for learning the machines, e.g., Lex-net, Inception, etc.

6.12 Conclusion

Conventionally in manufacturing industries, the quality control procedures are performed when the product completes its manufacturing process phases. This leads to the likelihood of producing low quality and less cost-effective product. Also, with this, the overall cost of the manufacturing process increases. This research aims to implement in-process measurement technique for developing of real-time metrology inspection system in which the quality control check of the product is performed during the manufacturing process. Using machine vision technology, the manufacturing process and quality control work simultaneously. This will improve the quality of the product and increase the productivity.

References

1. Y. F. Zhao and X. Xu, "Enabling cognitive manufacturing through automated on-machine measurement planning and feedback," *Advanced Engineering Informatics*, vol. 24, pp. 269–284, 2010.
2. W. Barkman, *In-process Quality Control for Manufacturing*: CRC Press, New York, 1989.
3. J. A. Sładek, "Analysis of the Accuracy of Coordinate Measuring Systems," in *Coordinate Metrology*, S.-B. Choi, H. Duan, Y. Fu, C. Guardiola, J.-Q. Sun (Eds.): Springer-Verlag, Berlin Heidelberg, pp. 131–225, 2016.
4. H. Nielsen, "ISO 14253-1 Decision Rules–Good or Bad," in *NCSL-I Workshop and Symposium*, Charlotte, NC, p. 7, 1999.
5. B. Ivanov, "Evaluation of accuracy of in-process gauging of diameters by rolling method," *Measurement Techniques*, vol. 43, pp. 123–128, 2000.

6. Y. Zhang, Z. Li, and X. Weng, "Measurement of the outer diameter of rotator in operation," in *International Conference on Sensors and Control Techniques (ICSC 2000)*, Wuhan, China, pp. 317–320, 2000.

7. L. L. CIRP and G. Reinhart, *CIRP Encyclopedia of Production Engineering*: Springer-Verlag, Berlin, Heidelberg, 2014.

8. R. Schmitt, G. Mallmann, and P. Peterka, "Development of a FD-OCT for the inline process metrology in laser structuring systems," in *Optical Measurement Systems for Industrial Inspection VII*, Munich, Germany, p. 808228, 2011.

9. F. Neumann, T. Holtermann, D. Schneider, A. Kulczycki, T. Gries, and T. Aach, "In-process fault detection for textile fabric production: onloom imaging," in *Optical Measurement Systems for Industrial Inspection VII*, Munich, Germany, p. 808240, 2011.

10. N. Jayaweera and P. Webb, "Robotic edge profiling of complex components," *Industrial Robot: An International Journal*, vol. 38, pp. 38–47, 2011.

11. S. Mekid and K. Vacharanukul, "In-process out-of-roundness measurement probe for turned workpieces," *Measurement*, vol. 44, pp. 762–766, 2011.

12. S. Addepalli, R. Roy, D. Axinte, and J. Mehnen, "'In-situ' inspection technologies: Trends in degradation assessment and associated technologies," *Procedia CIRP*, vol. 59, pp. 35–40, 2017.

13. H.-D. Lin, "Automated visual inspection of ripple defects using wavelet characteristic based multivariate statistical approach," *Image and Vision Computing*, vol. 25, pp. 1785–1801, 2007.

14. F. Torres, L. Jiménez, F. Candelas, J. M. Azorín, and R. Agulló, "Automatic inspection for phase-shift reflection defects in aluminum web production," *Journal of Intelligent Manufacturing*, vol. 13, pp. 151–156, 2002.

15. F. Lahajnar, F. Pernuš, and S. Kovačič, "Machine vision system for inspecting electric plates," *Computers in Industry*, vol. 47, pp. 113–122, 2002.

16. M.-J. J. Wang, W.-Y. Wu, and C.-C. Hsu, "Automated post bonding inspection by using machine vision techniques," *International Journal of Production Research*, vol. 40, pp. 2835–2848, 2002.

17. L. Su, Z. Zha, X. Lu, T. Shi, and G. Liao, "Using BP network for ultrasonic inspection of flip chip solder joints," *Mechanical Systems and Signal Processing*, vol. 34, pp. 183–190, 2013.

18. X. Zhang, W.-M. Tsang, K. Yamazaki, and M. Mori, "A study on automatic on-machine inspection system for 3D modeling and measurement of cutting tools," *Journal of Intelligent Manufacturing*, vol. 24, pp. 71–86, 2013.

19. Z. Zhang, Y. Wang, and K. Wang, "Fault diagnosis and prognosis using wavelet packet decomposition, Fourier transform and artificial neural network," *Journal of Intelligent Manufacturing*, vol. 24, pp. 1213–1227, 2013.

20. H. Zheng, L. X. Kong, and S. Nahavandi, "Automatic inspection of metallic surface defects using genetic algorithms," *Journal of Materials Processing Technology*, vol. 125, pp. 427–433, 2002.

21. W. Wu and C. Hou, "Automated metal surface inspection through machine vision," *The Imaging Science Journal*, vol. 51, pp. 79–88, 2003.

22. D. Steiner and R. Katz, "Measurement techniques for the inspection of porosity flaws on machined surfaces," *Journal of Computing and Information Science in Engineering*, vol. 7, pp. 85–94, 2007.

23. T.-H. Sun, C.-C. Tseng, and M.-S. Chen, "Electric contacts inspection using machine vision," *Image and Vision Computing*, vol. 28, pp. 890–901, 2010.

24. S. Ghorai, A. Mukherjee, M. Gangadaran, and P. K. Dutta, "Automatic defect detection on hot-rolled flat steel products," *IEEE Transactions on Instrumentation and Measurement,* vol. 62, pp. 612–621, 2013.

25. T.-H. Sun, C.-H. Tang, and F.-C. Tien, "Post-slicing inspection of silicon wafers using the HJ-PSO algorithm under machine vision," *IEEE Transactions on Semiconductor Manufacturing,* vol. 24, pp. 80–88, 2011.

26. W.-C. Li and D.-M. Tsai, "Wavelet-based defect detection in solar wafer images with inhomogeneous texture," *Pattern Recognition,* vol. 45, pp. 742–756, 2012.

27. R. T. Chin and C. A. Harlow, "Automated visual inspection: A survey," *IEEE Transactions on Pattern Analysis and Machine Intelligence,* vol. PAMI-4, pp. 557–573, 1982.

28. R. T. Chin, "Automated visual inspection: 1981 to 1987," *Computer Vision, Graphics, and Image Processing,* vol. 41, pp. 346–381, 1988.

29. T. S. Newman and A. K. Jain, "A survey of automated visual inspection," *Computer Vision and Image Understanding,* vol. 61, pp. 231–262, 1995.

30. E. N. Malamas, E. G. Petrakis, M. Zervakis, L. Petit, and J.-D. Legat, "A survey on industrial vision systems, applications and tools," *Image and Vision Computing,* vol. 21, pp. 171–188, 2003.

31. F.-C. Tien, C.-H. Yeh, and K.-H. Hsieh, "Automated visual inspection for micro-drills in printed circuit board production," *International Journal of Production Research,* vol. 42, pp. 2477–2495, 2004.

32. C.-Y. Chang, C.-H. Li, Y.-C. Chang, and M. Jeng, "Wafer defect inspection by neural analysis of region features," *Journal of Intelligent Manufacturing,* vol. 22, pp. 953–964, 2011.

33. S. Chen, B. Lin, X. Han, and X. Liang, "Automated inspection of engineering ceramic grinding surface damage based on image recognition," *The International Journal of Advanced Manufacturing Technology,* vol. 66, pp. 431–443, 2013.

34. Y. C. Liang, J. Gao, C. X. Jian, and X. Chen, "Online visual inspection system for OLED defects," in *Applied Mechanics and Materials,* vol. 241–244, pp. 3153–3158, 2013.

35. B. Scholz-Reiter, D. Weimer, and H. Thamer, "Automated surface inspection of cold-formed micro-parts," *CIRP Annals-Manufacturing Technology,* vol. 61, pp. 531–534, 2012.

36. H. Shen, S. Li, D. Gu, and H. Chang, "Bearing defect inspection based on machine vision," *Measurement,* vol. 45, pp. 719–733, 2012.

37. Y. Gan and Q. Zhao, "An effective defect inspection method for LCD using active contour model," *IEEE Transactions on Instrumentation and Measurement,* vol. 62, pp. 2438–2445, 2013.

38. C. Benedek, O. Krammer, M. Janóczki, and L. Jakab, "Solder paste scooping detection by multilevel visual inspection of printed circuit boards," *IEEE Transactions Industrial Electronics,* vol. 60, pp. 2318–2331, 2013.

39. W. Jianming, G. Biao, Z. Xiao, D. Xiaojie, and L. Xiuyan, "Error correction for high-precision measurement of cylindrical objects diameter based on machine vision," *2015 12th IEEE International Conference on Electronic Measurement & Instruments (ICEMI),* Qingdao, China, , pp. 1113–1117, 2015.

40. Y. Chen and C. Liu, "Radius and orientation measurement for cylindrical objects by a light section sensor," *Sensors,* vol. 16, p. 1981, 2016.

41. O. Fabio, K. Fabjan, L. R. Dario, A. Jacopo, and C. Stefano, "Performance evaluation of a low-cost stereo vision system for underwater object detection," *IFAC Proceedings Volumes,* vol. 47, pp. 3388–3394, 2014.

42. H. Tamura, T. Sasaki, H. Hashimoto, and F. Inoue, "Position measurement system for cylindrical objects using laser range finder," *Proceedings of SICE Annual Conference 2010*, Taipei, Taiwan, pp. 291–296, 2010.

43. M. Richtsfeld, R. Schwarz, and M. Vincze, "Detection of cylindrical objects in tabletop scenes," 2010 *IEEE* 19th *International Workshop on Robotics in Alpe-Adria-Danube Region (RAAD), Budapest and Balatonfüred*, Hungary, pp. 363–369, 2010.

44. M. A. Ayub, A. B. Mohamed, and A. H. Esa, "In-line inspection of roundness using machine vision," *Procedia Technology*, vol. 15, pp. 807–816, 2014.

Part 5

Design of Mechanisms and Machines

Components, Structures, Assemblies, Mechanisms, and Machines are discrete in nature and are designed as per system approach using contemporary modeling and analysis tools. The modeling and analysis spans over concept design, embodiment design, and detail design. The techniques are commonly known as Computer Aided Design (CAD), Computer Aided Engineering (CAE), Computer Aided Production Planning (CAPP), and Computer Aided Manufacturing (CAM). The state of the art in the field are CAD/CAM packages with modeling using wireframes, surfaces, and solids where neutral geometry is available in initial graphics exchange specification (IGES), STEP, DXF, etc. formats, STEP being the next-generation CAD/CAM scheme. In manufacturing, the surfaces are of prime importance.

In CAD, surfaces are represented by Bezier Surfaces, NURBS, Coons' patches, and the likes.

The first chapter in this part deals with surface manipulation by providing convex hull of two closed implicit surfaces allowing computing developable transition surface of two genus-0 free-form surfaces.

The second chapter deals with the micro texturing on free-form surfaces with high precision for aerospace, micro-electronics industry, and biological engineering.

The third chapter in this part deals with the technology, structure, and engineering design of metal forming presses—strategic machine tools.

7

The Convex Hull of Two Closed Implicit Surfaces

Xiaoping Wang
Nanjing University of Aeronautics and Astronautics

CONTENTS

7.1 Introduction 137
7.2 Mathematical Fundamentals 138
7.3 Convex Hull of Surfaces 141
    7.3.1 Computation of Initial Values 143
7.4 Convex Hull of a Genus-0 Surface and Curve 144
    7.4.1 The Case for Parametric Surface 144
    7.4.2 The Case for Implicit Surface 145
7.5 Numerical Integration 146
7.6 Implementation Examples 147
7.7 Conclusions 149
7.8 Acknowledgement 149
References 149

7.1 Introduction

Algorithms for computing convex hulls of various objects have a broad range of applications in pattern recognition, collision detection [1,2], image processing, statistics, and geographical information system (GIS). Computing the convex hull is to construct a non-ambiguous and efficient representation of the required convex shape. In computational geometry [3], numerous algorithms are proposed for computing the convex hull of a finite set of points, polygons, and polyhedra with various computational complexities [3–5]. For spline surface, we first take the convex hull of its control points as an approximation to its actual convex hull and subdivide the surface into smaller pieces. Then the union created by the control points of these smaller pieces will be a better approximation to its actual convex hull. Repeat the process until the approximating convex hull converges to the exact convex hull within a certain reasonable bound. Seong et al. [6] once presented an algorithm for computing the convex hull of a rational surface without using the subdivisions, usually a troublesome process, in which the problem of computing convex hull is reduced to one of finding the zero-sets of polynomial equations. Considering a special case, Geismann et al. [7] developed an algorithm by reducing the convex hull problem of a set of ellipsoids to that of computing a cell in an arrangement of quadrics.

Moreover, implicit functions are also used to represent volumes. Accordingly, algorithm for computing the hull of implicit surfaces should also be investigated. Here we develop a numerical method of computing the convex hull of genus-0 free-form implicit surfaces (or computing the developable transition surface of two genus-0 free-form surfaces).

The basic idea is to determine the curves where the developable transition surface is tangent to the two free-form surfaces under the condition of the ruled surface generated by connecting line between corresponding points on the two intersection curves being a developable surface. With techniques in classical differential geometry, we derive the differential equations of the two curves where the developable surface is tangent to two genus-0 free-form surfaces defined parametrically or implicitly. The two curves naturally share the same parameter. Moreover, they are independent of the two base surfaces' geometry and their parameterization and are obtained by numerically solving the initial-value problem for a system of first-order ordinary differential equations (ODEs) in the four-dimensional parametric domain associated to the surface representation for parametric case, or in R^6 for implicit case, or in the five-dimensional space for the case that one surface is in parametric form and another is in implicit form. Based on this, we finally construct the transition surface and hence the convex hull of the two free-form surfaces.

7.2 Mathematical Fundamentals

This section reviews some basic concepts concerning curves and surfaces, which are needed in the following discussion. In addition, we establish some terminologies to be used in subsequent discussions. For more detail and rigorous introductions, readers are referred to any textbooks on differential geometry such as [8].

Let's begin with a differentiable parametric surface that is described by a vector-valued function of two variables as follows:

$$S(u,v) = (x(u,v), y(u,v), z(u,v)), \quad (u,v) \in D \subset R^2, \tag{7.1}$$

where $x(u,v)$, $y(u,v)$, $z(u,v)$ are differentiable functions of bivariate u and v which are called the surface parameters and D denotes the surface domain. Usually equation 7.1 is called a parameterization of the surface S. The partial derivatives of the vector valued function S with respect to u and v are $S_u(u,v)$, $S_v(u,v)$, $(u,v) \in D$.

From the classical differential geometry, we know that the vectors S_u and S_v are tangent to the surface S at the point (u,v) while the vector $S_u \times S_v$ is normal to the surface at the same point, where "$\times$" denotes the cross product (same as below). The vector $N = \dfrac{S_u \times S_v}{|S_u \times S_v|}$ is called the unit normal vector (or just normal vector for short) of the surface S at the corresponding point. We assume the surface S is regular, i.e. $S_u \times S_v \neq 0$ for any $(u,v) \in D$. A curve on the surface can be described by parametric equations $u = u(t)$, $v = v(t)$, $t \in [a,b]$ in the surface domain. Its equation in R^3 can be written as the vector-valued function

$$P(t) = S(u(t), v(t)), \ t \in [a,b]. \tag{7.2}$$

Taking the derivative of $P(t)$ with respect to t to linearize equation 7.2, we get the following corresponding equation in the tangent space of the surface:

$$P'(t) = S_u \frac{du}{dt} + S_v \frac{dv}{dt}, \ t \in [a,b]. \tag{7.3}$$

Along the curve $P(t)$, the normal vector N is

$$N(t) = N(u(t), v(t)).$$

Let S be a C^2 regular parametric surface. It follows from classical differential geometry [18] that

$$I(du, dv) = (dP)^2 = E(du)^2 + 2Fdudv + G(dv)^2,$$

$$II(du, dv) = N \cdot d^2P = e(du)^2 + 2fdudv + g(dv)^2.$$

I and II are known as the first and the second fundamental forms respectively, while "$\cdot$" indicates the scalar product (same as below).

The Gauss curvature K of surface is defined as

$$K = \frac{eg - f^2}{EG - F^2}. \tag{7.4}$$

In addition, we have the following equations:

$$N_u = a_{11}S_u + a_{21}S_v, \ N_v = a_{12}S_u + a_{22}S_v,$$

where

$$a_{11} = (fF - eG)/(EG - F^2), \quad a_{12} = (gF - fG)/(EG - F^2),$$

$$a_{21} = (eF - fE)/(EG - F^2), \quad a_{22} = (fF - gE)/(EG - F^2).$$

So

$$N'(t) = (a_{11}S_u + a_{21}S_v)\frac{du}{dt} + (a_{12}S_u + a_{22}S_v)\frac{dv}{dt}. \tag{7.5}$$

Now let's consider implicit surfaces that are represented by the implicit equation $f(x,y,z) = 0$. Obviously implicit surfaces differ in appearance, and always differ in expression, from the parametric surfaces. So do the curve on a parametric surface and the counterpart on an implicit one. Here we demand that the function f must be continuous and differentiable. That is, the first partial derivatives $f_x = \partial f/\partial x$, $f_y = \partial f/\partial y$, $f_z = \partial f/\partial z$ must be continuous and not all zero, everywhere on the surface, i.e. the surface is regular. The vector $\nabla f = (f_x, f_y, f_z)$ is called as the gradient of the implicit surface at the point (x, y, z). The vector $N = \nabla f/|\nabla f|$ is the unit normal vector (or just normal vector for short) of the implicit surface at the point (x, y, z). The Jacobian of the gradient ∇f is called the Hessian of f. Write it as $H(f) = J(\nabla f)$. A curve on the implicit surface is described by parametric form $\bar{P}(t) = (x(t), y(t), z(t))$. It is characterized by the following equation:

$$f(x(t), y(t), z(t)) = 0.$$

Linearizing it, we get

$$f_x \frac{dx}{dt} + f_y \frac{dy}{dt} + f_z \frac{dz}{dt} = 0. \tag{7.6}$$

The gradient ∇f is normal (*i.e.*, perpendicular) to the surface at the point (x, y, z). Finally, one vector identity should be mentioned here for applications in discussion below. For any three vectors a, b, c, it follows that

$$(a \times b) \times c = (a \cdot c)b - (b \cdot c)a. \tag{7.7}$$

Developable surfaces [9,10] shall briefly be introduced as special cases of ruled surfaces. A ruled surface R carries a one-parameter family of straight lines L. These lines are called generators or generating lines. The general parameterization of a ruled surface R is

$$R(u,v) = C(u) + vA(u) \quad u \in I, v \in R, \tag{7.8}$$

where $C(u)$ is called directrix curve and $A(u)$ denotes a vector field (also called an indicatrix curve). For fixed values u, this parameterization represents the straight lines $L(u)$ on R. The normal vector $N(u, v)$ of the ruled surface $R(u, v)$ is computed as cross product of the partial derivative vectors R_u and R_v, and we obtain

$$N(u,v) = C'(u) \times A(u) + vA'(u) \times A(u). \qquad (7.9)$$

For fixed $u = u_0$, the normal vectors $N(u_0, v)$ along $L(u_0)$ are linear combinations of the vectors $C'(u) \times A(u)$ and $A'(u) \times A(u)$. The parameterization $R(u, v)$ represents a developable surface D if for each generator L all points $x \in L$ have the same tangent plane (with exception of the *singular point* on L). This implies that the vectors $C'(u) \times A(u)$ and $A'(u) \times A(u)$ are linearly dependent. Thus from (7.7) a ruled surface is said to be developable if

$$\det(C'(u), A(u), A'(u)) = 0. \qquad (7.10)$$

There exists a unique *singular point* $s(u)$ at each generating line $L(u)$, and it is determined by the parameter value

$$v_s = -(C'(u) \times A(u)) \cdot (A'(u) \times A(u))/(A'(u) \times A(u))^2. \qquad (7.11)$$

If $A(u)$ and $A'(u)$ are linearly dependent, the singular point s is at infinity; otherwise it is a proper point.

7.3 Convex Hull of Surfaces

A closed surface is one that is boundaryless and compact. The two-dimensional sphere, the two-dimensional torus, and the real projection plane are the relevant examples. Here we consider only closed C^2 surfaces of genus 0 detailed as follows:

Let R and S be regular C^2-continuous surfaces. Two surfaces R and S are defined by implicit form:

$$f(x,y,z) = 0 \text{ and } g(u,v,w) = 0,$$

where $(x,y,z) \in D_1 = [a_{11},b_{11}] \times [a_{12},b_{12}] \times [a_{13},b_{13}] \subset R^3$, $(u,v,w) \in D_2 = [a_{21},b_{21}] \times [a_{22},b_{22}] \times [a_{23},b_{23}] \subset R^3$. We find a developable surface T such that it is tangent to both surfaces R and S at the same time, i.e., T bounds the convex hull of the two surfaces. In fact, surface T is the envelope surface of the common tangent planes of surfaces R and S. Thus, finding surface T is reduced to finding two curves where surface T is tangent to R and S, respectively.

Let the two corresponding curves be $C_1(s)$ and $C_2(s)$, where s is taken as the arc-length parameter of C_1. Then s can be an ordinary parameter induced by C_1 for curve C_2. The vector $\nabla f = (f_x, f_y, f_z)$ is called the gradient of the implicit surface at the point (x, y, z). The Jacobian of the gradient ∇f is called the Hessian of f, written as $H(f) = J(\nabla f)$. Similarly, we have ∇g and $H(g)$. Then, from T being a developable surface, we further have the following equations that characterize C_1 and C_2, where N_1 and N_2 are the unit normal vectors of R and S, respectively:

$$\begin{cases} (C_1(s) - C_2(s)) \cdot N_1(s) = 0 \\ (C_1(s) - C_2(s)) \cdot N_2(s) = 0 \\ (C_1'(s), (C_1(s) - C_2(s)), (C_1'(s) - C_2'(s))) = 0 \end{cases} \quad (7.12)$$

From (7.12) we have *the following independent equations*:

$$\begin{cases} \left[(C_1(s) - C_2(s)) H(f) \right] \cdot C_1'(s) = 0 \\ (C_1'(s) H(g)) \cdot C_2'(s) = |\nabla g| (C_1'(s) H(f)) \cdot C_1'(s) / |\nabla f|. \\ \left[(C_1(s) - C_2(s)) H(g) \right] \cdot C_2'(s) = 0 \end{cases} \quad (7.13)$$

Supplementing additional equations $[C_1'(s)]^2 = 1$, $\nabla f(s) \cdot C_1'(s) = 0, \nabla g(s) \cdot C_2'(s) = 0$ to equation 7.13 and solving the resulting system of equations, we get

$$\begin{cases} \left(\dfrac{dx}{ds}, \dfrac{dy}{ds}, \dfrac{dz}{ds} \right) = \dfrac{\pm [(x-u, y-v, z-w) H(f)] \times \nabla(f)}{\sqrt{\{[(x-u, y-v, z-w) H(f)] \times \nabla(f)\}^2}} \\ \left(\dfrac{du}{ds}, \dfrac{dv}{ds}, \dfrac{dw}{ds} \right) = \dfrac{|\nabla g| (C_1'(s) H(f)) \cdot C_1'(s)\{[(x-u, y-v, z-w) H(g)] \times \nabla g\}}{|\nabla f| C_1'(s) \cdot \{H(g)[(x-u, y-v, z-w) H(g)] \times \nabla g\}} \end{cases}$$

$$(7.14a)$$

Adding initial value conditions which can be computed with the similar methods as described above to system (7.14a) as follows:

$$\begin{cases} x(s_0) = x_0, y(s_0) = y_0, z(s_0) = z_0 \\ u(s_0) = x_0, v(s_0) = y_0, w(s_0) = z_0 \end{cases}, \quad (7.14b)$$

we get an initial value problem for a first-order system of ODEs. By numerically integrating system (7.14), curves C_1 and C_2 and hence surface T can be obtained.

7.3.1 Computation of Initial Values

Since T is a developable surface, the tangent plane of surface R at (x, y, z) on curve C_1 is also tangent to surface S at (u, v, w) on curve C_2. Thus, the points on two curves C_1 and C_2 can be characterized by the following equations:

$$\begin{cases} ((x,y,z)-(u,v,w)) \cdot \nabla f = 0 \\ ((x,y,z)-(u,v,w)) \cdot \nabla g = 0. \\ \nabla f \times \nabla g = 0 \end{cases} \qquad (7.15)$$

Rewriting the equations gives

$$\begin{cases} g_1(x,y,z,u,v,w) = 0 \\ g_2(x,y,z,u,v,w) = 0 \\ \quad \cdots\cdots\cdots \\ g_5(x,y,z,u,v,w) = 0 \end{cases} \qquad (7.16)$$

Then the computation of initial values is reduced to finding a point on the intersection curve of three hypersurfaces $g_1 = 0, g_2 = 0, \ldots, g_5 = 0$ in $D_1 \times D_2 \subset R^6$. We define a six-variate function $G: D_1 \times D_2 \rightarrow R$ as follows:

$$G(x) = \sum_{i=1}^{5} g_i^{2}(x), \ x = (x,y,z,u,v,w) \in R^6. \qquad (7.17)$$

This function is positive definite and has a global minimum at each of the roots to the equation; that is, a solution x to the system (7.16) is precisely a global minimizer of (7.17) that makes $G(x) = 0$. Thus, the root-finding problem is transformed to the one of finding global minimizer where $G(x)$ is evaluated to be 0. Several methods [11–14] for non-polynomial cases and other methods [15–19] for polynomial cases can be used to deal with this new issue. Moreover, if three vectors $\nabla g_1, \nabla g_2, \ldots, \nabla g_5$ are *linearly independent in $D_1 \times D_2$, we give another method. Actually if three vectors $\nabla g_1, \nabla g_2, \ldots, \nabla g_5$ are linearly independent in $D_1 \times D_2$ and the function $G(x)$ has extreme value in $D_1 \times D_2$, it must be unique minimum zero.* Then the real-valued function $G(x)$ decreases *fastest* if one goes from $x \in D_1 \times D_2$ in the opposite direction of the gradient of $G(x)$ at x, i.e., toward the vector "$-\nabla G(x)$". Therefore, it creates a one-dimensional gradient vector field "$-\nabla G(x(t))$", where t is just the arc-length parameter of the integral curve $q(t)$ over the vector field in $D_1 \times D_2$. The integral curve can be obtained by numerically integrating the following equation with any point in $D_1 \times D_2$ as the starting point:

$$q'(t) = \left(\frac{dx}{dt}, \frac{dy}{dt}, \frac{dz}{dt}, \frac{du}{dt}, \frac{dv}{dt}, \frac{dw}{dt} \right) = -\frac{\nabla G(x)}{|\nabla G(x)|}.$$

We trace from the starting point along the curve until we arrive at the point $x^* = (x^*, y^*, z^*, u^*, v^*, w^*)$ which makes $G(x)$ reach its minimum value 0, i.e., $g_i(x^*) = 0$, $i = 1, 2, \ldots, 5$. And thus the initial values (7.14b) are obtained.

Remark: The presented methods can be used in computing the convex hull of a free-form surface. Here what we need to do is to take the equation of the surface S as $f(u, v, w) = 0$.

7.4 Convex Hull of a Genus-0 Surface and Curve

7.4.1 The Case for Parametric Surface

This section presents a case study for computing convex hull of a close surface of genus 0 and a curve (or computing the developable transition surface between a curve and a genus-0 free-form surface). Let S_1 be regular C^2-continuous surfaces. Let $C(t)$ be a C^2-continuous curve. We find a developable surface T such that it is tangent to both surface S and curve C at the same time; that is, T bounds the convex hull of the surface and curve. In fact, surface T is the envelope surface of the common tangent planes of surface S and curve C. Thus, finding surface T is reduced to finding a curve where surface T is tangent to S. Let the curve be P. Actually, the curve P can be regarded as tangential projection of curve $C(t)$ onto S. The projection curve thus inherits the parameter of the space curve $C(t)$. So, it can be described in a parametric form $P(t)$.

Firstly we consider the case of parametric surface; that is, surface S is defined by parametric form $S(u, v)$, where $(u, v) \in D$. Then we further have the following equations which characterize $P(t)$:

$$\begin{cases} [C(t) - P(t)] \cdot N(t) = 0 \\ (C'(t), C(t) - P(t), C'(t) - P'(t)) = 0, \\ P'(t) \cdot N(t) = 0 \end{cases} \tag{7.18}$$

where N is the unit normal vector of S.

From equation 7.18 *we can obtain the following two independent equations:*

$$\begin{cases} [C(t) - P(t)] \cdot N(t) = 0 \\ C'(t) \cdot N(t) = 0 \end{cases}. \tag{7.19}$$

By taking the derivative of equation (7.19) with respect to t, it follows that

$$\begin{cases} [C(t) - P(t)] \cdot N'(t) = 0 \\ C'(t) \cdot N'(t) = -C''(t) \cdot N(t) \end{cases}. \tag{7.20}$$

Assuming that there is no flat point on surface S and

$$(C'(t), C(t) - P(t), N(t)) \neq 0 \tag{7.21}$$

along the projection curve $P(t)$, finally equation 7.20 gives

$$\begin{cases} \dfrac{du}{dt} = \dfrac{C''(t) \cdot N(u,v)\left(N_v \cdot [C(t) - S(u,v)]\right)}{K(S_u \times S_v) \cdot ((C(t) - S(u,v)) \times C'(t))} \\ \dfrac{dv}{dt} = \dfrac{-C''(t) \cdot N(u,v)\left(N_u \cdot [C(t) - S(u,v)]\right)}{K(S_u \times S_v) \cdot (C(t) - S(u,v)) \times C'(t))} \end{cases}, \tag{7.21a}$$

where K is the Gauss curvature of surface S at the point on the projection curve $P(t)$, defined with equation 7.4.

For system (7.21a) to be completely determined, the following initial-value conditions, which can be computed through equation 7.19 for a fixed parameter t, on the surface domain D, must be added

$$u(t_0) = u_0, \; v(t_0) = v_0. \tag{7.21b}$$

Once the system (7.21) is solved by numerically integrating for u and v, the space projection curve $P(t)$ can be obtained by substituting u and v into the equation of surface S. Hence, surface T can finally be obtained.

7.4.2 The Case for Implicit Surface

Now let us consider the case of implicit surface, i.e., surface S is defined by implicit form $f(x, y, z) = 0$. Here the developable surface T is characterized by the following independent equation system:

$$\begin{cases} [C(t) - P(t)] \cdot \nabla f = 0 \\ C'(t) \cdot \nabla f = 0 \\ f(P(t)) = 0 \end{cases}. \tag{7.22}$$

By taking the derivative of equation 7.22 with respect to t, it is concluded that

$$\begin{cases} [(C(t) - P(t))H(f)] \cdot P'(t) = 0 \\ (C'(t)H(f)) \cdot P'(t) = -C''(t) \cdot \nabla f. \\ P'(t) \cdot \nabla f = 0 \end{cases} \tag{7.23}$$

From equation 7.23 we obtain

$$P'(t) = \left(\frac{dx}{dt}, \frac{dy}{dt}, \frac{dz}{dt}\right) = \frac{-\nabla f \cdot C''(t)\{[(C(t) - P(t))H(f)] \times \nabla f\}}{C'(t) \cdot \{H(f)[(C(t) - P(t))H(f)] \times \nabla f\}}, \tag{7.24a}$$

where $P(t) = (x, y, z)$.

Analogous to the parametric case, adding an initial value condition, which can be computed through equation 7.22 for a fixed parameter t, to equation 7.24a as follows,

$$x(t_0) = x_0, \, y(t_0) = y_0, \, z(t_0) = z_0, \tag{7.24b}$$

we get an initial value problem for a first-order system of ODEs. By numerically integrating the system (7.24), projection curve $P(t)$ and hence surface T can be finally obtained.

7.5 Numerical Integration

The desired tangent surface T in convex hull for all cases can be obtained by *numerical integration techniques feasible* for first-order explicit ODE systems associated with initial value problems, since the analytical solutions of these ODEs *might not* exist except some special surfaces such as sphere, etc. For example, the ODE solver of MATLAB [20] or Numerical Recipes [21], based on Runge–Kutta, Adams–Bashforth, and other numerical methods can be used to solve these systems. In addition, these solvers provide user with good controls of both absolute and relative error tolerances. Moreover, the initial conditions can also be obtained by finding a point on the intersection curve of several hypersurfaces in $D_1 \times D_2 \in R^6$.

The presented method works well for any implicit surface. For instance, if the surface is composed of several piecewise continuous surface patches, some kind of continuity conditions must be considered to ensure the differential model is still valid in the neighborhood of each patch boundary. In all cases for convex hull for two surfaces, the initial conditions can also be obtained by finding a point on the intersection curve of several hypersurfaces in R^6.

In practical applications, we may demand that the curves C_1 and C_2 in Section 7.3, and the curve P in Section 7.4, should be described in the standard form such as B-spline or NURBS. The preceding numerical integration yields, however, an array of points in parametric domain and hence in the surface. Fortunately, using, for example, a cubic B-spline to interpolate those points on the surface, we can create a closed form B-spline approximation of projection curve. Here the particular approximation method developed by Wolter [22] or Qu [23] can be used to obtain good accuracy. Moreover, based on the method proposed in [24], we can also get the B-spline approximation of the resulting curve with the array of points in the parametric domain.

7.6 Implementation Examples

The newly proposed method can be applied to any implicit surface. Since our purpose is to test the correctness of the algorithms rather than to obtain diversified results, for the sake of simplicity, we only take ellipsoid as the given surface to demonstrate the effectiveness of the method. Figure 7.1 illustrates the convex hull of two ellipsoids R and S defined as follows:

$$18x^2 + (y-z)^2 + 9\left(y+z-6\sqrt{2}\right)^2 - 18 = 0,\ x \in [-1,1],$$

$$y \in \left[3\sqrt{2}-\sqrt{5}, \sqrt{2}+\sqrt{5}\right],\ z \in \left[3\sqrt{2}-\sqrt{5}, \sqrt{2}+\sqrt{5}\right],$$

$$4u^2 + v^2 + 4w^2 - 16 = 0,\ u \in [-2,2],\ v \in [-4,4],\ u \in [-2,2],$$

where C_1 and C_2 are two curves and developable surface T is tangent to the given ellipsoids. The coordinates (0.98, 4.20, 4.52, 1.94, 0.79, 0.26) corresponding to (x, y, z, u, v, w) represent the initial point which is obtained via the

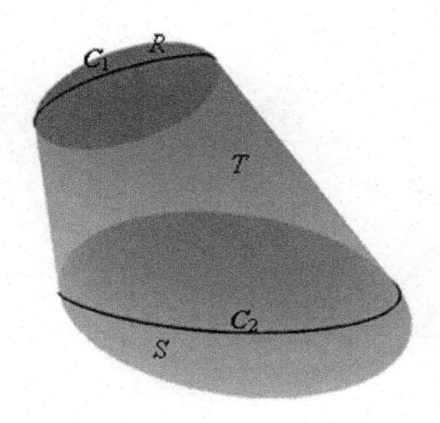

FIGURE 7.1
Convex hull of two ellipsoids.

gradient descent method with $(1, 3\sqrt{2}, 3\sqrt{2}, 2, 0, 0)$ as starting point, where $(0.98, 4.20, 4.52)$ is used for tracing C_1 and $(1.94, 0.79, 0.26)$ is used for tracing C_2.

Figure 7.2 shows the partial convex hull of an implicit surface, given by

$$\frac{y^2}{9(\cos x + 2)^2} + \frac{z^2}{(\sin x + 2)^2} - 1 = 0,\ x \in [0, 3\pi],$$

$$y \in \left[-4.5\sqrt{2}, 4.5\sqrt{2}\right], u \in \left[\sqrt{2}/2, 3\right].$$

Figure 7.3 shows the convex hull of an ellipse $C(t)$ and an ellipsoid, in which $P(t)$ is the curve where developable surface T is tangent to the given ellipsoid, given by

$$C(t) = \left(4\cos t,\ \sqrt{2}\sin t - 4\sqrt{2},\ \sqrt{2}\sin t + 4\sqrt{2}\right),\ t \in [0, 2\pi],$$

$$S(u, v) = (3\cos u \cos v,\ 6\cos u \sin v,\ 3\sin u),\ u \in [-\pi/2, \pi/2],\ v \in [0, 2\pi].$$

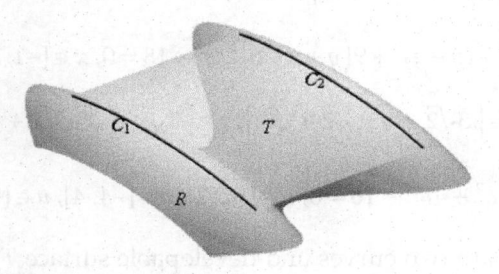

FIGURE 7.2
The partial convex hull surface.

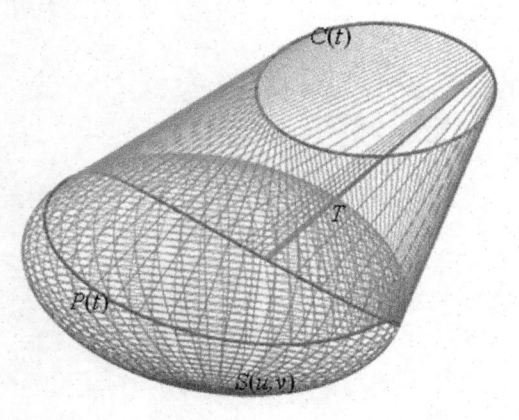

FIGURE 7.3
Convex hull of an ellipsoid and an ellipse.

All these above examples are implemented in MATLAB 6.5 on Windows PC with 2.6 GHz CPU and 512 M memory. CPU time shows that the proposed method has relatively high computational efficiency. In Figure 7.1, the computation of curves C_1 and C_2 (including the drawing of the curves C_1 and C_2 and the convex hull surface T) with the presented method needs only 0.206 s for absolute error tolerance 1e−6.

7.7 Conclusions

This paper proposes an approach for computing the convex hull surface of free-form surfaces defined implicitly. The basic idea is as follows. The convex hull problem is reduced to determining one or two curves on surface where developable surface is tangent to two given surfaces. Employing the approach of classical differential geometry, we deduce some differential equations that characterize the curves in convex hull computation of two given free-form surfaces. However, these equations are underdetermined. Taking arc-length of one such curve as the common parameter for both curves, the system of equations is made even-determined by supplementing additional equations. Finally, the system can be solved by numerical techniques.

Compared with the algebraic method [6] which only reduces the related computation to finding the zero-sets of polynomial equations but does not give further algorithms, the newly presented one offers a numerical computational method for the tangent curves and hence the convex hull surface. Moreover, analysis shows that the new method can be extended to the parametric surface case, even to the case of one parametric surface and one implicit surface without any difficulty.

7.8 Acknowledgement

The work reported in this paper was supported by National Natural Science Foundation of China under grant (No. 51575266).

References

1. T. Lozano-Pérez and M. A. Wesley, "An algorithm for planning collision-free paths among polyhedral obstacles," *Communications of the ACM*, vol. 22, pp. 560–570, 1979.

2. T. Perez-Lozano, "Spatial planning: A configuration space approach," *IEEE Transactions on Computers*, vol. 32, 1983.
3. F. P. Preparata and M. I. Shamos, "Convex hulls: Basic algorithms," in *Computational Geometry*, D. Gries, F. Schneider (Eds.): Springer, New York, pp. 95–149, 1985.
4. R. L. Graham and F. F. Yao, "Finding the convex hull of a simple polygon," *Journal of Algorithms*, vol. 4, pp. 324–331, 1983.
5. D.-T. Lee, "On finding the convex hull of a simple polygon," *International Journal of Computer & Information Sciences*, vol. 12, pp. 87–98, 1983.
6. J.-K. Seong, G. Elber, J. K. Johnstone, and M.-S. Kim, "The convex hull of free-form surfaces," in *Geometric Modelling*, S. Hahmann, G. Brunnett, G. Farin, and R. Goldman (Eds.): Springer, Vienna, Austria, 2004, pp. 171–183.
7. N. Geismann, M. Hemmer, and E. Schömer, "The convex hull of ellipsoids," in *Proceedings of the Seventeenth Annual Symposium on Computational Geometry*, Medford, MA, pp. 321–322, 2001.
8. M. P. do Carmo. *Differential Geometry of Curves and Surfaces*: Prentice Hall, Englewood, NJ, 1976.
9. H. Pottmann and J. Wallner, *Computational Line Geometry*: Springer Science & Business Media, 2009.
10. G. Aumann, "Interpolation with developable Bézier patches," *Computer Aided Geometric Design*, vol. 8, pp. 409–420, 1991.
11. S. Salhi and N. M. Queen, "A hybrid algorithm for identifying global and local minima when optimizing functions with many minima," *European Journal of Operational Research*, vol. 155, pp. 51–67, 2004.
12. M. J. Hirsch, P. M. Pardalos, and M. G. Resende, "Solving systems of nonlinear equations with continuous GRASP," *Nonlinear Analysis: Real World Applications*, vol. 10, pp. 2000–2006, 2009.
13. I. Tsoulos and A. Stavrakoudis, "On locating all roots of systems of nonlinear equations inside bounded domain using global optimization methods," *Nonlinear Analysis: Real World Applications*, vol. 11, pp. 2465–2471, 2010.
14. N. E. Mastorakis, "Solving non-linear equations via genetic algorithms," in *Proceedings of the 6th WSEAS International Conference on Evolutionary Computing*, Lisbon, Portugal, June, pp. 16–18, 2005.
15. G. Elber and M.-S. Kim, "Geometric constraint solver using multivariate rational spline functions," in *Proceedings of the sixth ACM symposium on Solid Modeling and Applications*, Ann Arbor, MI, pp. 1–10, 2001.
16. B. Mourrain and J.-P. Pavone, "Subdivision methods for solving polynomial equations," *Journal of Symbolic Computation*, vol. 44, pp. 292–306, 2009.
17. G. Elber and T. Grandine, "An efficient solution to systems of multivariate polynomial using expression trees," *IEEE Transactions on Visualization and Computer Graphics*, vol. 15, pp. 596–604, 2009.
18. M. Reuter, T. S. Mikkelsen, E. C. Sherbrooke, T. Maekawa, and N. M. Patrikalakis, "Solving nonlinear polynomial systems in the barycentric Bernstein basis," *The Visual Computer*, vol. 24, pp. 187–200, 2008.
19. P. S. Nataraj and M. Arounassalame, "A new subdivision algorithm for the Bernstein polynomial approach to global optimization," *International Journal of Automation and Computing*, vol. 4, pp. 342–352, 2007.
20. The MATH WORKS INC, Using MATLAB, Version 6, The Math Works, Natick, MA, 2000.

21. W. T. Vetterling, S. A. Teukolsky, and W. H. Press, *Numerical Recipes: Example Book (FORTRAN)*: Cambridge University Press, 1992.
22. F.-E. Wolter and S. T. Tuohy, "Approximation of high-degree and procedural curves," *Engineering with Computers*, vol. 8, pp. 61–80, 1992.
23. J. Qu and R. Sarma, "The continuous non-linear approximation of procedurally defined curves using integral B-splines," *Engineering with Computers*, vol. 20, pp. 22–30, 2004.
24. G. Renner and V. Weiss, "Exact and approximate computation of B-spline curves on surfaces," *Computer-Aided Design*, vol. 36, pp. 351–362, 2004.

8

Design of Multi-DOF Micro-Feed Platform Based on Hybrid Compliant Mechanism

Jianqiang Huo, Chen Zhang, and Yun Song
Nanjing University of Aeronautics and Astronautics

CONTENTS

8.1 Introduction 153
8.2 Design of Multi-DOF Micro-Feed Platform 155
    8.2.1 Design Idea and Description of Multi-DOF Micro-Feed
        Platform 155
    8.2.2 Design of Nested Bottom Platform 156
    8.2.3 Design of Parallel Top Platform 157
8.3 FEA Simulation of Multi-DOF Micro-Feed Platform 159
    8.3.1 Mesh Generation 159
    8.3.2 FEA of the Parallel Bottom Platform 159
    8.3.3 FEA of the Parallel Top Platform 162
    8.3.4 FEA of the Micro-Feed Platform 163
8.4 Conclusions 164
8.5 Acknowledgement 165
References 165

8.1 Introduction

Structured surface with micro/nano feature can provide advanced function such as self-cleaning action, anti-microbial effect, rapid integration ability, and optimal friction behavior. To achieve the maximum benefit from the structured surface, fabrication technologies of micro/nano-structured surfaces have been a fascinating research topic in the last few decades [1]. The typical manufacturing methods include laser scribing, micro-electrical discharge machining (μ-EDM), micro-electrochemical machining (μ-ECM), micro-end milling, micro-fly cutting, atomic force microscope (AFM) scribing, and elliptical vibration-assisted cutting (EVC) [2–8]. Among all the above-mentioned methods, mechanical micro-machining offers great advantages for its flexibility, given that it can process all types of work materials and generate different types of complex micro-structure features. Mechanical micro-machining provides a potential process for precision machining of sophisticated micro/nano-structured surfaces by using multi-axis machine tool.

Currently, it is still difficult to fabricate micro-textures on free-form surfaces by using mechanical micro-machining method because of the limited motion accuracy of multi-axis machine tool. In order to solve the problem, a multi-DOF (degree of freedom) micro-feed mechanism is proposed to provide the capability of fabricating micro-textures on free-form surfaces for the mechanical micro-machining method in this chapter.

One of the best approaches is to utilize flexure-based compliant mechanisms combining piezoactuators in order to construct multi-DOF micro-feed mechanism. The type of joints has a number of advantages including no backlash, zero friction, and negligible hysteresis [9]. In addition, combination of different types of flexure hinges based on amplification structure design will provide the amplification of output displacement [10–12]. The compliant mechanism with flexure hinges is widely used in ultra-precision machining, positioning stage, etc. Ni [13] designed an ultra-stiff nano-precision linear piezomotor, which is composed of three high voltage piezoelectric actuators and a monolithic flexure frame. Chen [14] presented a novel precision motion stage, which combines a flexural stage, electromagnetic actuation, capacitance position measurement, and nonlinear digital feedback control. In addition, the incorporation of the rotational motion not only increases the DOF but also provides capability for correcting possible undesired coupling between major axes. As a result, there would be fewer requirements in degree of precision for stage machining and assembly. Gao [15] proposed a surface-motor-driven XY planar motion stage equipped with a newly developed $XY\theta Z$ surface encoder for submicron position. The surface motor consists of four linear motors placed on the same surface, two pairs in the X and Y axes. The magnetic array and the stator winding of the linear motor are mounted on the platen (the moving element) and the stage base, respectively. Kim [16] presents two novel six-axis magnetic-levitation (maglev) stages capable of nanoscale positioning, which have very simple and compact structures to meet the demanding positioning requirements of next-generation nano-manipulation and nano-manufacturing focuses on the design and precision construction of the actuator units, the moving platens, and the stationary base plates.

The above studies made a lot of efforts from the design of micro-feed mechanism, and the designed mechanisms have high precision revolution, but it is still difficult in the realization of multi-DOF micro-feed motion. In this chapter, a multi-DOF micro-feed platform based on hybrid compliant mechanism is constructed by combining the advantages of serial structures and parallel structures.

This chapter is organized as follows. First, design process of multi-DOF micro-feed platform is given in Section 8.2. Section 8.2.1 describes the basic idea and component of multi-DOF micro-feed platform. Section 8.2.2

introduces design of nested bottom platform. Section 8.2.3 gives design of parallel top platform. Section 8.3 is about the kinematics analysis of the established multi-DOF micro-feed platform. The finite element analysis (FEA) simulation of nested bottom platform and parallel top platform is given in Sections 8.3.1–8.3.3. Section 8.3.4 is the transient analysis of the micro-feed platform. Conclusions are given in Section 8.4.

8.2 Design of Multi-DOF Micro-Feed Platform

8.2.1 Design Idea and Description of Multi-DOF Micro-Feed Platform

In order to fabricate micro/nano textures on free-form surface, the designed micro-feed platform needs to have high precision feed motion in various directions. Thus, two-layer structure with compliant mechanism is proposed to form the multi-DOF micro-feed mechanism. Compliant mechanism in top layer is constructed in parallel, and in bottom layer it is connected in series. Various motions are generated by the combination of parallel and series forms in the multi-DOF micro-feed mechanism. In order to obtain fast response of micro-feed motion and high-precision output, piezoelectric actuators are used to drive compliant mechanism in different layers.

Based on the above idea, a 5-DOF micro-feed platform based on hybrid compliant mechanism is designed as shown in Figure 8.1. The proposed 5-DOF micro-feed platform includes parallel top platform, nested bottom platform, and base platform. Parallel top platform is composed of

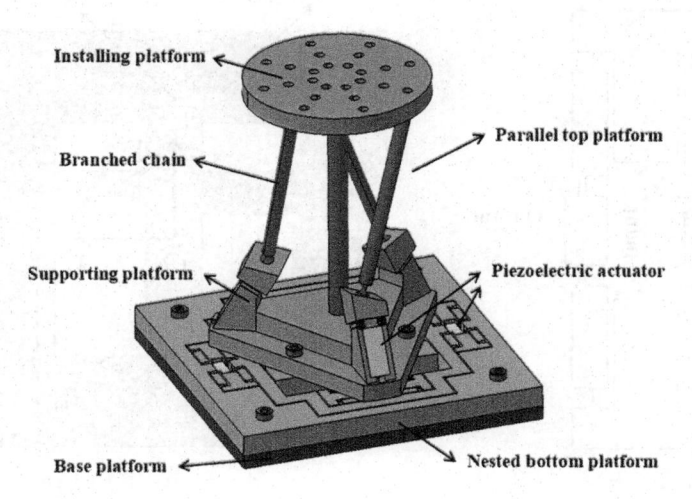

FIGURE 8.1
Model of 5-DOF micro-feed platform.

supporting platform, installing platform, three groups of symmetrical parallel branched chain and one middle branched chain, which is perpendicular to the *XOY* plane. The branched chain is designed by combining the series of multi-axis and single-axis flexure hinges. The rotations along three axes are generated in parallel top platform by three groups of piezoelectric transducers setup in three groups of symmetrical parallel branched chain. The nested bottom platform is designed with two groups of symmetrical-distribution-type shape to form two parts of *XOY* coordinate plane, which is used to generate movements along *X* and *Y* direction by four groups of piezoelectric transducers set up in two groups of symmetrical-distribution-type shape. The parallel top platform and nested bottom platform are constructed and connected in series to form the multi-DOF micro-feed mechanism. Base platform is used to support parallel top platform and nested bottom platform.

8.2.2 Design of Nested Bottom Platform

In this research, a bridge-type displacement amplifier is designed to increase the travel stroke of piezoelectric actuator in the design of nested bottom platform. Figure 8.2a depicts the schematic diagram of bridge-type displacement amplifier which has a compact structure and a large amplification ratio. Once driven with an input displacement, the amplifier produces a vertical output displacement along the backward direction. In order to achieve the micro-feed motion along the *X* and *Y* axes directions and avoid coupling effect, a compliant mechanism with this kind of

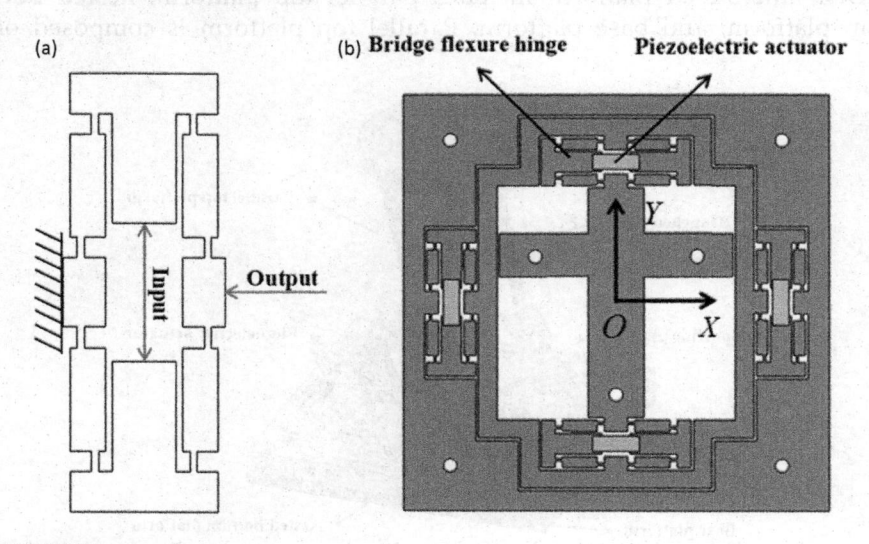

FIGURE 8.2
Nested bottom platform. (a) Schematic of bridge-type amplifier and (b) model of nested bottom platform.

flexure-hinge structure is designed to generate the decoupled micro-motion along the X and Y axis direction. Four groups of bridge-type displacement amplifier with symmetry to X or Y axis direction are combined together in series to generate motion along the X or Y direction, which form a nested structure shown in Figure 8.2b. Two bridge-type amplifiers along X axis direction, which is symmetrical to X axis, and two bridge-type amplifiers along Y axis direction, which is symmetrical to Y axis, are constructed to nested bottom platform. The structure containing amplifiers along Y axis is nested within the structure containing amplifiers along X axis, which forms decoupled mechanism. Piezoactuators are installed in the middle of bridge-type amplifiers to drive the nested bottom platform along X axis direction and Y axis direction. The structures form a nested mechanism in XOY coordinate plane for generating micro-feed motion along X and Y axes directions.

8.2.3 Design of Parallel Top Platform

In order to obtain the rotational motions of multi-DOF micro-feed platform, the configuration named 3-PSS/S (P – prismatic pair; S – spherical pair) is considered to be a good idea. Three flexure-hinge active legs and one middle flexure-hinge supporting leg are used to construct a spherical parallel platform. Each flexure-hinge active leg is a subchain, and its displacement is obtained by the driving of piezoactuator. Middle flexure-hinge supporting leg limits the translational movement of rotary platform. Figure 8.3 shows the schematic of the parallel top platform. In Figure 8.3, points A_1, A_2, A_3 are

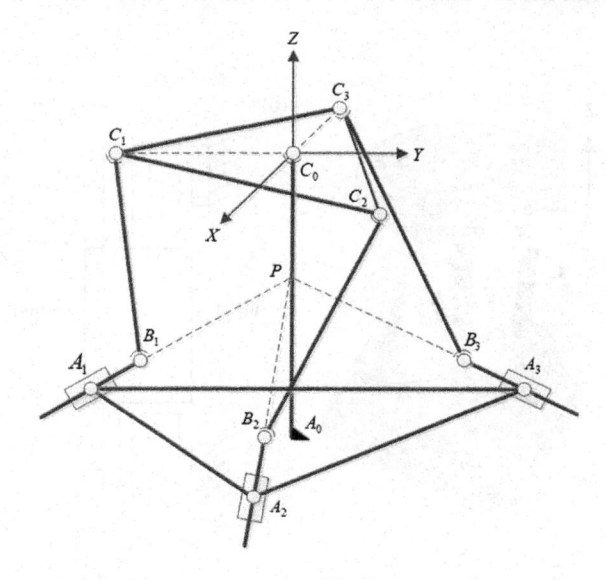

FIGURE 8.3
The schematic of the parallel top platform.

sliding pair and these value used to calculate displacement, and points B_1, B_2, B_3, C_1, C_2, C_3, C_0 are joints. Joints B_1, B_2, B_3, C_1, C_2, C_3 are multi-axis flexure hinges for realizing the plane rotations, and C_0 is a multi-axis flexure hinge for limiting the space translational motions.

Based on the above analysis, a parallel top platform with compliant mechanism is designed, and the established parallel top platform is shown in Figure 8.4a. The established parallel top platform model includes supporting platform, four groups of subchain structure, installing platform, and piezoelectric actuator. Each parallel subchain structure is constructed by two multi-axis spherical joints and one single-axis flexure-hinge structure in series. Single-axis flexure-hinge structure shown in Figure 8.4b is designed to output translational motions of piezoactuator. This structure not only reduces stress but also has a good output performance. Three groups of subchain structures are symmetrically placed along Z axis direction and are connected with supporting platform and also with the installing platform to form parallel subchain structure. One subchain placed in the middle of the supporting platform is connected with the installing platform with a spherical flexure hinge. Three piezoelectric actuators drive three groups of subchain structures to generate micro-feed rotary motion of the installing platform along three directions. When three piezoelectric actuators are exerted with the same sine voltage signals, the parallel top platform will generate rotation along Z axis direction. The rotations along other axes directions are generated when three piezoelectric actuators are exerted with the different sine voltage signals. Thus, the parallel top platform can realize the rotational movement of 3 DOF around the midpoint of the installing platform.

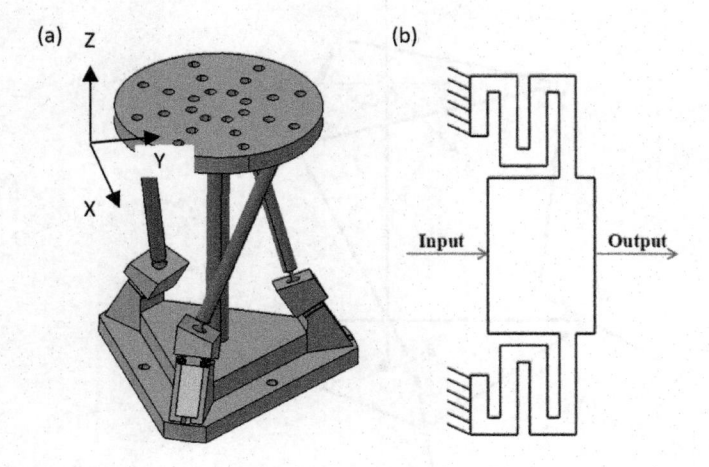

FIGURE 8.4
(a) Model of the parallel top platform and (b) single-axis flexure-hinge structure.

8.3 FEA Simulation of Multi-DOF Micro-Feed Platform

8.3.1 Mesh Generation

To investigate the dynamic characteristics of the multi-DOF micro-feed platform, FEA was performed by using ANSYS software. FEA process is divided into three steps: (1) the analysis of nested bottom platform, (2) the analysis of parallel top platform, and (3) the analysis of the multi-DOF micro-feed platform. A spring steel with the following physical material parameters is assumed: density ($7.81E{-}9\,t/mm^3$), Young's modulus ($210\,GPa$), Poisson's ratio (0.288).

Based on the geometric model (Figure 8.1), a finite element model of the designed multi-DOF micro-feed platform is established. The finite element models of nested bottom platform, parallel top platform, and micro-feed platform are established by using tetrahedral free grid division shown in Figure 8.6.

8.3.2 FEA of the Parallel Bottom Platform

In order to obtain the output performance of the nested bottom platform, the transient analysis of the nested bottom platform is conducted by the finite element method. The initial displacement load $10\,\mu m$ is exerted on piezoelectric transducer in the X axis direction shown in Figure 8.2b, and the fixed constraints are set. The finite element model of the nested bottom platform in Figure 8.5a is analyzed, and the simulation results such as stress distribution and deformation are obtained as shown in Figure 8.6.

From the simulation results of the nested bottom platform, it can be seen that the maximum stress that occurred on the flexible hinge structure is $67.474\,MPa$, which is much smaller than the allowable stress of spring steel.

(a) 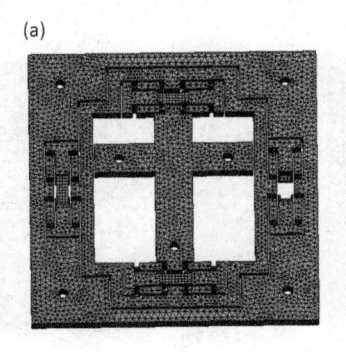 (b)

FIGURE 8.5
Finite element model. (a) Nested bottom platform and (b) parallel top platform.

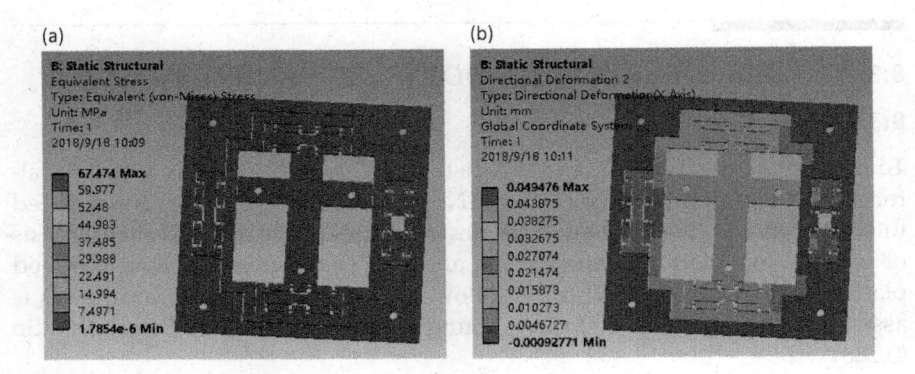

FIGURE 8.6
The simulation results of the parallel bottom platform in X axis direction. (a) Stress distribution and (b) deformation.

The maximum deformation of the middle part of the nested bottom platform is 37.61 μm in X direction from the Figure 8.6b, and the output of the bridge-type amplifier is 49.45 μm. Thus, the amplification ratio α_1 is 3.76, and the reduction ratio λ_1 of deformation from amplifier to middle part in X direction is calculated as

$$\lambda_1 = \frac{49.45 - 37.61}{49.45} \times 100\% = 23.9\% \tag{8.1}$$

The maximum displacement load 10 μm is applied on the Y direction piezoelectric transducer which is contacted with the nested bottom platform, and the analyzed result is shown in Figure 8.7. From Figure 8.7, it can be concluded that the maximum stress that occurred on the flexible hinge structure is 65.047 MPa, which is much smaller than allowable stress of spring steel. The maximum deformation of the middle part of the nested

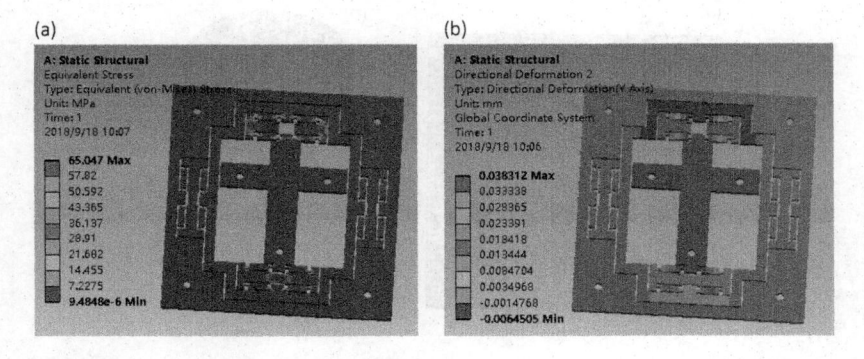

FIGURE 8.7
Results of the parallel bottom platforms along Y axis direction. (a) Stress distribution and (b) total deformation in Y direction.

bottom platform is 38.24 μm in Y direction from the Figure 8.7b, and output of the bridge-type amplifier is 38.27 μm. The amplification ratio α_2 is 3.82, and the reduction rate λ_2 of deformation from amplifier to middle part in Y direction is calculated as

$$\lambda_2 = \frac{38.27 - 38.24}{38.24} \times 100\% = 0.08\% \tag{8.2}$$

From the comparison results, it shows that the nested bottom platform has a big amplification ratio and little energy loss. Coupling is the output influence when exerting different direction input motion, and coupling effect is an important index of the designed multi-DOF micro-feed platform. In order to evaluate the coupling effect of the designed multi-DOF micro-feed platform, coupling effect of nested bottom platform is analyzed by finite element method.

The maximum deformation of the middle part in the Y direction is 0.014 μm shown in Figure 8.8a when exerting displacement load 10 μm in the X direction. Thus, the coupling coefficient K_X in the X direction can be defined as

$$K_X = \frac{0.014}{10} \times 100\% = 0.14\% \tag{8.3}$$

The maximum displacement output in the X direction is 0.014 μm shown in Figure 8.8a when exerting the displacement load 10 μm in the Y direction. Thus, the coupling coefficient K_Y in the Y direction can be obtained as

$$K_X = \frac{0.008}{10} \times 100\% = 0.08\% \tag{8.4}$$

It can be seen from the above analysis that the coupling effects in X and Y axes directions are less than 0.2% for the nested bottom platform. Thus, the coupling effects can be completely ignored in actual application.

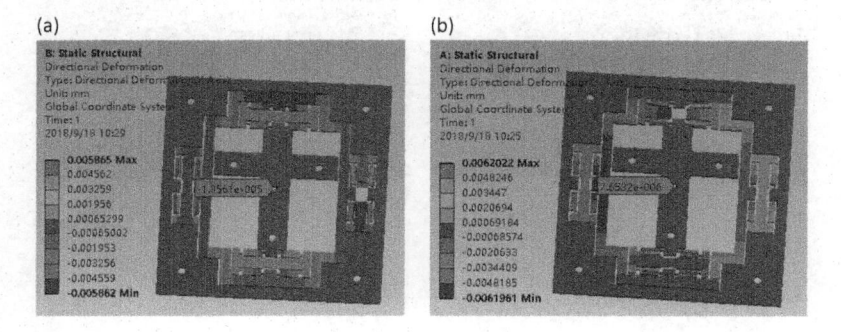

FIGURE 8.8
Coupling analysis of the parallel bottom platform. (a) Displacement in the X axis direction and (b) displacement in the Y axis direction.

8.3.3 FEA of the Parallel Top Platform

In order to obtain the output performance of the parallel top platform, the transient analysis of the parallel top platform is conducted by the finite element method. The initial displacement load 20 μm is exerted on three piezoelectric transducers along input direction as shown in Figure 8.4b, and the fixed constraints are set. The finite element model of parallel top platform in Figure 8.5b is analyzed, and the simulation results such as stress distribution and deformation are obtained shown in Figure 8.9.

The displacement load 20 μm is exerted on piezoelectric transducer in all subchains. The maximum stress of the parallel top platform depicted in Figure 8.9a is 168.27 MPa, which occurs in single-axis flexible-hinge structure and is much smaller than the allowable stress of spring steel. The maximum deformation is 69.14 μm along the edge of installing platform, so the rotation angle θ_1 along Z axis can be computed as

$$\theta_1 = \frac{180}{\pi} ar \tan\left(\frac{69.14}{110 \times 10^3}\right) = 0.036° \tag{8.5}$$

The simulation results are shown in Figure 8.10 when the displacement load 20 μm is exerted on one piezoelectric transducer in one of the subchains. The simulation results show that maximum stress occurs in one of the single-axis flexible-hinge structure of the parallel top platform and the maximum stress is 162.62 MPa, which is much smaller than the allowable stress of spring steel. The maximum deformation is 17.33 μm, and the minimum deformation is −15.98 μm in Z axis direction. Thus, the rotation angle θ_2 is obtained as

$$\theta_2 = \frac{180}{\pi} ar \tan\left(\frac{17.33 - (-15.98)}{110 \times 10^3}\right) = 0.017° \tag{8.6}$$

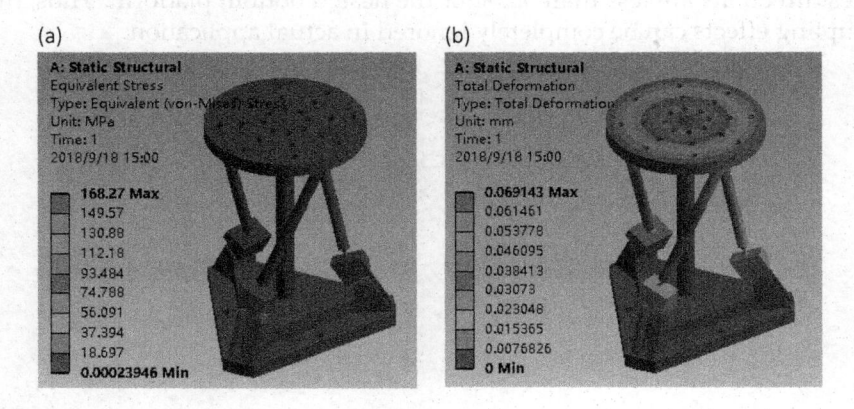

(a)   (b)

FIGURE 8.9
The input displacement of single subchain. (a) Stress distribution and (b) deformation in Z direction.

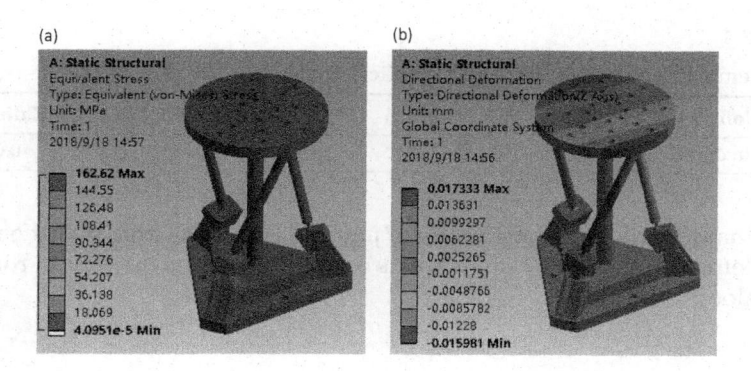

FIGURE 8.10
Simulation results of all the subchains with the same displacement. (a) Stress distribution and (b) deformation in Z axis direction.

8.3.4 FEA of the Micro-Feed Platform

In order to obtain the output performance of the micro-feed platform, the transient analysis of the micro-feed platform is conducted by the finite element method. The initial displacement load 10 µm is exerted on piezoelectric transducer in the nested bottom platform shown in Figure 8.2b, and the fixed constraints are set. The finite element model of parallel top platform in Figure 8.5c is analyzed, and the simulation results such as stress distribution and deformation are obtained as shown in Figure 8.11.

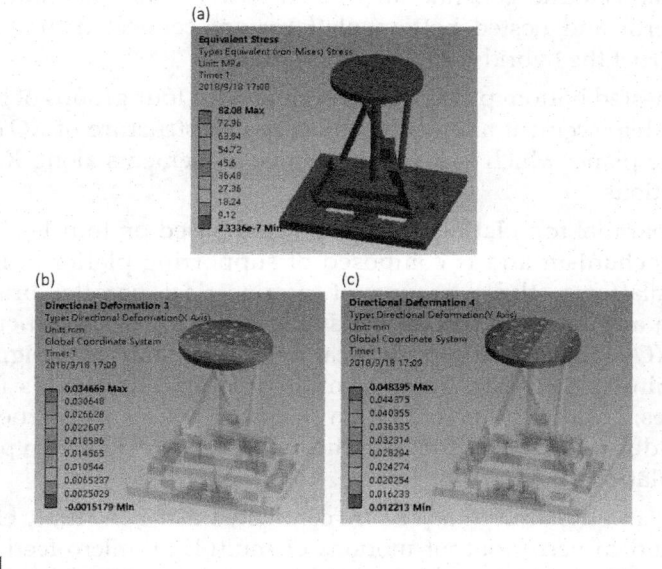

FIGURE 8.11
The results of the multi-DOF micro-feed platform. (a) Stress distribution, (b) deformation in X direction, and (c) deformation in Y direction.

TABLE 8.1

Displacement Performance Analysis of Micro-Platforms

Simulation	X direction	Y direction	Z direction
Maximum deformation	34.67 μm	48.40 μm	0.036°

The maximum deformation is 34.67 μm in X axis direction, and the maximum deformation is 48.40 μm in Y axis direction, and the maximum rotation angle along Z axis is 0.036° (Table 8.1).

8.4 Conclusions

A multi-DOF micro-feed mechanism was proposed to provide the capability of fabricating micro-textures on free-form surface for the mechanical micro-machining method. Structure design, kinematics modeling, and finite element simulation for the proposed multi-DOF micro-feed mechanism are performed. The following is a summary of findings from the present work:

1. A multi-DOF micro-feed mechanism with two-layer structure, which includes parallel top platform and nested bottom platform, is established to generate micro-feed movements. The parallel top platform and nested bottom platform are connected in series to construct the hybrid compliant mechanism.

2. The nested bottom platform is designed with four groups of bridge-type displacement amplifier to form nested structure of XOY coordinate plane, which is used to generate movements along X and Y directions.

3. The parallel top platform is constructed based on four legs parallel mechanism and is composed of supporting platform, installing platform, three groups of symmetrical parallel branched chain and one middle branched chain which is perpendicular to the XOY plane. The symmetrical branched chain is designed by combining the series form of multi-axis and single-axis flexure hinges; parallel branched chain is established; and piezoelectric transducer is used to generate the movements of 3 DOF in parallel top platform.

4. Static and dynamic analyses are conducted using the finite element method to verify output motions of multi-DOF micro-feed mechanism. The results show that the designed multi-DOF micro-feed mechanism is satisfactory.

8.5 Acknowledgement

This work was funded by the National Natural Science Foundation of China under grant nos. 51675277 and by China Postdoctoral Science Foundation under grant no. 2015M570447. The authors want to express their sincere gratitude to the Selection Committee for the Natural Science Foundation of China Grant and for the financial support that made this research possible.

References

1. J. Zhang, T. Cui, C. Ge, Y. Sui, and H. Yang, "Review of micro/nano machining by utilizing elliptical vibration cutting," *International Journal of Machine Tools and Manufacture*, vol. 106, pp. 109–126, 2016.
2. D. Dhupal, B. Doloi, and B. Bhattacharyya, "Modeling and optimization on Nd:YAG laser turned micro-grooving of cylindrical ceramic material," *Optics and Lasers in Engineering*, vol. 47, pp. 917–925, 2009.
3. H. Lim, Y. Wong, M. Rahman, and M. E. Lee, "A study on the machining of high-aspect ratio micro-structures using micro-EDM," *Journal of Materials Processing Technology*, vol. 140, pp. 318–325, 2003.
4. C. H. Jo, B. H. Kim, and C. N. Chu, "Micro electrochemical machining for complex internal micro features," *CIRP Annals*, vol. 58, pp. 181–184, 2009.
5. T. Schaller, L. Bohn, J. Mayer, and K. Schubert, "Microstructure grooves with a width of less than 50 μm cut with ground hard metal micro end mills," *Precision Engineering*, vol. 23, pp. 229–235, 1999.
6. D. P. Adams, M. J. Vasile, and A. Krishnan, "Microgrooving and microthreading tools for fabricating curvilinear features," *Precision Engineering*, vol. 24, pp. 347–356, 2000.
7. K. A. Bourne, Development of a *High-Speed High-Precision Micro-Groove Cutting Process*: University of Illinois, Urbana-Champaign, 2010.
8. E. Shamoto and T. Moriwaki, "Study on elliptical vibration cutting," *CIRP Annals-Manufacturing Technology*, vol. 43, pp. 35–38, 1994.
9. D. Zhang, D. Chetwynd, X. Liu, and Y. Tian, "Investigation of a 3-DOF micro-positioning table for surface grinding," *International Journal of Mechanical Sciences*, vol. 48, pp. 1401–1408, 2006.
10. H.-W. Ma, S.-M. Yao, L.-Q. Wang, and Z. Zhong, "Analysis of the displacement amplification ratio of bridge-type flexure hinge," *Sensors and Actuators A: Physical*, vol. 132, pp. 730–736, 2006.
11. Q. Xu and Y. Li, "Analytical modeling, optimization and testing of a compound bridge-type compliant displacement amplifier," *Mechanism and Machine Theory*, vol. 46, pp. 183–200, 2011.
12. H. Zhou and B. Henson, "Analysis of a diamond-shaped mechanical amplifier for a piezo actuator," *The International Journal of Advanced Manufacturing Technology*, vol. 32, pp. 1–7, 2007.

13. J. Ni and Z. Zhu, "Design of a linear piezomotor with ultra-high stiffness and nanoprecision," *IEEE/ASME Transactions on Mechatronics*, vol. 5, pp. 441–443, 2000.

14. K.-S. Chen, D. Trumper, and S. Smith, "Design and control for an electromagnetically driven X–Y–θ stage," *Precision Engineering*, vol. 26, pp. 355–369, 2002.

15. W. Gao, S. Dejima, H. Yanai, K. Katakura, S. Kiyono, and Y. Tomita, "A surface motor-driven planar motion stage integrated with an XYθZ surface encoder for precision positioning," *Precision Engineering*, vol. 28, pp. 329–337, 2004.

16. W.-j. Kim, S. Verma, and H. Shakir, "Design and precision construction of novel magnetic-levitation-based multi-axis nanoscale positioning systems," *Precision Engineering*, vol. 31, pp. 337–350, 2007.

9

Metal Forming Presses: Technology, Structure, and Engineering Design

Volkan Esat

Middle East Technical University

CONTENTS

9.1 Introduction 167
9.2 Technology and Structure 168
    9.2.1 State-of-the-Art Servo Presses 170
    9.2.2 A Case Study: FEA of Deep Drawing by Use of Servo Press 173
9.3 Engineering Design through Reverse Engineering 175
    9.3.1 Functional Reverse Engineering 176
9.4 Conclusions 177
9.5 Acknowledgments 177
References 178

9.1 Introduction

Metal forming processes constitute the backbone of contemporary manufacturing engineering along with machining. In metal forming, workpieces in solid state are shaped via mechanical plastic deformation. Metal forming processes range from bulk forming processes including forging, extrusion, wire drawing, and rolling to sheet metal processes such as bending, stamping, deep drawing, and shearing. All of the aforementioned processes require applications of varying types and often high levels of loads through generally strong, rigid, and expensive machine tools, and technology. Metal forming presses are the <u>machine tools</u> that induce permanent deformation to metals and/or shear them while transforming the workpiece into a functional product.

It is predicted that press technology was first used for olive oil and grape juice extraction around 600 BC. Sheet metal armor seemed to emerge in 1400s as a product of metal working. Slitting mills were used in Europe as a sheet metal shaping and shearing machine tool toward the end of 16th century. Joseph Bramah fabricated the hydraulic press, which was utilized as a compressive machine tool for shaping metals by the end of 18th century.

Hydraulic press technology further advanced with the addition of electrical motors aiding to generate the compression required for metal forming processes (Halicioglu et al., 2016a).

This chapter is dedicated to reviewing the technology, structure, and state-of-the-art of metal forming presses with an emphasis on the contemporary servo types. Current reverse engineering (RE) practices are addressed, acknowledging some of the functional reverse engineering (FRE) approaches. A case study embodying finite element analysis (FEA) of a typical metal forming operation with servo press, in the form of deep drawing, is presented.

9.2 Technology and Structure

Presses are grouped according to their working mechanism as hydraulic, mechanical, or pneumatic and according to their control strategy as conventional or servo type. In order to apply the desired load for a specific period of time and to have significant control over the application and dwelling durations, the structural mechanism utilized such as knuckle-joint or screw types and the drive system chosen such as servo motor play a vital role.

Hydraulic presses possess programmable strokes attaining full capacity at various speeds and strokes during the operation. In a hydraulic press, the slide can conduct various motions such as reversing or dwelling throughout the stroke. As a shortcoming, hydraulic presses lack accuracy and reliability when compared to the mechanical types. Mechanical presses are more advantageous than hydraulic presses in forming speed; however, they are relatively poor in terms of being controllable and adapting to a constant motion data (Halicioglu et al., 2016a). Sheet metal parts in mass production are currently manufactured predominantly through metal forming presses. One of the frequently used mechanical press constructions embodies an eccentric mechanism actuated by an induction motor coupled with a flywheel (Behrens et al., 2016).

A certain classification of conventional presses based on the means of control strategy categorizes them as follows:

    I. Force-based (hydraulic actuation)

    II. Stroke-based (knuckle-joint mechanism, linkage press, crank mechanism)

    III. Energy-based (screw mechanism).

Stroke-based control and screw mechanism possess the common characteristic where the power from the rotational motion is transformed into reciprocating translational motion of the punch plate. Motion can be converted through various mechanisms as mentioned above. Amongst those, the most frequently employed type utilizes the crank mechanism, whereas other mechanisms such as the knuckle-joint and the linkage types are preferred to mitigate the drawbacks of the crank type such as optimizing unfavorable motions at the extremum positions (Osakada et al., 2011).

Servo presses to be utilized in sheet metal forming emerged as another contemporary option when compared to the traditional presses. They embody superior features such as higher manufacturing paces, exceeding those of traditional counterparts. Servo press consumes energy only when it is moved, which appears as another advantage over a conventional press. On the other hand, they require significant electric power particularly in deep drawing operations. Typical conventional mechanical presses contain an electric motor coupled with a gear system connected to a flywheel with all necessary transmission elements. However, for servo presses, the motor is directly connected to pinion, sustaining essential control throughout the operation (Halicioglu et al., 2018). Some contemporary commercial servo presses can be seen in Figure 9.1.

Servo-driven presses can embody different working mechanisms. Mechanical servo machine tools possess greater production rates, with superior efficiency and increased trustworthiness when compared to the hydraulic servo versions, while also decreasing noise levels due to the essential mechanical parts of the fluid-based systems. Arguably, flexible slide movement is the most essential and distinguishing aspect of a servo machine tool. One of the strengths of servo presses is that they can achieve a great variety of motions (Osakada et al., 2011; Bias Makina, 2018).

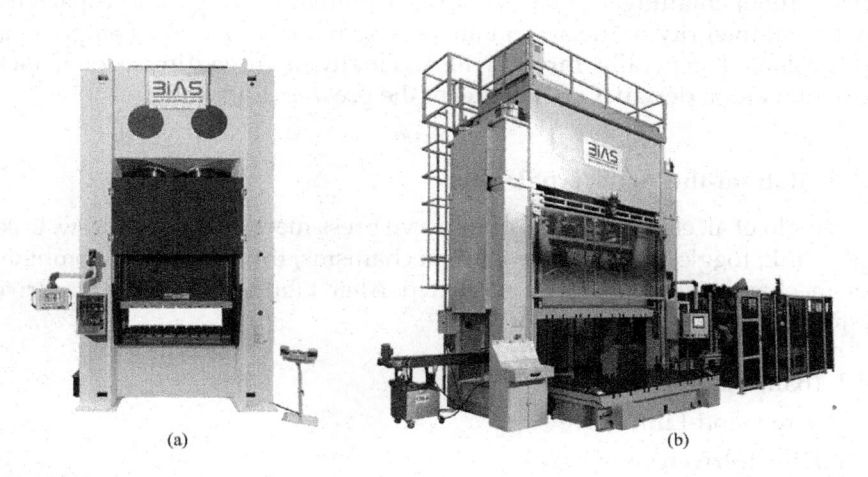

(a)   (b)

FIGURE 9.1
Servo presses: (a) 250 ton-force capacity press for white goods industry and (b) 500 ton-force capacity press for automotive industry. (Courtesy of Bias Makina, Turkey, with permission.)

Servo presses are constructed retaining the capabilities of established mechanisms employed in conventional mechanical presses. Keeping in mind the capabilities brought by a servo motor, this fact brings in an operational edge in processing rates and end product condition. Halicioglu et al. (2016a) tabulates the strengths and advantages of servo presses as follows:

I. Lubrication on demand
II. Reduced impact load through controlling touching speed
III. Decreased noise via adequate tribological and kinematic conditions
IV. Flexibility in operating confines due to desired slide motions such as pulsating or oscillating
V. Reduced vibration via optimizing the punch movements
VI. Accurate manufacturing of the product through obtaining adequate kinematics of the punch plates
VII. Smaller press dimensions within the processing space
VIII. Synchronization of additional similar presses for particular product lines
IX. Extended tool life owing to adequate tribological and kinematic conditions
X. Decreased waste due to the accuracy in the product
XI. Improved productivity through least possible processing times
XII. Controlled energy consumption due to start–stop technology.

Halicioglu et al. came up with a design guide for developing a servo crank press (2016b) entailing system dynamics, different loadings, and capacities. Structural integrity of the system and strength analyses of the components are evaluated generally through FEA. Finally, a three-dimensional (3D) computer aided design (CAD) model of the press is produced.

9.2.1 State-of-the-Art Servo Presses

Halicioglu et al. classified (2016a) the servo press mechanisms as screw, linkage, crank, toggle, and knuckle joint mechanisms, through which combination mechanisms may also be constructed. Altan tabulated (2007) servo-drive systems as follows:

I. (Ball) Screw Drive
II. Screw-and-Link Drive
III. Direct-driven Gear Drive
IV. Direct-driven Spindle Drive
V. Linear Drive.

Typically, high-torque slow-angular-speed servo motors are utilized in servo-driven presses. The servo motor is engaged to the crank shaft directly or through a pinion-and-gear transmission. In order to achieve a considerable level of dynamics in the system, the machine tool pressing slide is aimed to attain the minimum moment of inertia possible. Due to this fact, flywheels, defined with their capability of storing energy, are not used in servo versions (Behrens et al., 2016). Schematics of a typical gear drive servo press can be seen in Figure 9.2.

Nowadays servo motors can be manufactured as AC or DC types. In 1980s, they were predominantly constructed as DC type due to their large currents being controlled by silicon-controlled rectifiers, whereas AC servo motor became widespread after the use of transistors that can control and switch greater currents occurring at increased frequencies. Most motor manufacturers currently produce a great variety of electrical prime movers to be used in operations which employ servo motors (Osakada et al., 2011).

Presses with servo motors can impart superior resilience in one-dimensional reciprocating motion capable of following various paths of ram/punch motion. Figure 9.3 depicts comparison of punch motion and cycle characteristics of mechanical eccentric press versus servo press used for forming a similar product. This capability of servo presses also leads to shorter cycles, potentially increasing production rates. Groche et al. present (2010) fundamental conventions in devising 3D servo presses which can satisfy the provisions of various production systems. As the name of the press refers, tool movements in 3D space can be efficiently manipulated. Through other machine elements such as cams or gears, it is possible to add additional degrees of freedom to the system, yielding increased flexibility.

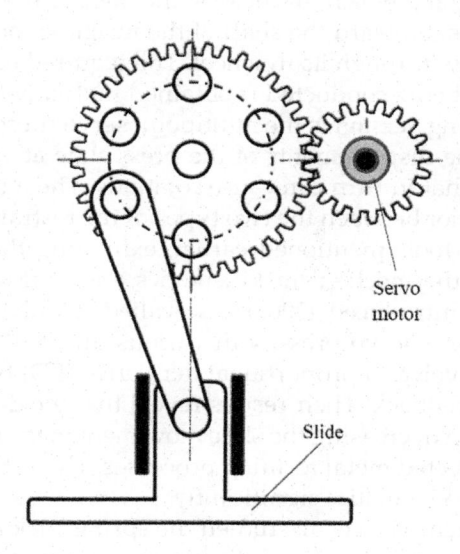

FIGURE 9.2
Schematics of a typical gear drive servo press.

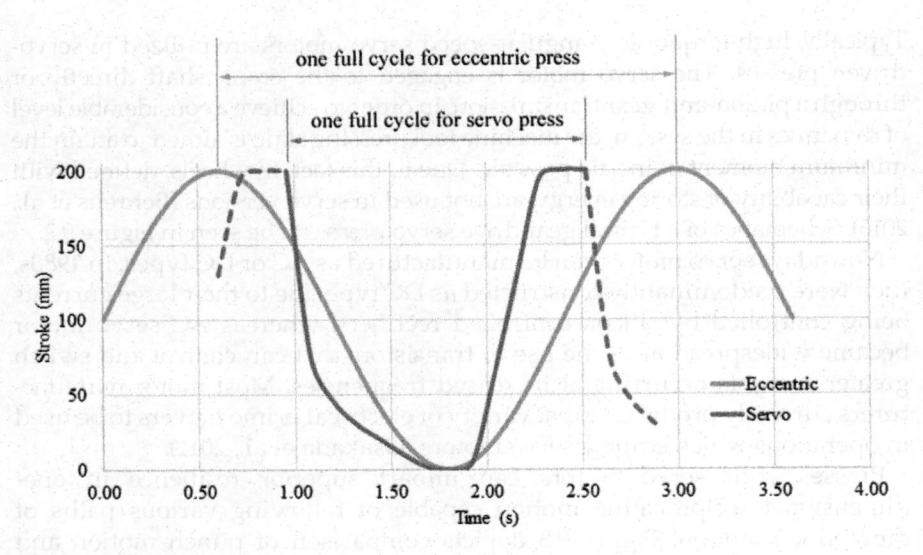

FIGURE 9.3
Comparison of punch motion and cycle characteristics of mechanical eccentric press versus servo press used for forming a similar product.

Behrens et al. generated (2016) a novel design for eccentric presses, which decreases costs while benefiting from the strengths of a servo drive. The main objective is achieving low-cost controllable ram/punch motion in comparison to current technology. Powersplit gearbox constitutes a so-called powersplit subsystem, which harnesses the mechanical energy of both actuators and steers it toward the shaft of the machine tool. The main drive system stores energy through its flywheel. The required energy in the metal forming operation being conducted is obtained mutually from both drives.

Another interesting version is the multipoint servo press, which is capable of manipulating the displacements of the press slide at various points in a rigorous manner; that in turn can cause rotations. The authors investigated the coupling behavior between the two types of the restrained/unrestrained 3D servo machine tools mentioned earlier, exhibiting the dependent relationship amongst the mechanism kinematics and stiffness (Groche et al., 2017). Li and Tso introduced (2008) a so-called iterative learning control (ILC) procedure for a servo press with various different combined drives. Their procedure involves a proportional derivative (PD) type ILC controller with closed-loop feedback. Their results reveal that the designed procedure proves competent in decreasing the slide movement inaccuracies of the servo machine tool in selected metal forming processes. The authors claim to have improved the slide kinematics significantly.

In another research, Mori et al. studied the spring back responses of ultra-high-strength steel sheet metal in V-die bending, formed meticulously via a computer numerical control (CNC) servo press (2007). Predicting spring back responses of sheet metals constitutes one of the vital manufacturing challenges in determining the required metal forming process parameters that leads to the desired end product. In this work, the authors researched the effects of various process parameters such as the punch speed and change of thickness as well as the effects of material properties on spring back (Mori et al., 2007).

Matsumoto et al. examined ram/punch kinematics in extrusion for avoiding forming problems such as galling through a novel forming technique that utilized a servo machine tool (2013). The quality of the end product is assessed experimentally as well as numerically in terms of tribological and thermal variations. Authors claim that their suggested forming technique can live up to the expectations with regard to the targeted tribological and thermal conditions.

Researchers applied load pulsation mode in a servo press on to stainless steel parts during plate forging process (Maeno et al., 2014). They investigated the effects of loading conditions on the shape change response using various steel and carbide dies. Authors suggest that loading pulses applied in forging aid ease of operation via decreasing friction. As another metal forming method, extrusion conducted with a double-axis servo machine tool is put forward by Osakada et al. in order to revamp the geometrical deformities of the workpieces (2005). Experiments and FEAs were both carried out on gear parts as case studies, through which optimal conditions for extrusion were figured out and their effectiveness was demonstrated. Cao et al. designed (2014) a servo drive utilizing a universal sheet forming performance test system.

9.2.2 A Case Study: FEA of Deep Drawing by Use of Servo Press

Research was conducted by an author-led project in order to investigate various deep drawing process parameters through FEA. The force profile was selected so that it could be produced by a servo press. The project team, Boga and Calayir, generated finite element (FE) models in the commercial software MSC Marc 2013 with an emphasis on the prevalent die profiles: in the form of cylindrical, tractrix, and conical dies (2014). Using the aforementioned die profiles, author and the researchers investigated the punch and die process parameters and their effects on force levels and the final products. The nonlinear elastic–plastic material properties of the selected steel and the nominal geometric properties of the generated FE models are provided in Figure 9.4.

The generated FE models of the tractrix and conical dies at the start of the simulations can be seen in Figure 9.5. The FE model of the cylindrical die is given in Figure 9.6 along with the total equivalent plastic strain distributions in the final fully loaded state of the deep drawn product. From the figure, it can be observed that highest of strains, and therefore stresses, accumulate around the die shoulder bent-up region of the formed workpiece.

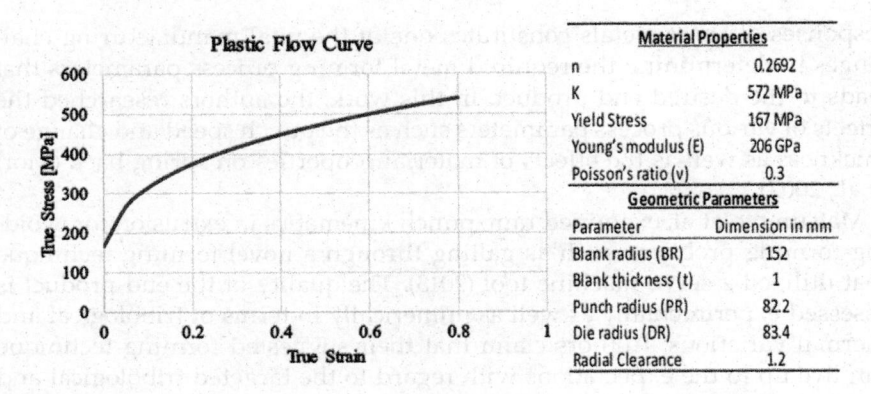

Material Properties	
n	0.2692
K	572 MPa
Yield Stress	167 MPa
Young's modulus (E)	206 GPa
Poisson's ratio (v)	0.3
Geometric Parameters	
Parameter	Dimension in mm
Blank radius (BR)	152
Blank thickness (t)	1
Punch radius (PR)	82.2
Die radius (DR)	83.4
Radial Clearance	1.2

FIGURE 9.4
Material properties of the selected steel and the nominal geometric properties of the generated FE models.

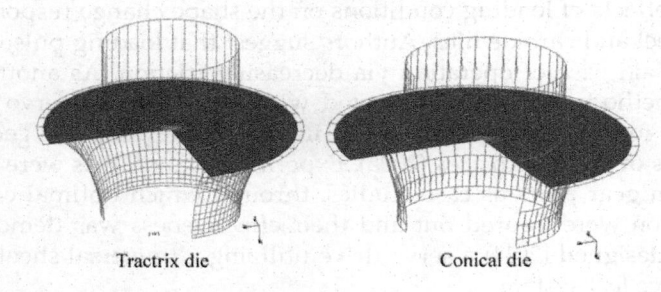

FIGURE 9.5
FE models of the tractrix and conical dies at the start of the simulations.

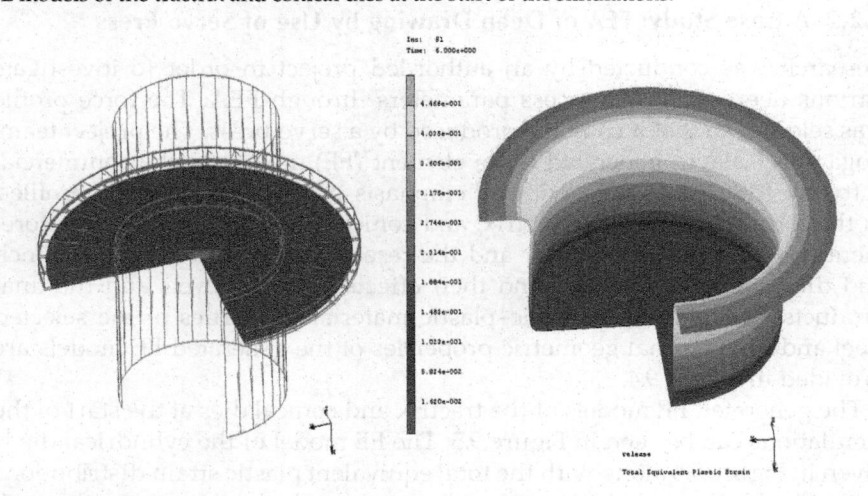

FIGURE 9.6
FE model of the cylindrical die at the start of the simulation and the total equivalent plastic strain distributions in the final state of the deep drawn product.

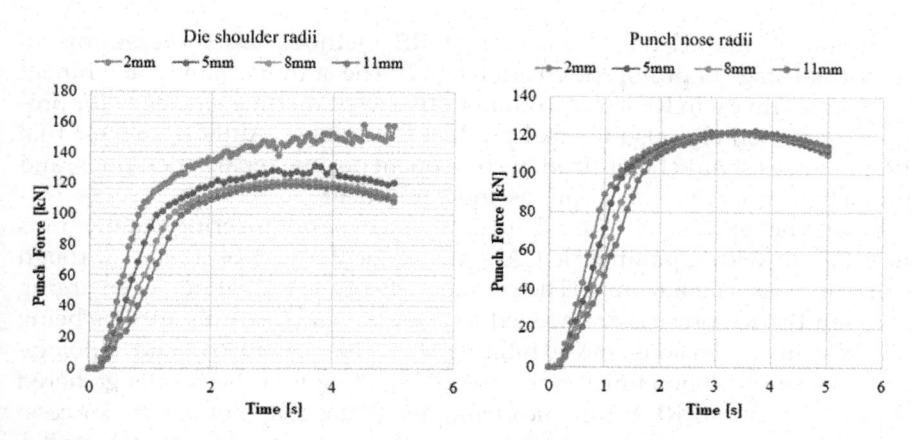

FIGURE 9.7
Punch force versus time with respect to varying die shoulder and punch nose radii.

Figure 9.7 shows the punch force variations versus time with varying die shoulder and punch nose radii. Die shoulder and punch nose radii values were selected as 2 mm, 5 mm, 8 mm, and 11 mm, with 3 mm increments. The predicted results clearly show that the required punch forces decrease as both die shoulder and punch nose radii values change from a tight radius (i.e. 2 mm) to a generous radius (i.e. 11 mm). The force profiles appear to be achievable by a typical servo press. The overall interpretation of results show that die shoulder geometry has the most significant effect on deep drawing. In order to draw a longer cup with a smaller diameter, use of tractrix or conical die appears to be a more viable option.

9.3 Engineering Design through Reverse Engineering

Ramnath et al. describe reverse engineering as a novel idea which indicates the methodology of developing design knowledge through already produced machines or systems (2018). RE is a methodology where a converse order is pursued in manufacturing a product. Anwer and Mathieu highlight the actual importance of reverse engineering in designing end products via addressing various inverse and deductive methodologies (2016). Authors point out that RE in mechanical engineering has grown from collecting technical specifications of a workpiece or machine part, and commencing manual procedure to reconstruct it, to a highly sophisticated practice utilizing state-of-the-art numerical modeling and contemporary digitizing. Anwer and Mathieu claim that RE of mechanical products can be comprehended from either a geometric or a shape perspective (2016).

Dubravcik and Kender explain that RE methods are representing an important stage of prototype creation (2012). The authors claim that eminent companies invest in RE to aid in competitiveness and time efficiency for prototype creation and also for real product production. Authors suggest that RE processes should be utilized in component design, repair operations, and maintenance services of machines and equipment.

Buonamici et al. state (2018) that RE practice for engineering mainly aims to come up with a parametric CAD model for the part of interest through collecting basic data from it. The components of the digital model emerging through the RE process are needed to possess some features such as being accurate in dimensions and exhibiting the actual geometric and topological associations. Buonamici et al. also claim (2018) that the results gathered through advanced RE techniques using templates may not ensure successfully generated attribute associations with an amendable digital model. According to the authors, advanced RE techniques may not warrant proper administration of real-life geometric limitations in line with designer needs while preserving the alterability of the constructed model. Authors propose a reverse engineering technique adequate for rebuilding the parts of interest via embedding a CAD template into 3D scanned data. The template embodies the a priori information on the part to be regenerated. In brief, authors introduce an innovative methodology in order to regenerate digital models commencing from the initial mesh structure into a completely amendable realistic CAD model (Buonamici et al., 2018).

Klein et al. make use of a combination of functional reasoning, RE, and measurement methods in order to incorporate new technologies, such as carbon fiber reinforced plastics, into existing products addressing the sustainability benefits (2016). Authors assess their methodology in various examples: inner-city transport (e.g. a trailer for bicycles), parts of a sport aerial platform (e.g. very light helicopters), and typical cultural objects (e.g. pipe organs).

9.3.1 Functional Reverse Engineering

Jain et al. (2014) describe FRE as a collection of protocols that incorporate detailed functional description of the whole object as well as the functional specification of its system structure including the subsystems. Through this approach, any existing functions of the object might not be overlooked during the ongoing RE process. FRE also helps in identifying any over designed aspects of the object, which may be eliminated during reengineering due to redundancy.

Corbo et al. propose (2004) an FRE analysis methodology involving practical disintegration and identification of varying portions gathered through the aspects of the constructed digital model. They claim that their methodology may be utilized to verify the model as soon as the faces are knit and the necessary engineering calculations are carried out, helping out with making important decisions and accordingly decreasing number of design trials and diminishing cost and time.

Main steps of FRE are suggested by Jain et al. as follows (2014):

1. *Subtract and Operate*: Subtraction of each component one by one and analysis of the effect of the removed part
2. *Force Flow Diagram*: Aims to demonstrate the transfer of force/energy amongst components.

Author of this chapter suggests use of FRE in order to model a servo press of interest. This effort would require a step-by-step orderly subtraction of its components; conducting RE of the subtracted components; identifying the working mechanisms, structural requirements, available degrees of freedom, and drive and control systems; and eventually generating a force flow diagram to gain a complete understanding of the machine tool. Generated CAD models of the components should be meshed and converted to FE models in order to investigate the structural integrity of the subsystems subjected to various static and dynamic mechanical loads.

9.4 Conclusions

This chapter aims to compile up-to-date information on metal forming presses with an emphasis on the contemporary servo press types. Technology, structure, and mechanisms of these presses are reviewed, highlighting their strengths and shortcomings. A case study by the author and colleagues is presented to demonstrate the effects of the use of servo presses in metal forming applications through computational methods. Possibility of feasible use of FRE for servo presses is also highlighted.

9.5 Acknowledgments

Consideration and invitation for participation in the 1st International Workshop on Functional Reverse Engineering of Machine Tools (WRE) 2018 and financial support from the Ghulam Ishaq Khan Institute of Engineering Sciences and Technology (GIKI), Pakistan; financial and in-kind support from the Middle East Technical University—Northern Cyprus Campus (METU NCC), Turkey; and efforts of METU NCC Mechanical Engineering students in their graduation design projects as referenced are gratefully acknowledged.

References

Altan, T. (2007) R&D Update: Servo press forming applications - Part II, Stamping Journal, April 2007, www.thefabricator.com/article/stamping/servo-press-forming-applications-part-ii.

Anwer, N., and Mathieu, L. (2016) From reverse engineering to shape engineering in mechanical design, *CIRP Annals - Manufacturing Technology*, Vol. 65, pp. 165–168.

Behrens, B. A., Bouguecha, A., Krimm, R., Teichrib, S., and Nitschke, T. (2016) Energy-efficient drive concepts in metal-forming production, *Procedia CIRP*, Vol. 50, pp. 707–712.

BIAS Makina Inc. (2018) Bias Multi Servo Press, **catalogue**, www.biasmakina.com/katalog/Bias%20Multi%20Servo%20Press%20Cataloque-English.pdf.

Boga, H. C., and Calayir, H. (2014) Assessment of Process Parameters in Designing Deep Drawing Dies by Using FEM, METU NCC Mechanical Engineering Graduation Design Project, Project generation and supervision by Dr Volkan Esat.

Buonamici, F., Carfagni, M., Furferi, R., Governi, L., Lapini, A., and Volpe, Y. (2018) Reverse engineering of mechanical parts: A template-based approach, *Journal of Computational Design and Engineering*, Vol. 5, pp. 145–159.

Cao, Y., Du, X., Su, Y., Dong, W., Deng, P., and Ruan, Q. (2014) Sheet stamping formability test system based servo crank press, *Procedia Engineering*, Vol. 81, pp. 1061–1066.

Corbo, P., Germani, M., and Mandorli, F. (2004) Aesthetic and functional analysis for product model validation in reverse engineering applications, *Computer-Aided Design*, Vol. 36, pp. 65–74.

Dubravcik, M., and Kender, S. (2012) Application of reverse engineering techniques in mechanics system services, *Procedia Engineering*, Vol. 48, pp. 96–104.

Groche, P., Hoppe, F., and Sinz, J. (2017) Stiffness of multipoint servo presses: Mechanics vs. control, *CIRP Annals - Manufacturing Technology*, Vol. 66, pp. 373–376.

Groche, P., Scheitza, M., Kraft, M., and Schmitt, S. (2010) Increased total flexibility by 3D servo presses, *CIRP Annals - Manufacturing Technology*, Vol. 59, pp. 267–270.

Halicioglu, R., Dulger, L. C., and Bozdana, A. T. (2016a) Mechanisms, classifications, and applications of servo presses: A review with comparisons, *Proceedings of the Institution of Mechanical Engineers, Part B: Journal of Engineering Manufacture*, Vol. 230, No. 7, pp. 1177–1194.

Halicioglu, R., Dulger, L. C., and Bozdana, A. T. (2016b) Structural design and analysis of a servo crank press, *Engineering Science and Technology, an International Journal*, Vol. 19, pp. 2060–2072.

Halicioglu, R., Dulger, L. C., and Bozdana, A. T. (2018) Improvement of metal forming quality by motion design, *Robotics and Computer-Integrated Manufacturing*, Vol. 51, pp. 112–120.

Jain, V., Jain, S., and Yadav, K. (2014) FRE: Functional reverse engineering for mechanical components, *International Journal of Advanced Mechanical Engineering*, Vol. 4, No. 2, pp. 139–144.

Klein, M., Thorenz, B., Lehmann, C., Boehner, J., and Steinhilper, R. (2016) Integrating new technologies and materials by reengineering: Selected case study results, *Procedia CIRP*, Vol. 50, pp. 147–152.

Li, C.-H., and Tso, P.-L. (2008) Experimental study on a hybrid driven servo press using iterative learning control, *International Journal of Machine Tools & Manufacture*, Vol. 48, pp. 209–219.

Maeno, T., Mori, K., and Hori, A. (2014) Application of load pulsation using servo press to plate forging of stainless steel parts, *Journal of Materials Processing Technology*, Vol. 214, pp. 1379–1387.

Matsumoto, R., Jeon, J.-Y., and Utsunomiya, H. (2013) Shape accuracy in the forming of deep holes with retreat and advance pulse ram motion on a servo press, *Journal of Materials Processing Technology*, Vol. 213, pp. 770–778.

Mori, K., Akita, K., and Abe, Y. (2007) Springback behaviour in bending of ultra-high-strength steel sheets using CNC servo press, *International Journal of Machine Tools & Manufacture*, Vol. 47, pp. 321–325.

Osakada, K., Matsumoto, R., Otsu, M., and Hanami, S. (2005) Precision extrusion methods with double axis servo-press using counter pressure, *CIRP Annals*, Vol. 54, No. 1, pp. 245–248.

Osakada, K., Mori, K., Altan, T., and Groche, P. (2011) Mechanical servo press technology for metal forming, *CIRP Annals - Manufacturing Technology*, Vol. 60, No. 2, pp. 651–672.

Ramnath, B. V., Elanchezhian, C., Jeykrishnan, J., Ragavendar, R., Rakesh, P. K., Dhamodar, J. S., and Danasekar, A. (2018) Implementation of reverse engineering for crankshaft manufacturing industry, *Materials Today: Proceedings*, Vol. 5, pp. 994–999.

Part 6

Decision Making

Decision making is a common exercise performed by individuals and group of people in different situations. Industrial decision making relies on human capabilities as well as data collected from manufacturing processes and the manufacturing system. Sensors and transducers play an important role in collecting data from factory floor for single parameter as well as for collection of parameters to optimally run the Small to Medium Enterprises (SMEs). Theoretical formulation of decision-making process requires abstraction from the real situation. First chapter in this part devises this scheme through sequential Markov Decision Process. The second chapter provides physical data from sensors and transducers for the sake of industrial decisions in a hierarchical industrial management pyramid. The third chapter uses the concept of multi-formalism to optimize an enterprise's underlying process model in the form of an executable workflow, thus affecting many decision parameters.

10

Modular Approach of Writing MATLAB Code for a Markov Decision Process

Ali Nasir

University of Central Punjab

CONTENTS

10.1   Introduction 183
10.2   Main Components of the Code 184
10.3   Basic Assumptions 185
10.4   Reward Function File 185
10.5   State to Index Function File 186
10.6   Index to State Function File 187
10.7   State Transition Function File 188
10.8   Transition Probability Function File 189
10.9   Value Iteration Main File 191
10.10 Value Calculation Function 192
10.11 Policy Calculation Function File 193
10.12 Analysis of the Optimal Policy 194
10.13 Conclusions 212
References 213

10.1 Introduction

Markov decision process (MDP) is a tool that is used to model the problems involving decision making in the presence of uncertainty [1]. MDP has been used extensively by three communities namely computer science community, control systems community, and operations research community. A wide range of decision-making problems have been and are being formulated as MDPs. There are many variations of MDP that have been developed for modeling specific types of problems [2,3]. Once a problem is formulated as an MDP, it can be solved for calculation of optimal decision policy using any of many techniques available in the literature under the umbrella of stochastic dynamic programming. Most common techniques used for solving an MDP are value iteration and policy iteration.

An MDP consists of five elements namely (i) a set of states representing all possible situations (in the form of variable values) in the problem, (ii) a set of actions representing all possible decisions that can be made under a given situation or state, (iii) a reward or cost function mapping each state to a number representing its desirability or undesirability (sometimes cost is also associated with decision or a decision–state pair e.g. a decision may be expensive in one state while cheap in another state), (iv) a transition probability function that assigns a probability to each state–decision–state trio (this function is used to depict the probability of reaching a next state from current state under current decision), and (v) a discount factor ranging between zero and one that depicts the depreciation of reward or cost with the passage of time. It is important to mention here that representation of time in an MDP is relative and not absolute. Specifically, time is measured in terms of decisions made i.e. one decision is made in one unit of time. This unit of time is called decision epoch, and it may be a millisecond or an hour or a day depending upon the actual process for which MDP has been formulated. Discount factor depreciates reward or cost with respect to decision epochs rather than absolute time. On the other hand, for processes where time is a critical element of decision making, it can be used as part of the state; that is, a state may include value of time. In such cases, the decision epoch is not a representation of time because some decisions may take longer than the others, whereas the depreciation is equal for all decisions; hence in such cases, depreciation is with respect to number of decision epochs rather than time.

MATLAB is a computing software that is widely used among researchers and practitioners for calculations, simulations, and analysis. MATLAB has many tool boxes dedicated for computations and simulations in specific fields of engineering and computer science. It is easy to use and allows for both script-based code and block-based constructions (SIMULINK). The coding methodology discussed here is in script form and can be used with C++ or JAVA or other similar languages with appropriate syntax modifications. All the examples provided here are in MATLAB.

10.2 Main Components of the Code

Coding of MDP involves seven major function files and one main script file. Besides these files, some auxiliary functions may be required depending upon the structure (mapping of the state transitions under various decisions) of the MDP. A list of files is presented below, and details are discussed in later sections. It is recommended (not necessary) that the files may be written in the order they are discussed here:

1. Reward function file
2. State to index function file
3. Index to state function file
4. State transition function file
5. Transition probability function file
6. Value iteration main file
7. Value calculation function file
8. Policy calculation function file.

10.3 Basic Assumptions

Before discussing the details of how to write each module of an MDP code, it is important to understand the underlying assumptions. First, it has been assumed that the states and actions are discrete and finite. Second, the reward function and transition probabilities do not change as the decision process is being executed although variable reward function, and variable transition probabilities can be accommodated in finite horizon solution of an MDP. Third, each variable in an MDP takes on integer values (fractional values are not discussed). However, fractional values (if finite) can be converted into integer values using scaling or any predefined mapping.

10.4 Reward Function File

A state of an MDP may consist of several variables and each variable has its own range of values. Therefore, in order to calculate the reward function, all possible combinations of the values of state variables have to be considered. Typically, a set of state can be written as

$$S = \{s_1, s_2, \ldots, s_n\}$$
$$s_i = \{v_1, v_2, \ldots, v_m\}, i \in \{1, 2, \ldots, n\} \tag{10.1}$$
$$v_j \in V_j, j \in \{1, 2, \ldots, m\}$$

Equation 10.1 indicates that there are n states where each state is a combination of values of m variables. Each variable v_j can take values from a discrete and finite set V_j of real numbers. Therefore, the total number of states $n = |V_1| \times |V_2| \times \cdots \times |V_m|$, where $|V_j|$ represents the set cardinality of V_j.

Now, in order to write the reward function file, the nested for-loop structure is required. This is demonstrated via an example.

Example 1

Suppose that a person is rolling four dice simultaneously at a casino. The reward of the outcome is sum of the values that turn up at each dice. Write the code for reward function file.

Solution

```
function R = rewardEx1()
k = 1;
R = -ones(6^4,1);
for d1 = 1:6
    for d2 = 1:6
        for d3 = 1:6
            for d4 = 1:6
                R(k) = d1+d2+d3+d4;
                k = k+1;
            end
        end
    end
end
```

Note that there is an indexing variable k that is required to be defined. Also declaring the size of the vector R prior to the start of the nested for-loops saves computational time. Furthermore, the way code is written defines the indexing pattern of R vector with respect to four state variables. This pattern can be changed by rearranging the loops. For example, $d2$ can be set in the outermost loop, and $d1$ can be set in the second loop (next to the outermost).

10.5 State to Index Function File

Indexing is defined by the reward function file. Therefore, a function can be written for calculation of the index for any given state. Indexing function is very important because implementation of MDP requires identification of each state with minimum computational burden, and indexing function provides it. In order to define the indexing function, we need three types of information. First is the number of values of each variable; second is the difference between each consecutive value for each variable; and third is the difference between the minimum value of each state variable and zero. The following example demonstrates how to write state to index function file.

Example 2

For the experiment in Example 1, write state to index function file.

Solution

```
function i = S2indexEx2(S)
i = S(4) + 6*(S(3)-1) + 36*(S(2)-1) + 216*(S(1)-1);
```

Note that the last (inner most) variable has no multiplication factor. Observe that each of the four variables have subtraction of one, i.e. the difference between the actual minimum value of a variable and zero (except for the innermost variable). Also, the multiplication factor of each variable is the product of the sizes of variables that have for-loops inside the loop of that variable in reward function file. Calculation of the index does not involve any for-loop or iterations, and hence it is computationally cheap.

10.6 Index to State Function File

Index to state function file may not be required for writing the MDP code, but it is used when analyzing the MDP policy. Another purpose of this file is to check the accuracy of state to index function file. In order to write this file, one needs to define a small function that can be thought of as modified form of floor function defined as

```
function f = floor2(i,j)
f = floor(i/j);
if rem(i,j)==0
  f = f - 1;
end
```

The function returns floored quotient of its two inputs, and if the quotient is a perfect integer, then it is decreased by one. This function is used in index to state function to determine the contribution of an index to the value of a state variable. Other information used for calculation of the state variable values is the difference between minimum value of a state variable and zero, the number of possible values of a variable, and the difference between two consecutive values. The following example shows how to write index to state function file.

Example 3

Write the index to state function file corresponding to the state to index file in Example 2.

Solution

```
function S = index2SEx3(i)
S(1) = rem(floor2(i,216),6) + 1;
```

```
S(2) = rem(floor2(i,36),6) + 1;
S(3) = rem(floor2(i,6),6) + 1;
S(4) = rem(floor2(i,1),6) + 1;
```

It is relatively easier to understand this file if it is read from the end to the beginning. Value of the last state is simplest to calculate. It is determined by two nested functions which basically extract the contribution of this variable in the index. The inside function floor2 calculates the quotient of the index and the location of the variable in the state (location is one for the inner most variable and for subsequent variables; it is the product of the sizes of variables that have for-loops inside the loop of that variable in reward function file. Once the quotient is calculated, its remainder is computed from the number of possible values of the state variable. This value is then added with the difference between the minimum value of the state variable and zero.

10.7 State Transition Function File

State transition function is used to arrange the states in two dimensions. The result is a matrix with number of rows equal to the number of states and number of columns equal to sum of all possible outcomes with all possible actions. For example, if an MDP has ten states and two possible actions where one action has two possible outcomes and the other action has four possible outcomes, then the transition matrix will have ten rows and six columns. The following example demonstrates how to write the transition function file.

Example 4

Consider the dice experiment in Example 1, and assume that the player has the option of rolling the dice again or retaining the result of the previous dice role. Write the transition function file for this situation.

Solution

```
function N = NextStateEx4()
N = -ones(6^4,6^4+1);
k = 1;
for d1 = 1:6
    for d2 = 1:6
        for d3 = 1:6
            for d4 = 1:6
                j = 1;
                for d1n = 1:6
                    for d2n = 1:6
                        for dn3 = 1:6
                            for dn4 = 1:6
                                N(k,j) = j;
```

```
                                        j = j+1;
                            end
                     end
                 end
             end
             N(k,j) = k;
             k = k + 1;
         end
      end
    end
end
Alternate Solution
function N = NextStateEx4()
N = -ones(6^4,6^4+1);
n1 = 1:6^4;
for k = 1:6^4
    N(k,:) = [n1,k];
end
```

Two solutions have been provided here. One is computationally complex with eight nested loops. This solution is generic and can be used for many types of transition functions. Second solution makes use of the structure of the problem and saves a lot of computations. In general, the transition function file can be written in many ways, and mostly state to index function file is used in the transition function as shall be demonstrated later in the case studies.

10.8 Transition Probability Function File

Similar to the state transition file, transition probability function file generates a matrix of probabilities. The size of the matrix is exactly the same as that of state transition file. However, the contents include probabilities of possible outcomes rather than the indices of possible outcomes. The following example demonstrates how to write transition probability function file.

Example 5

Suppose that for the experiment in Example 1 with transitions as in Example 4, the transition probabilities are uniformly distributed. Write the transition probability function file.

Solution

```
function T = TransitionProbEx5()
T = -ones(6^4,6^4+1);
k = 1;
for d1 = 1:6
    for d2 = 1:6
        for d3 = 1:6
```

```
            for d4 = 1:6
                j = 1;
                for d1n = 1:6
                    for d2n = 1:6
                        for dn3 = 1:6
                            for dn4 = 1:6
                                T(k,j) = 1/(6^4);
                                j = j+1;
                            end
                        end
                    end
                end
                T(k,j) = 1;
                k = k + 1;
            end
        end
    end
end
Alternate Solution
function T = TransitionProbEx5()
T = [ones(6^4,6^4)* 1/(6^4),ones(6^4,1)];
```

Note that the sum of probabilities of all possible outcomes due to a single action is one. Second action has only one outcome; therefore its probability is 1. Due to similarity between the state transition function file and transition probability function file, both matrices can be generated using a single file. In fact, the reward functions, the transition matrix, and the probability matrix all can be generated using a single file. This is depicted in Example 6.

Example 6

Write a single function file that calculates rewards, transitions, and probabilities of the experiment in Examples 1 through 5.

Solution

```
function [R,N,T] = AllPurposeEx6()
R = -ones(6^4,1);
N = -ones(6^4,6^4+1);;
T = N;
n1 = 1:6^4;
k = 1;
for d1 = 1:6
    for d2 = 1:6
        for d3 = 1:6
            for d4 = 1:6
                j = 1;
                for d1n = 1:6
                    for d2n = 1:6
                        for dn3 = 1:6
                            for dn4 = 1:6
                                T(k,j) = 1/(6^4);
                                    N(k,j) = j;
```

```
                                                j = j+1;
                                      end
                                end
                          end
                    end
                    N(k,j) = k;
                                     T(k,j) = 1;
                                     R(k)  = d1+d2+d3+d4;
                          k = k + 1;
                    end
              end
         end
end
Alternate Solution
function [R,N,T] = AllPurposeEx6()
R = -ones(6^4,1);
N = -ones(6^4,6^4+1);
T = [ones(6^4,6^4)* 1/(6^4),ones(6^4,1)];
k = 1;
for d1 = 1:6
    for d2 = 1:6
        for d3 = 1:6
            for d4 = 1:6
                         N(k,:) = [n1,k];
                   R(k) = d1+d2+d3+d4;
                   k = k+1;
            end
        end
    end
end
```

10.9 Value Iteration Main File

This file is the main script file that calls all the function files and calculates optimal policy and optimal values of the states. Although the author prefers to write it before writing the value calculation function and policy calculation function, this file may be written after all function files have been written. Example 7 demonstrates the coding involved in the main file.

Example 7

Write the value iteration algorithm file for the experiment in Examples 1 through 5.

Solution

```
clc;clear all;close all;
[R,N,T] = AllPurposeEx6();
V = R;
err = 1E1;
```

```
itr = 0;
P = -ones(length(R),1);
gamma = 0.95; % can set any value between 0 and 1
v = -ones(6^4+1);
while (err>1E-6)
    V_old = V;
    for i = 1:length(R)
        for k = 1:6^4+1
            v(k) = V(N(i,k));
        end
        V(i) = vcalcEx8(R(i),v,T(i,:),gamma);
    end
    err = max(abs(V-V_old));
    itr = itr+1;
end
for j = 1:length(R)
    for k = 1:6^4+1
        v(k) = V(N(j,k));
    end
    P(j) = pcalcEx9(R(j),v,T(j,:),gamma);
end
save('V_Ex7','V');
save('P_Ex7','P');
```

Note that the transition file used to calculate values of next possible states for each state variable "v" is used in the code for this purpose. The functions "vcalcEx8" and "pcalcEx9" shall be presented in later sections. It is important to save the optimal values and policy to avoid having to run the value iteration file repeatedly. The error limit in the code (for convergence test of state values) is 10^{-6}. This may be varied depending upon the problem. Also the calculation of error is calculated by maximum of the absolute values of the difference between state values in consecutive iterations. Some people may prefer error to be defined as Euclidian norm of the difference between state values in consecutive iterations.

10.10 Value Calculation Function

Value calculation function calculates the value of each state and is called iteratively by the value iteration file. If an action has a cost associated with it, that cost is included in the value calculation function file rather than the reward function file. Example 8 shows how to write the code for value calculation.

Example 8

Write the value calculation function for the experiment in Example 1, and assume that rolling the dice has a cost of 2 units.

```
function V = vcalcEx8(R,v,T,gamma)
u(1) = R + gamma*sum(v(1:length(v)-1).*T(1:length(T)-1)) - 2;
u(2) = R + gamma*v(length(v))*T(length(T));
V = max(u);
```

Note that the components of vector "u" represent the expected value of executing a particular action; hence maximum of those values is the maximum possible value that can be achieved by the state for which value is being calculated. Also the 2 unit cost of action has been included in "u(1)" which represents the dice rolling action. The other action (holding on the current numbers) has no cost. Cost is subtracted from state value whereas the reward is added to it.

10.11 Policy Calculation Function File

Policy calculation function file is very similar to the value calculation file. It uses a function called "findmax" that may not be available in MATLAB library and so is presented below:

```
function ind = findmax(n)
m = max(n);
l = length(n);
for i=1:l,
    if (n(i)==m),
        ind = i;
    end
end
```

This function is used to find the index of the maximum value of a vector. Use of "findmax" and calculation of policy are demonstrated in Example 9.

Example 9

Write the policy calculation function for the experiment in Example 1, and assume that rolling the dice has a cost of 2 units.

```
function P = pcalcEx9(R,v,T,gamma)
u(1) = R + gamma*sum(v(1:length(v)-1).*T(1:length(T)-1)) - 2;
u(2) = R + gamma*v(length(v))*T(length(T));
P = findmax(u);
```

With the development of policy calculation file, the MDP code is complete, and the optimal policy can be computed by running the value iteration file.

10.12 Analysis of the Optimal Policy

Once the policy is developed, its analysis can be done in multiple ways. A user-defined function that is useful for analyzing the optimal policy is "indexlist" function. This function returns the indices of the vector at which a particular value has occurred. It can be used to figure out which actions have been executed how many times by the optimal policy. The code for "indexlist" is given below:

```
function L = indexlist(V,i,tol)
if nargin<3
    tol = 0;
end
L = 0;
k = 1;
for j = 1:length(V)
    if V(j)==i || (V(j)<=i+tol && V(j)>=i-tol),
        L(k) = j;
        k = k+1;
    end
end
```

The following example demonstrates the use of indexlist.

Example 10

For the optimal policy calculated using the code in Example 7, determine how many times the optimal action is to roll the dice and how many times it is optimal to hold on to the current score.

Solution

```
>> Number_of_a1 = length(indexlist(P,1))
Number_of_a1 =
        1090
>> Number_of_a1 = length(indexlist(P,2))
Number_of_a1 =
        206
```

Note that an important information is obtained by just one line of code. Above example reveals that rolling the dice (*a*1) is optimal more often than holding on to the current score (*a*2). Such information is useful in practical problems. Another type of information that can be obtained from the optimal policy using indexlist is to determine the characteristics of the states at which a particular action is executed. This is shown in the following example.

Example 11

Given the results of Example 10, determine the highest and lowest scores of the states at which a1 and a2 are executed.

Solution

```
clc;clear all;close all;
load('P_Ex7');
l1 = indexlist(P,1);
l2 = indexlist(P,2);
s1 = zeros(length(l1),4);
s2 = zeros(length(l2),4);
score1 = zeros(length(l1),1);
score2 = zeros(length(l2),1);
for i = 1:length(l1)
    s1(i,:) = index2SEx3(l1(i));
    score1(i) = sum(s1(i,:));
end
for j = 1:length(l2)
    s2(j,:) = index2SEx3(l2(j));
    score2(j) = sum(s2(j,:));
end
n1 = max(score1)
n2 = max(score2)
n3 = min(score1)
n4 = min(score2)
Result
n1 = 17, n2 = 24, n3 = 4, n4 = 18
```

The above example reveals that it is optimal to hold on to the current score if the score is between 18 and 24, whereas it is optimal to roll the dice again if the score is less than 18. Such categorical information is useful for practical purposes. Note that the function "index2SEx3" has been used and the policy has been loaded from the file saved in Example 7.

Another useful way of evaluating the policy is to generate sample trajectory. The following example demonstrates how to write the trajectory generation code.

Example 12

Generate sample trajectories for the dice rolling experiment using the policy calculated in Example 7.

Solution

```
clc;clear all;close all
load('P_Ex7');
[R,N,T] = AllPurposeEx6();
k = 50;
S = [1 3 4 2]; % initial state
p = 1;
pt = 1;
```

```
nextstate = zeros(k,4);
V = zeros(k,1);
nextstate(1,:) = S;
for q = 1:k
    i = S2indexEx2(S);
    pt = pt*p;
    a = P(i);
    if a==1
        f = round(rand(1)*6^4);
        p = T(i,f);
    else
        f = i;
        p = 1;
    end
    S = index2SEx3(N(i,f));
    nextstate(q+1,:) = S;
end
plot(nextstate(:,1)+nextstate(:,2)+nextstate(:,3)+nextstate
(:,4),'linewidth',3)
xlabel('Number of decisions')
ylabel('score')
title('Sample Trajectory of Dice Experiment')
grid on
```

Three of the sample trajectories generated by using the code in Example 12 are presented in Figures 10.1–10.3. Note that in all cases, the optimal policy brings the score to 18 or above and keeps it there.

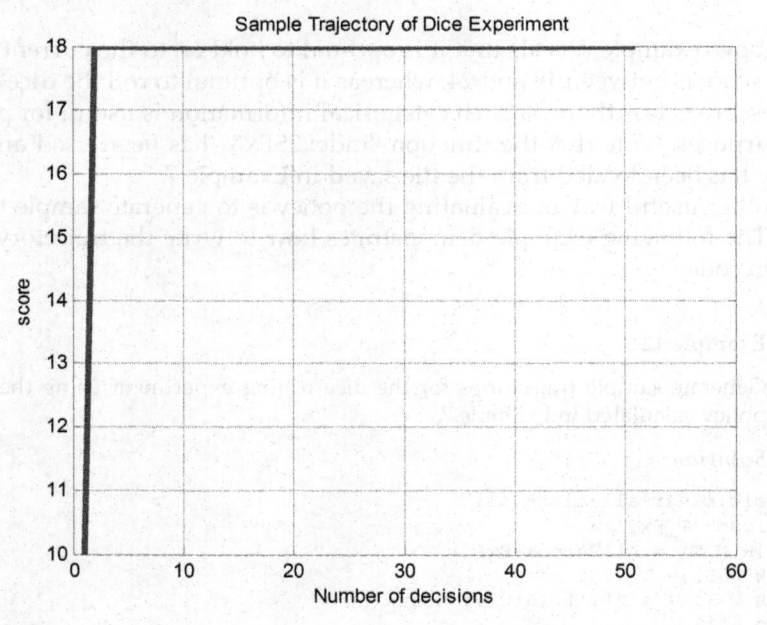

FIGURE 10.1
Sample trajectory 1 with score 18.

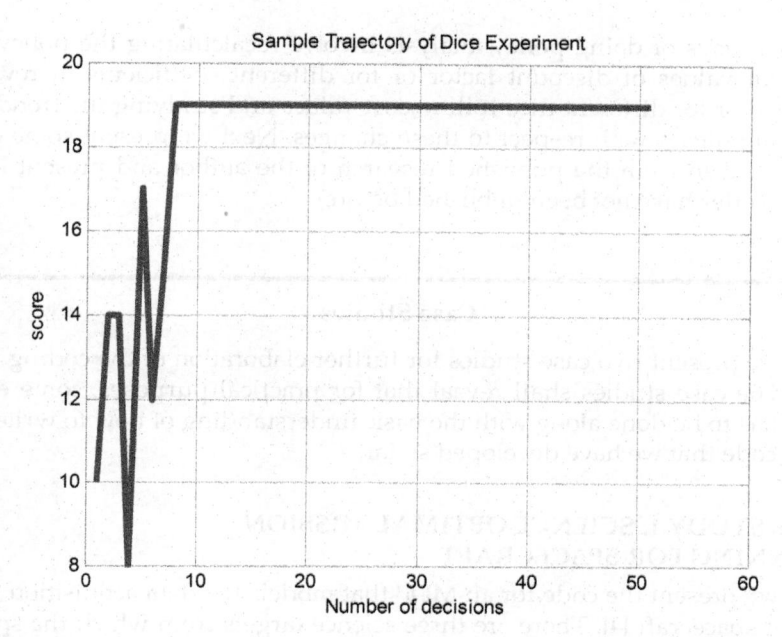

FIGURE 10.2
Sample trajectory 2 with score 19.

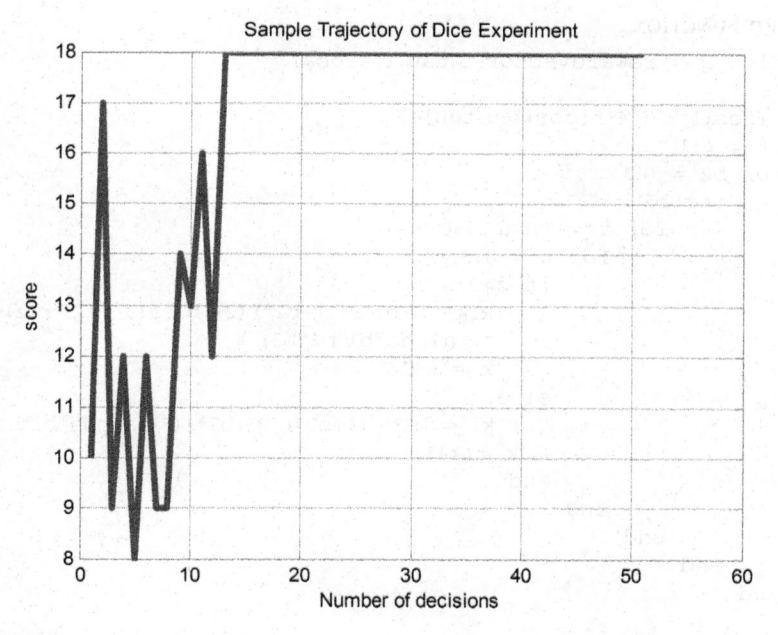

FIGURE 10.3
Sample trajectory 3 with score 18.

Other ways of doing policy analysis involve recalculating the policy for different values of discount factor or for different coefficients in reward function or for different dice-rolling cost values and studying the trends in optimal policies with respect to these changes. Next we present some case studies taken from the published research of the author and present their codes (codes have not been published before).

Case Studies

Now we present two case studies for further elaboration of the coding process. The case studies shall reveal that for practical purposes, some extra work has to be done along with the basic understanding of how to write the MDP code that we have developed so far.

CASE STUDY 1: SCIENCE OPTIMAL MISSION PLANNING FOR SPACECRAFT

Here we present the code for an MDP that models the data acquisition mission of spacecraft [4]. There are three science targets from which the spacecraft has to collect the data. For details of the symbols and other information, the reader is encouraged to read the research paper [4].

REWARD FUNCTION

```
function R = rewardvectorPL(tau,r,alpha)
k = 1;
R = zeros(1,2^3*4*floor(360/tau)+1);
for b1 = 0:1
    for b2 = 0:1
        for b3 = 0:1
            for v = 1:tau:360
                for z = 0:3
                    if z>0
                        R(k)  = b1*r(1)+b2*r(2)+b3*r(3)+0.5*r(z)+
                            alpha*b1*b2*b3;
                        k = k+1;
                    else
                        R(k)  = b1*r(1)+b2*r(2)+b3*r(3)+alpha*b1*b2*b3;
                        k = k+1;
                    end
                end
            end
        end
    end
end
R(k)  = 0;
```

Note that the reward function is assigned based on whether the variable $z > 0$ or not. In a similar way, innovation may be required for implementation of reward function.

COST MATRIX

```
function J = costmatrixPL(c,n)
%Given costs of actions in specific order, and number of
%targets, this function computes the cost matrix J
J = zeros(n+1);
delta = 1;%min(c)/10;
k = 1;
for i = 1:n,
    J(i,i+1:n+1) = c(k:k+n-i);
    k = k+n+1-i;
end
for i = 1:n+1,
    for j = 1:n+1,
        J(j,i) = J(i,j);
    end
end
for k=1:n+1,
    J(k,k) = delta;
end
J(1,1) = delta/2;
```

Cost matrix is not always required for an MDP code. But it is used here to simplify the value calculation function. Usually the calculation of cost is straightforward, but when it is not, then the recommended option is to write a separate cost function calculation file.

STATE TRANSITION FUNCTION

```
function N = nxtPL(prd,tau,t,tl,tu,tnop)
% data collection windows must avoid zero crossing
% elements of t must be multiples of tau
k = 1;
N = zeros(2305,8);
for b1 = 0:1
    for b2 = 0:1
        for b3 = 0:1
            for v = 1:tau:360
                for z = 0:3
                    N(k,2) = 2305;
                    N(k,4) = 2305;
                    N(k,6) = 2305;
                    N(k,8) = 2305;
                    N(k,9) = S2indexPL([(b1 - b1*(prd(1)==1)*
((v+tnop(1))>360)) (b2 - b2*(prd(2)==1)*((v+tnop(1))>360)) (b3 - b3*
(prd(3)==1)*((v+tnop(1))>360)) rem(v+tnop(1),360) z],tau);
```

```
            N(k,10) = S2indexPL([(b1 - b1*(prd(1)==1)*
((v+tnop(2))>360)) (b2 - b2*(prd(2)==1)*((v+tnop(2))>360)) (b3 - b3*
(prd(3)==1)*((v+tnop(2))>360)) rem(v+tnop(2),360) z],tau);
            N(k,11) = S2indexPL([(b1 - b1*(prd(1)==1)*
((v+tnop(3))>360)) (b2 - b2*(prd(2)==1)*((v+tnop(3))>360)) (b3 - b3*
(prd(3)==1)*((v+tnop(3))>360)) rem(v+tnop(3),360) z],tau);
            N(k,12) = S2indexPL([(b1 - b1*(prd(1)==1)*
((v+tnop(4))>360)) (b2 - b2*(prd(2)==1)*((v+tnop(4))>360)) (b3 - b3*
(prd(3)==1)*((v+tnop(4))>360)) rem(v+tnop(4),360) z],tau);
            N(k,13) = S2indexPL([(b1 - b1*(prd(1)==1)*
((v+tnop(5))>360)) (b2 - b2*(prd(2)==1)*((v+tnop(5))>360)) (b3 - b3*
(prd(3)==1)*((v+tnop(5))>360)) rem(v+tnop(5),360) z],tau);
            N(k,14) = S2indexPL([(b1 - b1*(prd(1)==1)*
((v+tnop(6))>360)) (b2 - b2*(prd(2)==1)*((v+tnop(6))>360)) (b3 - b3*
(prd(3)==1)*((v+tnop(6))>360)) rem(v+tnop(6),360) z],tau);
            N(k,15) = S2indexPL([(b1 - b1*(prd(1)==1)*
((v+tnop(7))>360)) (b2 - b2*(prd(2)==1)*((v+tnop(7))>360)) (b3 - b3*
(prd(3)==1)*((v+tnop(7))>360)) rem(v+tnop(7),360) z],tau);
            N(k,16) = S2indexPL([(b1 - b1*(prd(1)==1)*
((v+tnop(8))>360)) (b2 - b2*(prd(2)==1)*((v+tnop(8))>360)) (b3 - b3*
(prd(3)==1)*((v+tnop(8))>360)) rem(v+tnop(8),360) z],tau);
            N(k,17) = S2indexPL([(b1 - b1*(prd(1)==1)*
((v+tnop(9))>360)) (b2 - b2*(prd(2)==1)*((v+tnop(9))>360)) (b3 - b3*
(prd(3)==1)*((v+tnop(9))>360)) rem(v+tnop(9),360) z],tau);
            N(k,18) = S2indexPL([(b1 - b1*(prd(1)==1)*
((v+tnop(10))>360)) (b2 - b2*(prd(2)==1)*((v+tnop(10))>360)) (b3 - b3*
(prd(3)==1)*((v+tnop(10))>360)) rem(v+tnop(10),360) z],tau);
                    if z ==0
            N(k,1) = S2indexPL([(b1 - b1*(prd(1)==1)*
((v+t(1,1))>360))   (b2 - b2*(prd(2)==1)*((v+t(1,1))>360))   (b3 - b3*
(prd(3)==1)*((v+t(1,1))>360)) rem(v+t(1,1),360) 0],tau);
            N(k,3) = S2indexPL([(b1 - b1*(prd(1)==1)*
((v+t(1,2))>360))   (b2 - b2*(prd(2)==1)*((v+t(1,2))>360))   (b3 - b3*
(prd(3)==1)*((v+t(1,2))>360)) rem(v+t(1,2),360) 1],tau);
            N(k,5) = S2indexPL([(b1 - b1*(prd(1)==1)*
((v+t(1,3))>360))   (b2 - b2*(prd(2)==1)*((v+t(1,3))>360))   (b3 - b3*
(prd(3)==1)*((v+t(1,3))>360)) rem(v+t(1,3),360) 2],tau);
            N(k,7) = S2indexPL([(b1 - b1*(prd(1)==1)*
((v+t(1,4))>360))   (b2 - b2*(prd(2)==1)*((v+t(1,4))>360))   (b3 - b3*
(prd(3)==1)*((v+t(1,4))>360)) rem(v+t(1,4),360) 3],tau);
                    k = k+1;
                elseif z==1
            N(k,1) = S2indexPL([(b1 - b1*(prd(1)==1)*
((v+t(2,1))>360))   (b2 - b2*(prd(2)==1)*((v+t(2,1))>360))   (b3 - b3*
(prd(3)==1)*((v+t(2,1))>360)) rem(v+t(2,1),360) 0],tau);
            N(k,3) = S2indexPL([(b1 - b1*(prd(1)==1)*
((v+t(2,2))>360))   (b2 - b2*(prd(2)==1)*((v+t(2,2))>360))   (b3 - b3*
(prd(3)==1)*((v+t(2,2))>360)) rem(v+t(2,2),360) 1],tau);
```

```
                N(k,5)  =  S2indexPL([(b1  -  b1*(prd(1)==1)*
((v+t(2,3))>360))  (b2  -  b2*(prd(2)==1)*((v+t(2,3))>360))  (b3  -  b3*
(prd(3)==1)*((v+t(2,3))>360))  rem(v+t(2,3),360) 2],tau);
                N(k,7)  =  S2indexPL([(b1  -  b1*(prd(1)==1)*
((v+t(2,4))>360))  (b2  -  b2*(prd(2)==1)*((v+t(2,4))>360))  (b3  -  b3*
(prd(3)==1)*((v+t(2,4))>360))  rem(v+t(2,4),360) 3],tau);
                if v>tl(1) && v<tu(1)-t(2,2)
                N(k,3) = S2indexPL([1 b2 b3 rem(v+t(2,2),360)
1],tau);
                end
                k = k+1;
        elseif z==2
                N(k,1)  =  S2indexPL([(b1  -  b1*(prd(1)==1)*
((v+t(3,1))>360))  (b2  -  b2*(prd(2)==1)*((v+t(3,1))>360))  (b3  -  b3*
(prd(3)==1)*((v+t(3,1))>360))  rem(v+t(3,1),360) 0],tau);
                N(k,3)  =  S2indexPL([(b1  -  b1*(prd(1)==1)*
((v+t(3,2))>360))  (b2  -  b2*(prd(2)==1)*((v+t(3,2))>360))  (b3  -  b3*
(prd(3)==1)*((v+t(3,2))>360))  rem(v+t(3,2),360) 1],tau);
                N(k,5)  =  S2indexPL([(b1  -  b1*(prd(1)==1)*
((v+t(3,3))>360))  (b2  -  b2*(prd(2)==1)*((v+t(3,3))>360))  (b3  -  b3*
(prd(3)==1)*((v+t(3,3))>360))  rem(v+t(3,3),360) 2],tau);
                N(k,7)  =  S2indexPL([(b1  -  b1*(prd(1)==1)*
((v+t(3,4))>360))  (b2  -  b2*(prd(2)==1)*((v+t(3,4))>360))  (b3  -  b3*
(prd(3)==1)*((v+t(3,4))>360))  rem(v+t(3,4),360) 3],tau);
                if v>tl(2) && v<tu(2)-t(3,3)
                N(k,5) = S2indexPL([b1 1 b3 rem(v+t(3,3),360)
2],tau);
                end
                k = k+1;
        elseif z==3
                N(k,1)  =  S2indexPL([(b1  -  b1*(prd(1)==1)*
((v+t(4,1))>360))  (b2  -  b2*(prd(2)==1)*((v+t(4,1))>360))  (b3  -  b3*
(prd(3)==1)*((v+t(4,1))>360))  rem(v+t(4,1),360) 0],tau);
                N(k,3)  =  S2indexPL([(b1  -  b1*(prd(1)==1)*
((v+t(4,2))>360))  (b2  -  b2*(prd(2)==1)*((v+t(4,2))>360))  (b3  -  b3*
(prd(3)==1)*((v+t(4,2))>360))  rem(v+t(4,2),360) 1],tau);
                N(k,5)  =  S2indexPL([(b1  -  b1*(prd(1)==1)*
((v+t(4,3))>360))  (b2  -  b2*(prd(2)==1)*((v+t(4,3))>360))  (b3  -  b3*
(prd(3)==1)*((v+t(4,3))>360))  rem(v+t(4,3),360) 2],tau);
                N(k,7)  =  S2indexPL([(b1  -  b1*(prd(1)==1)*
((v+t(4,4))>360))  (b2  -  b2*(prd(2)==1)*((v+t(4,4))>360))  (b3  -  b3*
(prd(3)==1)*((v+t(4,4))>360))  rem(v+t(4,4),360) 3],tau);
                if v>tl(3) && v<tu(3)-t(4,4)
                N(k,7) = S2indexPL([b1 b2 1 rem(v+t(4,4),360)
3],tau);
                end
                k = k+1;
        end
```

```
            end
        end
    end
  end
end
N(2305,1:8)=2305*ones(1,8);
```

State transition function is often complicated for real life problems. As shown above, the use of logical operators inside the call to state to index function is useful for defining state transitions. Sometimes it may be required to define a separate function file that calculates the next state from any given state; then that function may be called inside the state transition function file.

TRANSITION PROBABILITY FUNCTION

```
function T = trasvectorPL(rho1,rho2,delta,tau)
k = 1;
T = zeros(2^3*4*floor(360/tau)+1,8);
for b1 = 0:1
    for b2 = 0:1
        for b3 = 0:1
            for v = 1:tau:360
                for z = 0:3
                    if z ==0
                        T(k,1) = 1 - rho2(1);
                        T(k,2) = rho2(1);
                        T(k,3) = 1 - rho1(1);
                        T(k,4) = rho1(1);
                        T(k,5) = 1 - rho1(2);
                        T(k,6) = rho1(2);
                        T(k,7) = 1 - rho1(3);
                        T(k,8) = rho1(3);
                        k = k+1;
                    elseif z==1
                        T(k,1) = 1 - rho1(1);
                        T(k,2) = rho1(1);
                        T(k,3) = 1 - rho2(2)*delta(1);
                        T(k,4) = rho2(2)*delta(1);
                        T(k,5) = 1 - rho1(4);
                        T(k,6) = rho1(4);
                        T(k,7) = 1 - rho1(5);
                        T(k,8) = rho1(5);
                        k = k+1;
                    elseif z==2
                        T(k,1) = 1 - rho1(2);
                        T(k,2) = rho1(2);
                        T(k,3) = 1 - rho1(4);
                        T(k,4) = rho1(4);
                        T(k,5) = 1 - rho2(3)*delta(2);
                        T(k,6) = rho2(3)*delta(2);
```

```
                            T(k,7) = 1 - rho1(6);
                            T(k,8) = rho1(6);
                            k = k+1;
                    elseif z==3
                            T(k,1) = 1 - rho1(3);
                            T(k,2) = rho1(3);
                            T(k,3) = 1 - rho1(5);
                            T(k,4) = rho1(5);
                            T(k,5) = 1 - rho1(6);
                            T(k,6) = rho1(6);
                            T(k,7) = 1 - rho2(4)*delta(3);
                            T(k,8) = rho2(4)*delta(3);
                            k = k+1;
                    end
                end
            end
        end
    end
end
T(2305,:) = 0.5*ones(1,8);
```

If the calculation of the probabilities involve complex operations, transition probability function may also require separate probability calculation function. Here it is relatively simple.

INDEX TO STATE CONVERSION

```
function S = index2SPL(i,tau)
S(1) = rem(floor2(i,4*72*4),2);
S(2) = rem(floor2(i,4*72*2),2);
S(3) = rem(floor2(i,4*72),2);
S(4) = rem(floor2(i,4),72)*tau + 1;
S(5) = rem(floor2(i,1),4);
```

Noticeable state variable in the above index to state function is the fourth state where "tau" is multiplied with the "rem" function output. This is because the consecutive values of this variable are "tau" integers apart.

STATE TO INDEX CONVERSION

```
function i = S2indexPL(S,tau)
i = S(5)+1 + 4*((S(4)-1)/tau) + 72*4*S(3) + 72*4*2*S(2) + 72*4*4*S(1);
```

State to index function involves division by "tau" to accommodate the gap between two consecutive values of fourth state variable.

VALUE ITERATION MAIN FILE

```
clc;clear all
% This file calculates Values of states and Policies 0 through 3
% Where 0 means go to initial state and 1,2,3 means go to star
number
```

```
% 1,2,3 respectively
% Functions used: 'costmatrix','rewardvector','nxt','trasvector', and
% 'vcalc'
prd = [0 0 0];td = [5 10 10 10];
rho2 = 1E-4*td;delta = [1 1 1];
alpha = 1000;tau = 5;
tnop = [5 50 70 90 20 105 135 20 50 20]; % Size = ((n+1)*(n+2))/2
cnop = tnop/20;
% tl = [70 280 180];tu = [120 330 250];
tl = [50 170 40];tu = [150 330 130];
%tl = [300 180 210];tu = [350 270 240];
% t = [5 50 70 90;50 20 105 135;70 105 20 50;90 135 50 20];
dlmwrite('tau.txt',tau);dlmwrite('tu.txt',tu);dlmwrite('tl.txt',tl);
dlmwrite('td.txt',td);dlmwrite('tnop.txt',tnop);
dlmwrite('prd.txt',prd);
vf = 0;
r = [30 50 70];% Science reward for each individual Star
c = [5 7 9 10.5 13.5 5]; % Fuel consumptions for various travels
Size = (n*(n+1))/2
rho1 = 0.01*c;
dlmwrite('cost.txt',c);
n = length(r);
J = costmatrixPL(c,n);
t = 10*J;dlmwrite('t.txt',t);
R = rewardvectorPL(tau,r,alpha);
V = zeros(1,length(R));
V(length(R)) = vf;
P = -ones(1,length(R));P1=P;
T = trasvectorPL(rho1,rho2,delta,tau);
N = nxtPL(prd,tau,t,tl,tu,tnop);
%s = length(R)-(n+2);
s = length(R)-1;
% gamma = 0.3;
% gamma = 0.5;
% gamma = 0.7;
% gamma = 0.8;
% gamma = 0.90;
 gamma = 0.99; % takes about 3 min and 30 sec 1449 itr
k = 1;itr = 0;
err = 1E1;
while (err>1E-4)
    V_old = V;
    for i = 1:s,
        v = [V(N(i,1)) V(N(i,2)) V(N(i,3)) V(N(i,4)) V(N(i,5)) V(N(i,6))
V(N(i,7)) V(N(i,8)) V(N(i,9)) V(N(i,10)) V(N(i,11)) V(N(i,12)) V(N(i,13))
V(N(i,14)) V(N(i,15)) V(N(i,16)) V(N(i,17)) V(N(i,18))];
        [V(i),P1(i)] = vcalcPL(R(i),J(k,:),T(i,:),v,gamma,cnop);
        k = k+1;
        if k>4
```

```
                k = 1;
            end
        end
err = norm(V-V _ old)^2
itr = itr+1;
end
k = 1;
for i = 1:s,
      v = [V(N(i,1)) V(N(i,2)) V(N(i,3)) V(N(i,4)) V(N(i,5)) V(N(i,6))
V(N(i,7)) V(N(i,8)) V(N(i,9)) V(N(i,10)) V(N(i,11)) V(N(i,12)) V(N(i,13))
V(N(i,14)) V(N(i,15)) V(N(i,16)) V(N(i,17)) V(N(i,18))];
      P(i) = pcalcPL(J(k,:),T(i,:),v,cnop,gamma);
      k = k+1;
      if k>4
          k = 1;
      end
end
save('PolicyPL','P');
save('nxtPL','N');
```

Value iteration function file here involves initialization of various parameters for the problem that are used in the calculation of state transitions, transition probabilities, cost, and reward functions. It is a good practice to initialize parameters in value iteration main file instead of defining within the transition and reward functions.

VALUE CALCULATION FUNCTION

```
function [V P] = vcalcPL(R,J,T,v,gamma,cnop)
% This file solves Belman equation
% Uses function 'findmax'
%tnop = dlmread('tnop.txt');
j = 1;u = zeros(1,length(J)+length(cnop));
for i=1:length(J)
    u(i) = R + (-J(i)+gamma*(T(j)*v(j)));
    j = j+2;
end
u(i+1) = R - cnop(1) + (gamma*v(j));
u(i+2) = R - cnop(2) + (gamma*v(j+1));
u(i+3) = R - cnop(3) + (gamma*v(j+2));
u(i+4) = R - cnop(4) + (gamma*v(j+3));
u(i+5) = R - cnop(5) + (gamma*v(j+4));
u(i+6) = R - cnop(6) + (gamma*v(j+5));
u(i+7) = R - cnop(7) + (gamma*v(j+6));
u(i+8) = R - cnop(8) + (gamma*v(j+7));
u(i+9) = R - cnop(9) + (gamma*v(j+8));
u(i+10) = R - cnop(10) + (gamma*v(j+9));
V = max(u);
P = findmax(u)-1;
```

Value calculation function involves calculation of the cost of two types of actions. One type involves action cost from cost function matrix, and the other type involves cost "cnop". Note that the value calculation function is being used to calculate the policy as well. This is useful when the policy error is to be determined for the value function that is being used to calculate the policy.

POLICY CALCULATION FUNCTION

```
function P = pcalcPL(J,T,v,cnop,gamma)
% This file solves Belman equation
% Uses function 'findmax'
%tnop = dlmread('tnop.txt');
j = 1;u = zeros(1,length(J)+length(cnop));
for i=1:length(J)
    u(i) = (-J(i)+gamma*(T(j)*v(j)));
    j = j+2;
end
u(i+1)  = -cnop(1)  + gamma*(v(j));
u(i+2)  = -cnop(2)  + gamma*(v(j+1));
u(i+3)  = -cnop(3)  + gamma*(v(j+2));
u(i+4)  = -cnop(4)  + gamma*(v(j+3));
u(i+5)  = -cnop(5)  + gamma*(v(j+4));
u(i+6)  = -cnop(6)  + gamma*(v(j+5));
u(i+7)  = -cnop(7)  + gamma*(v(j+6));
u(i+8)  = -cnop(8)  + gamma*(v(j+7));
u(i+9)  = -cnop(9)  + gamma*(v(j+8));
u(i+10) = -cnop(10) + gamma*(v(j+9));
P = findmax(u)-1;
```

Code for policy calculation is the same as that for value calculation. It is written separately here because in value iterations, policy is calculated after the values have converged. But there are many variations of value iteration and policy iteration, so the reader can choose to write a separate file or use the same file.

TRAJECTORY GENERATION

```
clc;clear all;
P = dlmread('PolicyPL _ 2.txt');
N = dlmread('nxtPL _ 2.txt');
k = 65;tau = 5;
S = [0 0 0 1 0]; % initial state
for i = 1:k
    i = S2indexPL(S,tau);
    a = P(i);
    if a<4
        S = index2SPL(N(i,2*(a+1)-1),tau);
        action = a
    else
        S = index2SPL(N(i,a+5),tau);
        display ('action = NOOP with value ')
        display (a+1)
```

```
        end
        nextstate = S
end
```

POLICY EVALUATION AND PLOTS

```
clc;clear all
P = dlmread('PolicyPL.txt');
a = 1;b = 1;c = 1;d = 1;
for i = 1:length(P)-1,
    if rem(i,4)==1,
        Pt0(a) = P(i);
        if Pt0(a)>4,
            Pt0(a) = 4;
        end
        t0(a) = (ceil((rem(i,288) + (rem(i,288)==0)*288)/4)*5)-4;
        a = a+1;
    end
    if rem(i,4)==2,
        Pt1(b) = P(i);
        if Pt1(b)>4,
            Pt1(b) = 4;
        end
        t1(b) = (ceil((rem(i,288) + (rem(i,288)==0)*288)/4)*5)-4;
        b = b+1;
    end
    if rem(i,4)==3,
        Pt2(c) = P(i);
        if Pt2(c)>4,
            Pt2(c) = 4;
        end
        t2(c) = (ceil((rem(i,288) + (rem(i,288)==0)*288)/4)*5)-4;
        c = c+1;
    end
    if rem(i,4)==0,
        Pt3(d) = P(i);
        if Pt3(d)>4,
            Pt3(d) = 4;
        end
        t3(d) = (ceil((rem(i,288) + (rem(i,288)==0)*288)/4)*5)-4;
        d = d+1;
    end
end
figure(1)
subplot(2,2,1)
plot(t0,Pt0,'r+')
xlabel('true anomaly (deg)');ylabel('Policy (z=0)')
title('mu _ 0 = 0, mu _ 1 = 1, mu _ 2 = 2, mu _ 3 = 3, NOOP = 4')
%figure(2)
subplot(2,2,2)
```

```
plot(t1,Pt1,'r+')
xlabel('true anomaly (deg)');ylabel('Policy (z=1)')
%title('mu_0 = 0, mu_1 = 1, mu_2 = 2, mu_3 = 3, NOOP = 4')
%figure(3)
subplot(2,2,3)
plot(t2,Pt2,'r+')
xlabel('true anomaly (deg)');ylabel('Policy (z=2)')
%title('mu_0 = 0, mu_1 = 1, mu_2 = 2, mu_3 = 3, NOOP = 4')
%figure(4)
subplot(2,2,4)
plot(t3,Pt3,'r+')
xlabel('true anomaly (deg)');ylabel('Policy (z=3)')
%title('mu_0 = 0, mu_1 = 1, mu_2 = 2, mu_3 = 3, NOOP = 4')
```

The trajectory generation and policy evaluation results are discussed in [4].

CASE STUDY 2: CONTROL OF EPIDEMIC DISEASES USING SIR MODEL

This case study presents the MDP for Susceptible-Infected-Recovered (SIR) model of epidemic infection [5]. This MDP is different in a sense that some states are unreachable because the sum of susceptible, infected, and recovered individuals has to be exactly equal to the total population. The details of the model (MDP) are presented in [5].

REWARD/COST FUNCTION

```
function R = rewardMDP4SIR()
R = zeros(6130251,1);
k = 1;
a0 = 100;a1 = 40;
for r = 0:3500
    for i = 0:(3500 - r)
        s = 3500 - i - r;
        R(k) = a0*i + a1*s;
        k = k+1;
    end
end
```

Note that there is no dedicated for-loop for the last state variable. This is because of the constraint on states. The reward function itself is straightforward.

STATE TRANSITION FUNCTION

```
function N = nxtMDP4SIR()
N = zeros(6130251,6);
k = 1;
for r = 0:3500
    for i = 0:(3500 - r)
        s = 3500 - i - r;
```

```
        N(k,1) = k;
        N(k,3) = k;
        N(k,5) = k;
        N(k,7) = k;
        if s>0
            N(k,2) = S2indexMDP4SIR([r+1,i,s-1]);
            N(k,4) = S2indexMDP4SIR([r+1,i,s-1]);
            N(k,8) = S2indexMDP4SIR([r,i+1,s-1]);
        else
            N(k,2) = k;
            N(k,4) = k;
            N(k,8) = k;
        end
        if i>0
            N(k,6) = S2indexMDP4SIR([r+1,i-1,s]);
        else
            N(k,6) = k;
        end
        k = k+1;
    end
end
```

Transition function has the same loop structure as the reward function, but there are some conditions that have been used to specify the allowable transitions.

TRANSITION PROBABILITY FUNCTION

```
function T = transMDP4SIR()
T = zeros(6130251,6);
k = 1;
mu1 = 0.8;
mu2 = 0.9;
mu3 = 0.7;
mu4 = 0.6;
for r = 0:3500
    for i = 0:(3500 - r)
        s = 3500 - i - r;
        T(k,1) = 1-mu1;
        T(k,2) = mu1;
        T(k,3) = 1-mu2;
        T(k,4) = mu2;
        T(k,5) = 1-mu3;
        T(k,6) = mu3;
        T(k,7) = 1-mu4;
        T(k,8) = mu4;
        k = k+1;
    end
end
```

Probability specification is straightforward in this case.

INDEX TO STATE CONVERSION

```
function s = index2SMDP4SIR(i)
k = 0;
for j = 1:3501
    if i<=((3501*j)-k-j+1)
        s(1) = j-1;
        rf = 3501*s(1)-k;
        break;
    end
    k = k+(j-1);
end
s(2) = i - rf - 1;
s(3) = 3500 - s(1) - s(2);
```

Index to state conversion presented above is computationally expensive in this case because the variables have constraints.

STATE TO INDEX CONVERSION

```
function i = S2indexMDP4SIR(s)
N = 3500;
i = (N+1)*s(1)+(s(2)+1)-(s(1)*(s(1)-1)/2)*(s(1)>1);
```

State to index conversion is also different from the one presented in the main text because of the constraints in state space. This case study shows that there could be MDPs for which the presented code may need modifications, but overall the structure of the modules remains the same.

VALUE ITERATION MAIN FILE

```
clc;clear all;close all;
tic
R = rewardMDP4SIR();
N = nxtMDP4SIR();
T = transMDP4SIR();
load VMDP4SIR;
err = 1E1;
itr = 0;
P = -ones(length(R),1);
gamma = 0.95;
while (err>1E-6)
    V _ old = V;
    for i = 1:length(R)
        v = [V(N(i,1)) V(N(i,2)) V(N(i,3)) V(N(i,4)) V(N(i,5)) V(N(i,6))
V(N(i,7)) V(N(i,8))];
        V(i) = vcalcMDP4SIR(R(i),v,T(i,:),gamma);
    end
    err = max(abs(V-V _ old))
    itr = itr+1
end
```

```
for j = 1:length(R)
    v = [V(N(j,1)) V(N(j,2)) V(N(j,3)) V(N(j,4)) V(N(j,5)) V(N(j,6))
V(N(j,7)) V(N(j,8))];
    P(j) = pcalcMDP4SIR(R(j),v,T(j,:),gamma);
end
save('VMDP4SIR','V');
save('PMDP4SIR','P');
t = toc;
```

Value iteration file is straightforward in this case.

VALUE CALCULATION

```
function V = vcalcMDP4SIR(R,v,T,gamma)
C = [20 50 70 0];
u = zeros(4,1);
k = 1;
for i = 1:2:8
    u(k) = -R - C(k) + gamma*(v(i)*T(i) + v(i+1)*T(i+1));
    k = k+1;
end
V = max(u);
```

Value calculation is simple, but the cost function in this case is defined inside the value calculation file, whereas in previous case study, such parameters were defined in value iteration main file (it is user's choice). Policy calculation file is not presented here because it is same as value calculation file.

POLICY EVALUATION

```
clc;clear all;close all;
load ('PMDP4SIR');
I1 = -ones(6130251,1);
I2 = -ones(6130251,1);
I3 = -ones(6130251,1);
I4 = -ones(6130251,1);
k = 1;
for r = 0:3500
    for i = 0:(3500 - r)
        s = 3500 - i - r;
        if P(k)==1
            I(k) = (s)
        I(k) = i;
        k = k+1;
    end
end
```

REWARD EVALUATION AND PLOTS

```
clc; clear all;
R1 = zeros(4,3501);
R2 = zeros(4,3501);
```

```
Ct = zeros(1,3501);
k = 1;
a0 = 100;a1 = 40;
for s = 0:875:3500
    for i = 0:3500
        R1(k,i+1) = a0*i + a1*s;
        R2(k,i+1) = exp(i*0.0025) + exp(s*0.002);
        Ct(i+1) = i*70;
    end
    k = k+1;
end
figure(1)
% a0 = 100, a1 = 40, a0exp = 1/400, a1exp = 1/500;
subplot 211
plot(R1')
xlabel('No. of Infected Individuals (I)');
ylabel('Linear Cost Function Value');
grid on
subplot 212
plot(R2')
xlabel('No. of Infected Individuals (I)');
ylabel('Exponential Cost Function Value');
legend('s = 0','s = 875','s = 1750','s = 2625','s = 3500')
grid on
figure(2)
plot(Ct,'LineWidth',2)
hold on
plot(R1(1,:),':g','LineWidth',2)
plot(R2(1,:),'-r','LineWidth',2)
grid on
xlabel('No. of Infected Individuals');
ylabel('Cost Function Value');
legend('Cost of Treatment','Infection Cost (linear)','Infection
Cost (exponential)')
axis([0 4000 0 400000])
```

The results are plots obtained from policy evaluation and reward function evaluation and are presented in [5].

10.13 Conclusions

A guideline for MATLAB coding for an MDP has been presented. Basic structure of the MDP code consists of eight modules. This modular approach can be used to code a vast variety of MDPs. Simple examples have also been presented to make the user understand the concept. On the other hand, case studies of real life problems have been presented to indicate that there is no straightforward way of coding all MDPs. The presented methodology covers the basics of coding an MDP under certain assumptions, but when implementing the methodology for real life problems, nontrivial innovation is often required. Nonetheless, this document shall serve well to anyone who is acquainted to the theoretical concept of an MDP and wants to learn how to write the code for the same.

References

1. Puterman, M. L. *Markov Decision Processes: Discrete Stochastic Dynamic Programming*, Wiley and Sons, Inc., Hoboken, NJ, 1994.
2. Russell, S. and Norvig, P. *Artificial Intelligence: A Modern Approach*, 2nd Edition, Prentice-Hall, Upper Saddle River, NJ, 2005.
3. Powell, W. B. *Approximate Dynamic Programming: Solving the Curses of Dimensionality*, Vol. 703, Wiley & Sons, Hoboken, NJ, 2007.
4. Nasir, A., Atkins, E. M. and Kolmanovsky, I. V. "Science-optimal Spacecraft Attitude Maneuvering While Accounting for the Failure Mode." *18th IFAC World Congress*, Milano, Italy, 2011.
5. Nasir, A. and Rehman, H. "Optimal control for stochastic model of epidemic infections," *2017 14th International Bhurban Conference on Applied Sciences and Technology (IBCAST)*, Islamabad, Pakistan, 2017, pp. 278–284. doi:10.1109/IBCAST.2017.7868065.

11

A Smart Microfactory Design: An Integrated Approach

Syed Osama bin Islam, Liaquat Ali Khan, Azfar Khalid, and Waqas Akbar Lughmani
Capital University of Science and Technology (CUST)

CONTENTS

11.1 Introduction 216
    11.1.1 Industry 4.0 216
    11.1.2 Internet of Things 217
    11.1.3 Smart Factories 218
    11.1.4 Human–Machine Interaction 219
    11.1.5 Cobots 219
    11.1.6 Microsystems 220
    11.1.7 Micromanufacturing Techniques 220
    11.1.8 Micromechanical Machining 221
    11.1.9 Microfactory 221
    11.1.10 Micromachines and Designs Based on Flexures 222
    11.1.11 Flexures 223
    11.1.12 Microactuation 225
11.2 A Smart Microfactory Approach 226
    11.2.1 Scenario Details 226
    11.2.2 Development on the Physical Domain 227
        11.2.2.1 Design of the Proposed Microstage Design 227
        11.2.2.2 Material Properties 228
        11.2.2.3 Methodology 230
        11.2.2.4 Mathematical Formulation 230
        11.2.2.5 Meshing 231
        11.2.2.6 FEM Analysis 232
        11.2.2.7 Results 235
    11.2.3 Development on Virtual Domain 238
        11.2.3.1 Social Safety of CPS 238
        11.2.3.2 Methodology 240
        11.2.3.3 UR3 Robot Components and Capability 240
        11.2.3.4 Object-Detection API 241
        11.2.3.5 Hazard Assessment 241

11.3 Conclusion 245
11.4 Appendix: Tutorial on Object-Detection API 246
References 251

11.1 Introduction

High productivity and high flexibility are the demands of digital manufacturing industry. The current trend in manufacturing came up with the fourth industrial revolution, i.e. Industry 4.0 [1]. The concept is taking its shape from automated manufacturing systems to intelligent manufacturing systems but is still in its nascent stage. One of the basic components of these systems is a cyber-physical system (CPS) [2], i.e. a mechanism controlled by computer-based algorithms integrated with users over a network. The CPS is the smart system that consists of physical and computational elements; these elements can be distributed into four-layered architecture, which is made up of a sensing layer, networking layer, analyzing layer, and application layer [3]. The benefits of these systems are that they are time saving and flexible, feasible for a demand of even one quantity placed by an individual customer, and do not require reconfiguration of the manufacturing system. The term CPPS (cyber-physical production system) was coined in Germany that proposed a complete automated system in the realm of Industry 4.0: a manufacturing system based on CPS that comprises of physical elements which are robots, conveyors, sensors, actuators, etc. and a cyber-layer based on computational elements [4]. The independent elements of CPPS can cooperate with each other through Internet of Things (IoT) [5], a concept in which components having unique identity can transfer data to each other over a network without requiring any human–computer interaction (HCI), thus creating smart factories [6]. Internet can be one such communication protocol in IoT. A similar case of smart-factory production system is presented in [7]. Though the robots and computers take a major share in the CPS, human presence is essential for productivity either for supervision or complicated jobs that robots cannot undertake. The smart-factory concept exists for large production systems; however, there is very little research that exists for manufacturing at microdomain which is deemed necessary due to the limitations of the macro devices, i.e. their large size, greater power consumption, large cost effect, higher susceptibility to environment conditions, and control loop that is believed to be significantly larger [8]. In this chapter, a smart factory is proposed; a collaboration is envisaged between a human, a cobot, and a multistaged micromilling machine. The related concepts are stated below.

11.1.1 Industry 4.0

The latest trend of automation in manufacturing technologies incorporating data exchange is referred to as Industry 4.0. The concept suggests the use of IoT,

CPSs, cognitive computing, and cloud computing [9]. The modernization of industry started with the use of steam for mechanization when first machines were built; that was the first era of modern industry. Then with the advent of electricity, the machines were built which came up with the concept of mass production, and later the assembly lines were built; that was the second era. Then the digital world came into being which brought logic and control in the industry; the incorporation of computers came up with the beginning of automation, where machines and later robots replaced human workers on the assembly lines; that was the third era of industrial modernization. And presently we are entering into an era known as Industry 4.0 (the fourth industrial revolution), in which remotely placed robots and machines are connected to artificial intelligence (AI) fed computers that can control them with very little human interference. The interaction between operators, robotics, and computers happens in an entirely new way in which machine learning algorithms are used to learn and then control the process [10]. The term Industry 4.0 was originally conceived in the context of manufacturing; however, the concept evolved with the passage of time. Different industrial, governmental, and academic collaborations now fall under the scope of Industry 4.0 which has led to a new term "Industrie 4.0". But still in the broader context, Industry 4.0 is only about manufacturing which incorporates smart factories and processes/activities/technologies related to production and the areas related to them. Also Industry 4.0 is not merely related to some group of technology like IoT. It can be related to production, servicing, consumer interaction/ feedback. This improves upon cost and quality which can be attained by acquiring real-time data, cutting the inefficiencies, and removing irrelevance in this customer-centric environment where the value is for speed, cost effectiveness, and value-added innovative services. The concept is also related to improvement in digital supply chain model. In the other sense, it means that this term actually benefits business models with the use of innovation while transforming business models and processes. The benefits are profit, decreased cost, enhanced customer relationship, and optimized lifetime value—in short, increase in customer loyalty. Another aspect is in terms of flexibility, i.e. to sell more and innovate products in order to grow and remain relevant; this would be due to customer demand or desire to be part of the top most service/product or low-margin commoditized services/products or the services/products/solution that will disappear shortly due to "digital disruption" [11].

11.1.2 Internet of Things

A system in which a large number of embedded devices communicate with each other through Internet protocols is termed as IoT. Because of the use of Internet, these devices are also called "smart objects". These devices spread over the environment and are not directly operated by humans, for example, some components in vehicles or buildings not necessarily taking commands through human operator. The IoT provides a concept in which network connectivity is extended up to everyday items, objects, and sensors. The computing capability embedded into these systems are not necessarily a central computer for devices to consume, generate, and exchange data; this all is done with minimal human intervention. The implementation of IoT is done through different models of communications, each having unique characteristics. Internet Architecture Board presented four common models: Device-to-Device, Device-to-Gateway, Device-to-Cloud, and Back-End Data-Sharing. The variance in these models provides the flexibility in terms of connectivity and value to the users provided by the IoT devices [12]. The devices in the IoT not only include traditional PCs and mainframes but also refer to a worldwide network of devices like smartphones, embedded sensors, appliances, wearables (e.g. health sensors, smart watches) all outfitted with Internet protocol (IP) connectivity. It can be the connectivity of machines and electronic devices via a network that could be Wi-Fi or Ethernet. The transmission/reception of data could be direct amongst each other or from the cloud. In a manufacturing industry scenario, the machines, the devices, and the actuators, embedded with sensors, exchange the data directly or through a central computer over a wired/wireless network using the same IP [13]. The examples of IoT technologies which can be used in manufacturing industry are wireless sensor networks (WSNs) and other sensor networks that can be used to give information on quantitative/physical properties like materials, work in progress, tooling, and finished products. RFID (Radio Frequency Identification) can be used to support production scheduling by capturing the status of the job, and the overall performance of the system can be evaluated [14].

11.1.3 Smart Factories

The smart factory adapts and learns from new demands in which a constant flow of data is coming real time from productions and operations. It represents a way ahead of the existing automated world, where the components are fully connected and the processes are flexible [15]. The integration of data is system wide which contains human operators, physical elements, and controlling elements. The aim is to accomplish manufacturing through digitization of processes, keeping track of inventory while providing in process maintenance, inspection, or any other type of activity that happens within the entire framework. The outcome expected from this is to provide an agile system that should be more efficient, can reduce the lead time, and must be able to adjust to the unforeseen from within/outside or even predict them, so that a better place is made in the competitive market [16]. The concept of smart factory states that, while being flexible, it can autonomously run processes of the entire production system; the system has the capability to optimize itself, even from a broader network and has a real-time/near-real-time capability of self-adaptation to changes. Although factories in the past have some degree of automation, even few had higher levels, "automation", the actual term for a smart factory, suggests that the process or task's performance should behave as a single/discrete entity. Old machines which were automated used to take decisions on the basis of linear logic, like turning on/off a motor or opening of valve based on predefined logic. With the advent of AI and then its use in CPSs where physical systems and cyber-systems are combined, complex decision-making processes are introduced in automation to increase optimization in business processes just like humans do. Hence, the concept of "smart factory" integrates the decisions taken on the shop floor with the supply chain in the context of a broad enterprise; all this is done through information technology (IT)/operations technology (OT) connections. This has an effect on the production by ultimately improving the interaction of customers with suppliers. As this connectivity has changed the manufacturing processes, the emergence of Industry 4.0 (fourth industrial revolution) which suggests the integration of physical and digital entities based on OT and IT has also altered the functioning of supply chain. The new concept of digital supply network has emerged that shifts from linear operations in sequence to open interconnected operations which has modified the way of competition among the companies. These new concepts demand different capabilities from manufacturers like connected manufacturing systems (i.e. vertical integration), myriad operational systems (i.e. horizontal integration), and end-to-end operations (i.e. holistic integration) which enhance the organization of complete supply chain [17]. Therefore, the new concept of smart factory is a way ahead of traditional automation that has shifted to a flexible and fully integrated system where constant flow of data is coming real time from production and operations and that can adapt to any unforeseen real time.

11.1.4 Human–Machine Interaction

The interaction between humans and the machines is known as "human–machine interaction" (HMI). It is a technical system which is dynamic in nature and accomplishes itself through human–machine interface [18]. The HMI is related to HCI which on the other hand is based on computer technology; it can be called as the interaction between human users and computers. This field is not only related to the ways and means by which humans interact with computers but also the novel designing of technologies to let that happen [19]. So in the above context HMI is a multidisciplinary field where research is done on interactions between humans and machines accommodating inputs from HCI, exoskeleton control, AI, robotics, human–robot interaction (HRI), and humanoid robots [20].

11.1.5 Cobots

"Cobot" is the abbreviation for a collaborative robot which works in collaboration with a human operator. The cobots manipulate the objects that in turn assist humans; this will be done in accordance with the constraints and guidelines set by the users. These guidelines and constraints can be in terms of virtual surfaces defined by user [21]. The difference between cobots and autonomous industrial robots is that they directly interact with a human operator, sharing the same workspace and even payload, whereas autonomous industrial robots remain isolated from humans due to safety issues [22]. This collaboration between the humans and robots is in the revolutionary stage; it is expected that these robots work as companions in line with the humans by reading their behaviors and adapting to any changes real time. This is also termed as human–robot collaboration (HRC), the efficiency of HRC depends on effective monitoring of human's actions and the environment, the use of AI to anticipate the actions and state of mind by processing previous knowledge so that likely contribution to the task by human can be ascertained. This type of learning requires robots (cobots) to adapt to variety of humans, different types of human behaviors experienced by them, and different human needs. These types of robots are also termed as "social-cobots", where such adaptation results in more efficient and synchronous working of both the partners; this in turn increases the overall yield of the process [23].

11.1.6 Microsystems

The miniaturization of mechanical microsystems is under research that promises to enhance quality of life, health care, and economic growth. Understanding of mechanical properties of materials at microscale level is a very important aspect in fabrication of microdevices. The behavior of microsystems depends on these properties, but another major aspect is the structural geometry of microsystems. Fabrication of microdevices involves special fabrication processes which are widely different from the practices involved in fabrication of macrodevices. These are mainly categorized as non-lithography-based micromanufacturing (NLBMM) and lithography-based micromanufacturing (LBMM) techniques. NLBMM is gaining popularity to make micro 3D artifacts with various engineering materials. Being in the nascent stage, this technology looks promising for future micromanufacturing trends. Applications of these devices are in aerospace, biomedical, consumer products, telecommunication industry, and sensors.

11.1.7 Micromanufacturing Techniques

Microelectromechanical systems (MEMS) technique is one of the most common methods used to manufacture microdevices. These techniques use silicon-based semiconductor processing technology for large batch production, where photo-etching is used to shape silicon wafers using chemical and dry processes. Various other commercially viable techniques are also researched for fabrication of microdevices like ultrasonic, microelectrodischarge machining methods, photo-lithography, laser, and ion beam. Majority of these processes are slow, only viable for materials based on silicon, cater for planar shapes, unable to manufacture in small batches (customization), and are less cost effective [24].

11.1.8 Micromechanical Machining

Micromechanical machining is one of the latest techniques for fabricating micro-devices. The size range of these components can vary from tens of micrometers to a few millimeters. The advantages of this technique are that it bridges the gap between macrodomain and the 3D structures of nanodomains/microdomains; lithographic methods that are very expensive are no more required; they are suitable for accommodating individual components and monitoring of in-process quality of components [25]. Two types of micromachine tools are found mainly, i.e. precision machines and miniature machines. The characteristics of precision machine tools are a large foot print, high rotational speed of the spindle to decrease chip removal rate, use of air bearings/air turbines that allow low-torque operations, linear drive motors, and a large control system. The characteristics of miniature machine tools are as follows: they are cost effective; they have higher natural frequencies due to substantially smaller mass; they produce low vibration amplitudes; and the portability of these systems is easy thus making them beneficial. The actuators used in micromachine actuators are either voice coil actuators or flexure-based piezoelectric designs [24].

11.1.9 Microfactory

A microfactory is a factory of miniature size whose products are also of small dimensions. This name was coined in 1990 by Mechanical Engineer Laboratory (MEL) of Japan [26]. Requirements emanating from agile and flexible manufacturing, cost effectiveness, technology, and environmental issues demand greater challenges and competition from manufacturing industry in borderless business. The parts used in latest gadgetries are becoming smaller, but still the machine tools in practice are of conventional size, lacking justification. Reduction in the size of manufacturing systems can accrue many benefits like reduction of space, cost effectiveness, decrease in energy consumption, smart solutions, better environmental conditions, and low initial investment. This will have overall effect on agility in manufacturing industry as the factories can be reconfigured easily. Furthermore, the portability of the machine tools will be very easy, eliminating their requirement of fixture at factory. They can be even placed at manufacturing laboratories, offices, classrooms, or even in living areas. One of the major advantages of the microfactory despite saving materials, space, and energy is saving time especially in reconfiguration [27]. To achieve this advantage, full automation is one of the major requirements of microfactories that demands fully automated machine tools, in-process automatic inspection, automated assembly lines, automated material feeding/waste removal systems, tool replacement and evaluation systems, etc. [28] (Figure 11.1).

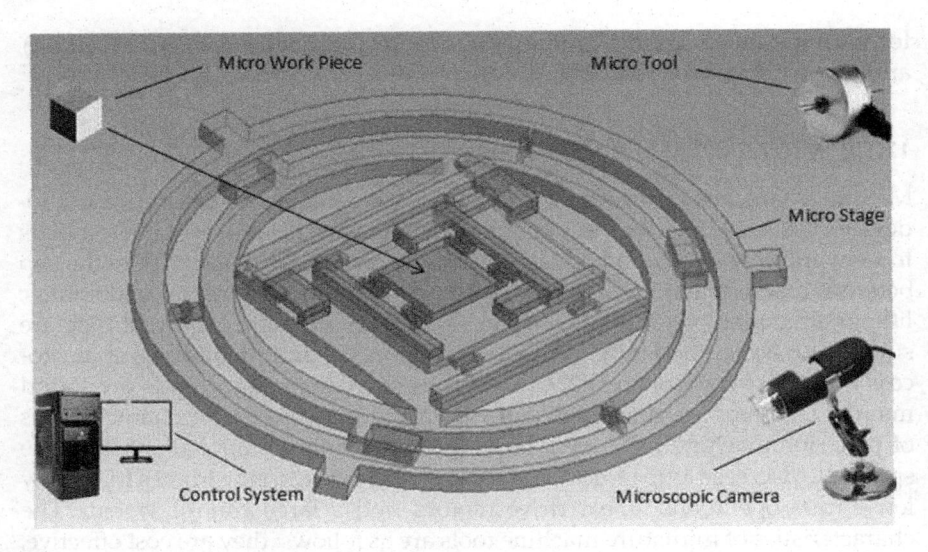

FIGURE 11.1
A conceptual microfactory.

11.1.10 Micromachines and Designs Based on Flexures

A five-axis micromilling machine based on PC control system is presented in [29], the machine is designed from microstages in market, control board that can be installed in PC, and available air spindle. Stepping motors drive each stage; therefore stages have high-speed resolution. Another five-axis micromilling machine based on PC control system is proposed in [30]; the machine tool is supported throughout with aerostatic bearings, and in addition these bearings are further assisted by squeeze oil-film. Diamond tool is proposed for cutting the job. There are shortcomings of these conventional technologies: high cost effect, low natural frequencies, friction, low control, and low accuracy, which can be overcome through use of flexure-based compliant mechanisms. Different advantages can be accrued by using these mechanisms like cost effectiveness, frictionless joints, removal of backlash as in case of gears, and compatibility to vacuum. A compliant mechanism can be described as a uniform shape structure whose working depends on its flexible material's deflection. It should be ensured that the compliant mechanism should work in elastic domain without inducing any plastic deformation by manipulating its structural parameters [31]. A 3-DOF compliant micropositioning stage was presented in [32], which is developed using notch flexures. Three piezoelectric (PZT) actuators are used for actuation and are placed at 120° apart in a symmetrical manner because of which large yaw motion can be achieved. A 2-DOF translational parallel micropositioning stage was presented in [33]. The degrees of freedom for each stage are achieved by serially connecting different types of compound flexures. PZT actuation is used for micropositioning/nanopositioning. The results showed good tracking and positioning performance. A simple idea of flexures to be used as control devices for linear stages was presented in [34] for the MEMS accelerometer design. A unique design where the flexures are used for controlling the rotation of rotary stage was presented in [35].

11.1.11 Flexures

Flexures are bearings that allow motion by bending load elements such as beams. In linkages, the major error in motion can be produced by pin joints. These joints can be traded off with flexures when there is a requirement of only small motions. When these flexural linkages are used as joints in mechanisms, they are referred to as compliant mechanisms. These can either be hourglass-shaped hinges or long thin blades that can flex throughout their length. The latter can have more deflection but have a constraint that it is having more compliance in out-of-plane directions. Advantages using flexures are as follows [33]:

- Good Control
- Motion devices having small range can be developed which are highly accurate
- Ideal to be implemented in precision machines
- Flexures are not affected by dirt.

Different types of flexure strips in use are generally categorized into parallel faces, cylindrical neck, and elliptical neck. Ones with the parallel section, i.e. rectangular shape, amongst them are advantageous; they are generally easy to manufacture as these micrometer structures used to be made by deposition of layers, though it is very difficult to make complex structures from the same process [36]. The motion of parallel face flexures is governed by lateral beam bending. The bending can be defined as a single-dimensional element where the axis of the beam is perpendicular to the load applied. The load under consideration can be distributed all along or can be a concentration on a specific point; it can also be a combined situation. Euler–Bernouli beam equation states the basic formulae for the lateral beam bending. The displacement in flexures is related to the force acting at some point and the spring constant of the beam. A flexure's bending stiffness K when not subjected to tensile load can be represented as

$$K = \frac{CEI}{L^3},$$
(11.1)

where I is moment of inertia,

$$I = \frac{wt^3}{12},$$
(11.2)

where

> w = Width of the flexible pivot
>
> t = Thickness of the beam
>
> E = Modulus of elasticity
>
> I = Moment of inertia
>
> L = Length of beam
>
> C = Constant determined by the end-to-end configuration.

The value of constant C can be depicted from Figure 11.2. With respect to end-to-end configuration, the recommended values of C are shown against different configurations. For example, the first figure shows a typical canti-lever beam which is fixed at one end; the force is applied perpendicular to

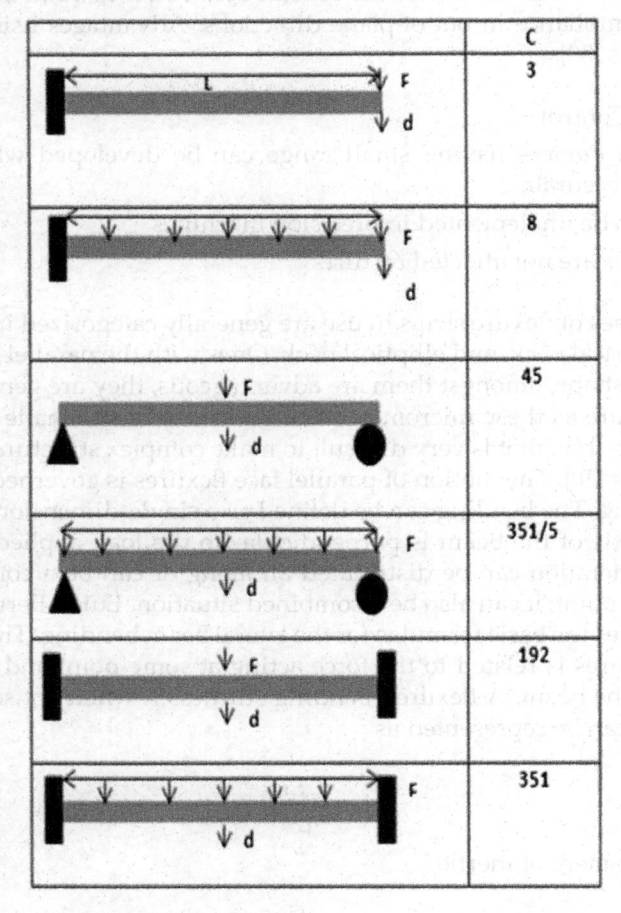

FIGURE 11.2
Values of C with respect to different configurations.

longitudinal axis at other end; the value of C is to be taken as 3. The second figure shows the same beam; however, the load is uniformly distributed all over the beam, the value of C is to be taken as 8. The third figure shows a pinned and a roller support at the ends and a point load at the center. The fourth figure shows pinned and roller supports at ends; however, the load is uniformly distributed all over the beam. The fifth and sixth figures show fixed supports at both ends with point load and distributed load, respectively; the respective values of C are shown against each.

The Euler–Bernouli beam model is based on different assumptions which are as follows:

- Isotropic material should be used that should maintain homogeneity and follow Hooke's law.
- It is considered that beam's cross section is constant and has initial straight orientation.
- Only pure bending is considered (i.e. no axial or torsional loads).
- Cross sections in y–z plane are considered to be unaffected during bending.
- Symmetry of axis is considered throughout the beam.

11.1.12 Microactuation

Applying force at microlevel is a special domain. Different types of actuating systems are available at microlevel that produce force on activation, whereas research is under progress on mechanisms and actuators of this range. This particular area of research is termed as "micromechatronics", in the microscopic world; it is the use of mechanics and electronics. Mostly the fabrication techniques used are integrated circuit (IC) manufacturing-based compatible processes [37]. Commonly available systems are capacitance devices (transverse comb drive devices, lateral comb drive actuators), thermal actuators, electrostatic actuators, and piezoactuators. Considering transverse comb actuators, the axis of action is orthogonal to the orientation of the fingers of the comb. The pros are that they are easy to fabricate and good for sensing the sensitivity of movement; however, they are difficult to be used as actuators because of the physical limit of distance. When considering lateral comb actuators, the force generated by them is proportional to the overlapped width and length of the fingers and inversely proportional to the separation among combs. They can be used where relatively long strokes are required from actuators. The output is mainly dependent on the thickness of fingers; the thicker they are, the larger the force will be. However, their foot print is relatively large. The working of thermal actuators is based on expansion of materials when subjected to heat; they can be solid, liquid, or gas, where coefficient of thermal expansion (CTE) can characterize the expansion of solids. When considering working of lateral thermal actuators, they consist of two legs: one hot and the other cold and wide; temperature difference is generated due to the different current flow densities in legs when heated by Joule heating. Joule/ ohmic/resistive heating is a process when electric current is passed through a resistance and converted into heat. Due to the difference in temperature of legs, the actuator deflects laterally; however, the actuator can have only one axis of action which cannot be in reverse direction. The electrostatic actuators are fabricated usually with metals and dielectrics. They are precise in movement; however, they suffer from short range and pull in phenomenon. The piezoactuators are made up of piezocrystals. Piezoelectric crystals are solid ceramic compounds that produce piezoelectric effects; that is, when mechanical force is applied on piezocrystals, electric voltage is produced, or when electric voltage is given to crystals, the mechanical deformation is induced. Natural piezoelectric crystals are quarts, tourmaline, and sodium potassium tartrate. A servo controller can be used to determine the input voltage given to the PZT ceramics that compares the signal from actual position sensor with a reference signal, which in turn will control the movement of the actuator. The main advantages of piezoactuators are their accuracy and repeatability; they are very stable and have linearity. The piezoactuators have unlimited lifetime, i.e. no wear and tear; it is proved that they can perform billions of cycles without any measurable wear. They have virtually infinite stiffness (within load limits), and there will be very little hysteresis and creep effects [36].

11.2 A Smart Microfactory Approach

11.2.1 Scenario Details

A microfactory approach is visualized which is operating in a socially safe environment whose operations are handled through a CPS. The micro-factory contains high-precision micropositioning/nanopositioning stage installed on a tabletop-size machine tool. A flexure-based, three-axis micro-positioning stage is considered that can be installed on a desktop-size mill-ing machine. A collaborative robot is envisaged to perform operations and handle microparts in the presence of human operator who is on a supervisory role. The microsize and delicate nature of the parts demand sensitive collision prevention, precise controlled operations, and safe handling. The smart microfactory is designed in two portions: initially a microstage is proposed for handling milling operations at microlevel that will be placed on a tabletop machine tool; secondly a collaborative robot is proposed for safe handling of microparts. For handling safety, a new technique is suggested based on virtual domain. A new concept of psychological safety of system is introduced while handling collaborative operations in the presence of a human supervisor (Figure 11.3).

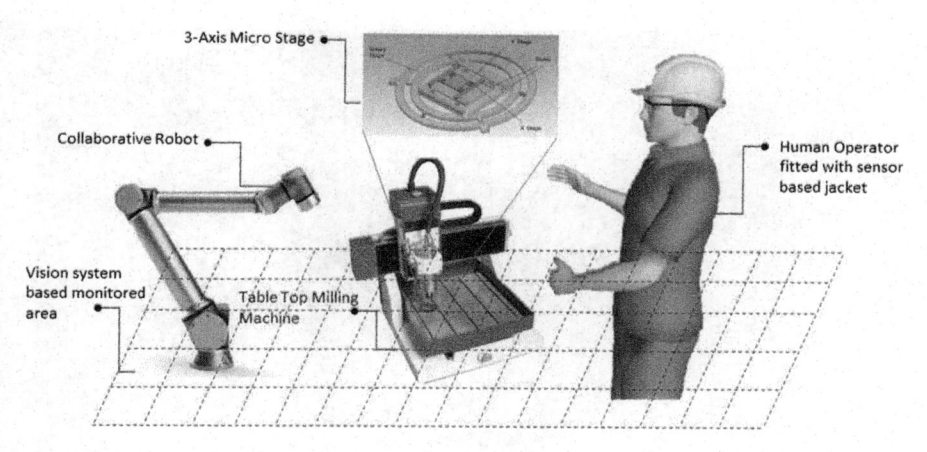

FIGURE 11.3
Proposed microfactory scenario.

11.2.2 Development on the Physical Domain

11.2.2.1 Design of the Proposed Microstage Design

A microstage is designed in SolidWorks software. Three stages were made overlapping each other. The prototype is specifically designed to achieve small range of motion through these stages, i.e. in micrometers, that can be used in a machine tool for microfabrication. Piezoactuators are used in each stage to produce lateral motion that will be converted into linear and rotary motion as per the design of each stage. The idea of linear stages was adopted from the MEMS accelerometer design in [34]. The idea of rotary stage was adopted from the design in which flexures are used for rotation [35]. Figure 11.4 shows the design of three-axis microstage in which the stages are clamped through bridges (flexures) connected with subsequent stages. When the force is applied through piezodevices, the flexures are bent, and motion is produced as per the stage design. The range of motion depends upon the stress limits of flexures that will be produced due to induced motion. The flexures used in the design can be considered as fixed–free beam configuration (Figure 11.5).

Actuation mechanism is shown in Figure 11.6 where piezoactuators are used for producing controlled range of motion.

Following design specifications were used for the prototype structure as shown in Figure 11.7.

The other specifications related to thickness are as follows:

Thickness 1st stage	6 μm
Thickness 2nd stage	6 μm
Thickness rotary stage	12 μm
Thickness for inner springs	6 μm
Thickness for outer springs	6 μm

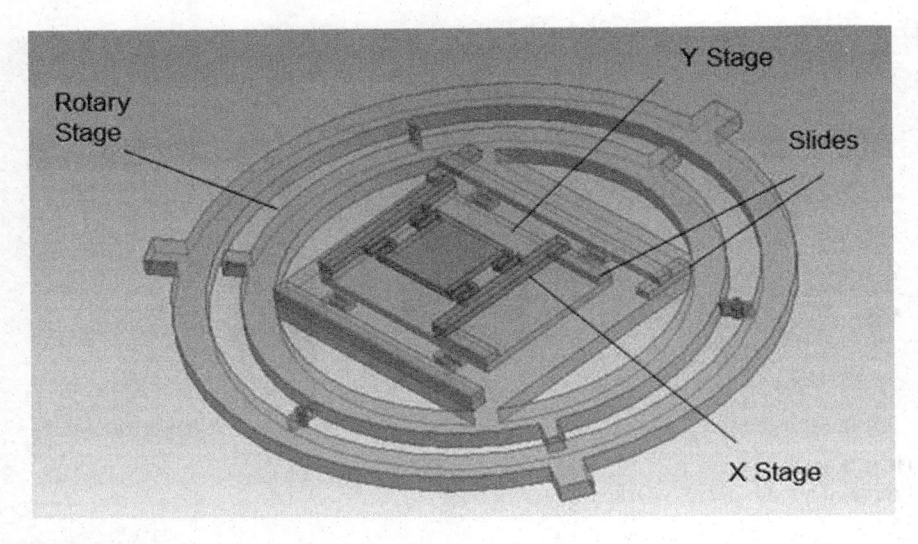

FIGURE 11.4
Design of microstage without piezoactuators.

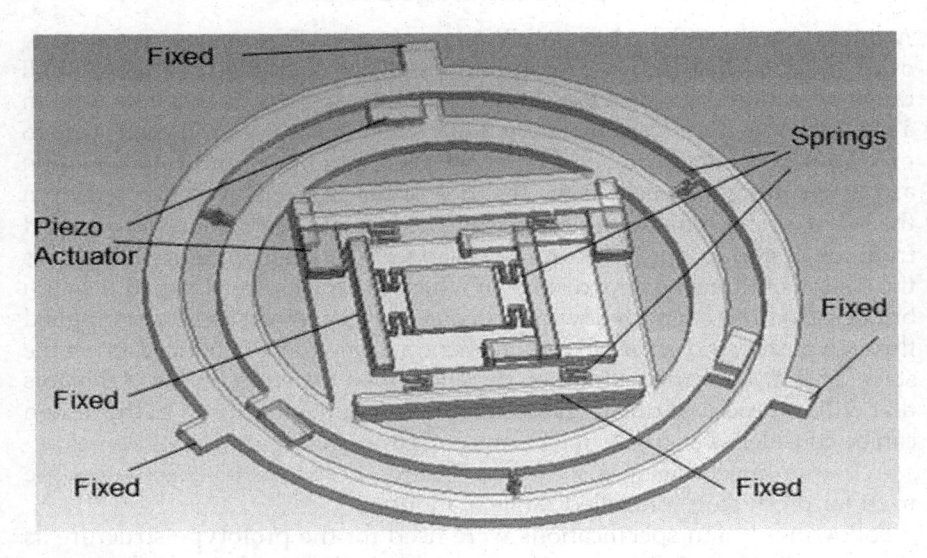

FIGURE 11.5
Design of microstage with piezoactuators and boundary conditions.

11.2.2.2 Material Properties

Three commonly used materials for micromachining were considered and later were analyzed for stress and deflection properties. A table showing material properties is given (Table 11.1).

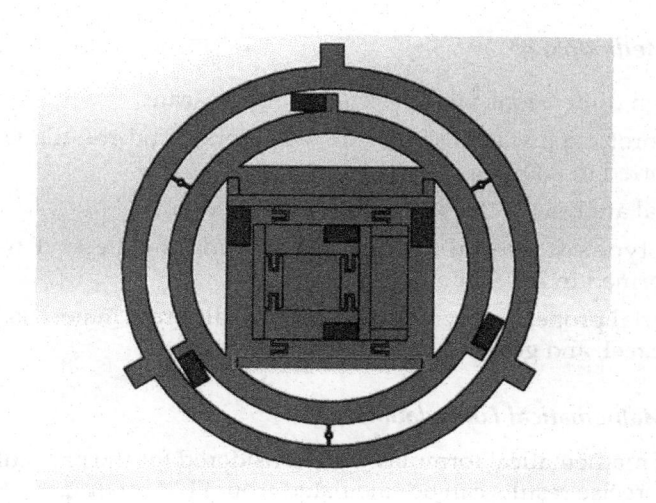

FIGURE 11.6
Piezoactuators shown in blue.

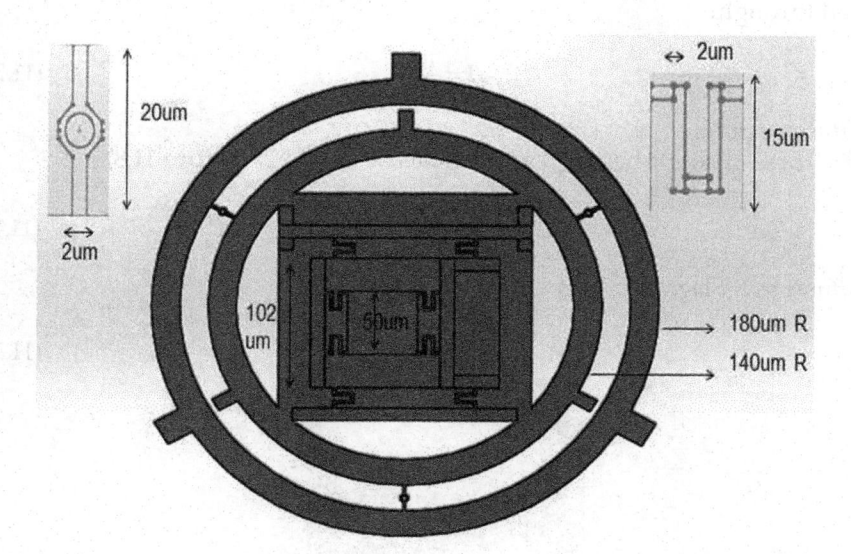

FIGURE 11.7 Specifications of prototype in μm.

TABLE 11.1

Material Properties

	Silicon	Gold	Steel
Density (kg/m3)	2,330	19,320	7,850
Young modulus (GPa)	165	98.5	200
Poisson ratio	0.22	0.42	0.3

11.2.2.3 Methodology

- Design dimensions were selected for microscale.
- The prototype was designed in SolidWorks, and the file was then imported in ANSYS.
- Modal analysis of the structure was carried out.
- Prototype's structural analysis for maximum deflection was then performed in ANSYS.
- Material properties were given for three different materials, i.e. silicon, steel, and gold.

11.2.2.4 Mathematical Formulation

Following mathematical formulas were considered for the particular design; however, precise results can be calculated from FEM analysis:

Moment of Inertia (I)

Moment of inertia (*I*) of the flexural beam used in the design can be calculated through

$$I = \frac{wt^3}{12}. \tag{11.3}$$

Stiffness of Springs (k)

Stiffness of single beam (*k*) as estimated from (11.2) (Figure 11.8):

$$k = \frac{3EI}{L^3}. \tag{11.4}$$

Stiffness of X Stage

$$K = 8 \times \frac{3EI}{L^3}. \tag{11.5}$$

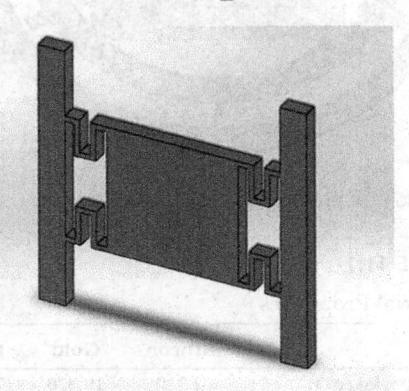

FIGURE 11.8
X stage.

For the eight springs in X stage, w was taken as 6 µm, L was taken as 15 µm, and t was taken as 2 µm (Figure 11.9).

Stiffness of Y Stage

$$K = 8 \times \frac{3EI}{L^3}. \tag{11.6}$$

For the eight springs in Y stage, w was taken as 6 µm, L was taken as 15 µm, and t was taken as 2 µm.

Stiffness, Total Force, and Rotation of Rotary Stage

$$K = 3 \times \frac{3EI}{L^3}, \tag{11.7}$$

$$3F = 3 \times Kd, \tag{11.8}$$

$$\tan theta = \frac{d}{R}, \tag{11.9}$$

where theta is the rotation, d is the deflection, and R is the radius of the stage (Figure 11.10).

For the three springs of rotary stage, w was taken as 12 µm, L was taken as 20 µm, and t was taken as 2 µm.

11.2.2.5 Meshing

Meshing of the structure in ANSYS is shown in Figure 11.11.

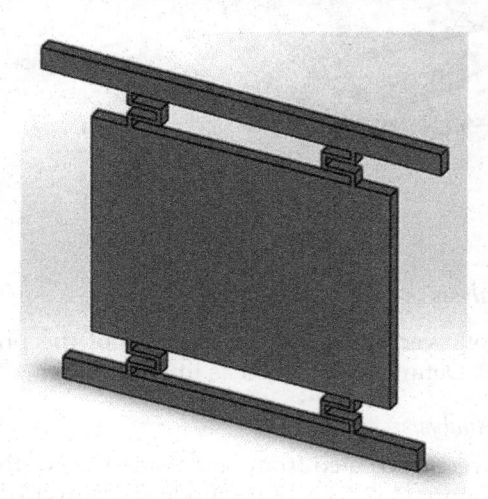

FIGURE 11.9
Y stage.

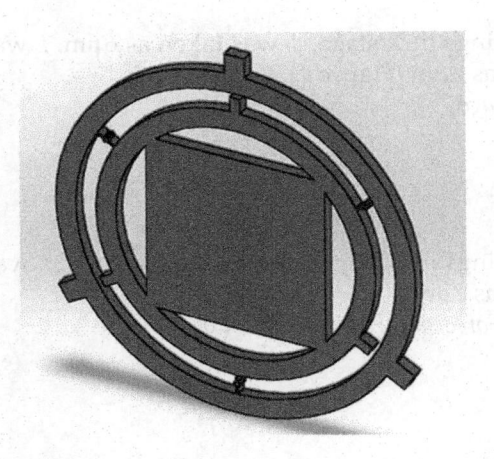

FIGURE 11.10
Rotary stage.

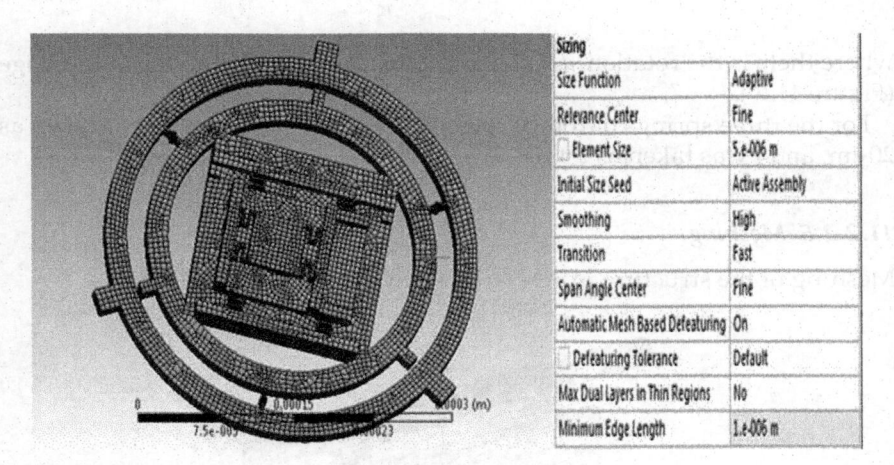

FIGURE 11.11
Meshing of structure.

11.2.2.6 FEM Analysis

Modal analysis followed by structural analysis of the prototype was performed in ANSYS. Details are covered as under.

11.2.2.6.1 Modal Analysis

Initial six modes were calculated from modal analysis for the designed prototype as shown in Figure 11.12. The fourth mode conforms to the desired motion.

Different modes, their frequencies, and mode shapes are given in tabulated form in Table 11.2.

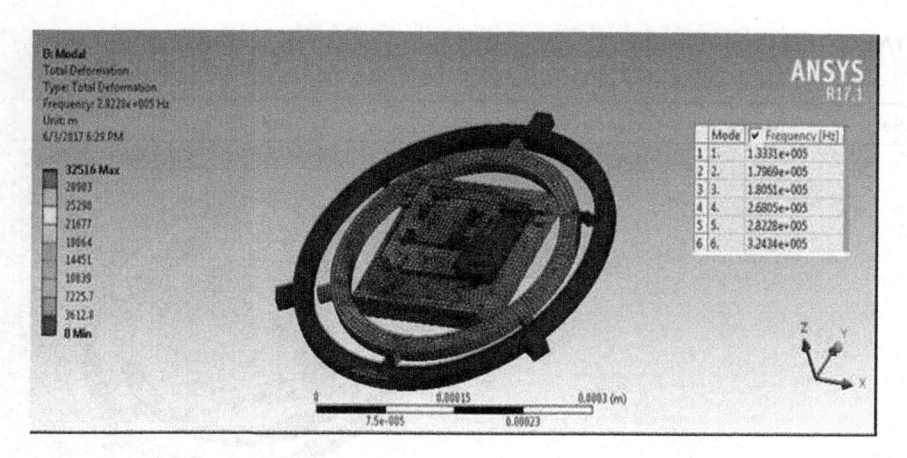

FIGURE 11.12
Modal analysis.

TABLE 11.2

First Six Frequency Modes of Prototype

Ser	Mode	Frequency (Hz)	Mode Shape
1.	1	1.3331e+005	
2.	2	1.7969e+005	

(Continued)

TABLE 11.2 (*Continued*)

First Six Frequency Modes of Prototype

Ser	Mode	Frequency (Hz)	Mode Shape
3.	3	1.8051e+005	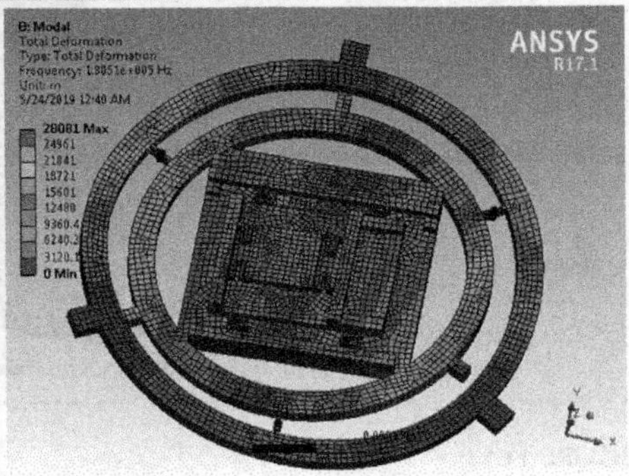
4.	4	2.6805e+005	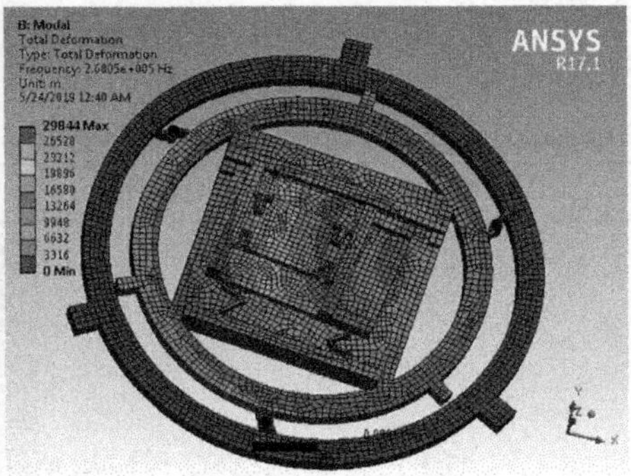
5.	5	2.8228e+005	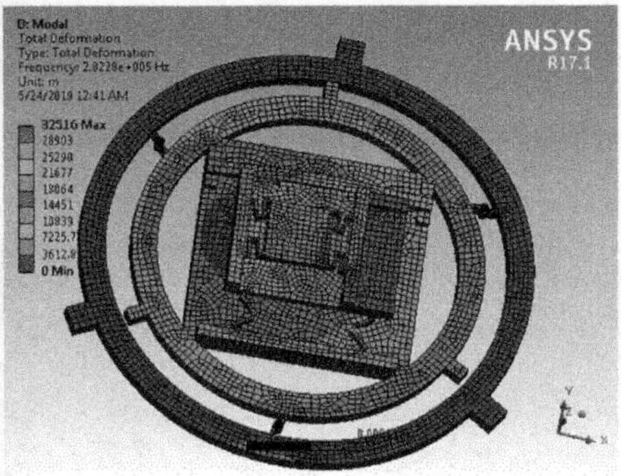

(*Continued*)

TABLE 11.2 (*Continued*)

First Six Frequency Modes of Prototype

Ser	Mode	Frequency (Hz)	Mode Shape
6.	6	3.2434e+005	

11.2.2.6.2 Structural Analysis

Structural analyses for *X*, *Y*, and rotary stages were carried out. A force ranging from 100 µN to 1 N was applied through each piezoactuator to get the finest resolution of each stage while using particular material. The maximum stress in the structure was obtained at each force level to ascertain the working of the structure below yield stress. Maximum deformations for each stage using all materials were calculated and compared to find the best material for the best design (Figures 11.13–11.15).

11.2.2.7 Results

The results show that 100 µN force on single actuator is the safe limit for this particular design as the obtained stress is quite below the yield

FIGURE 11.13
X stage static structural analysis (deformation of *X* stage).

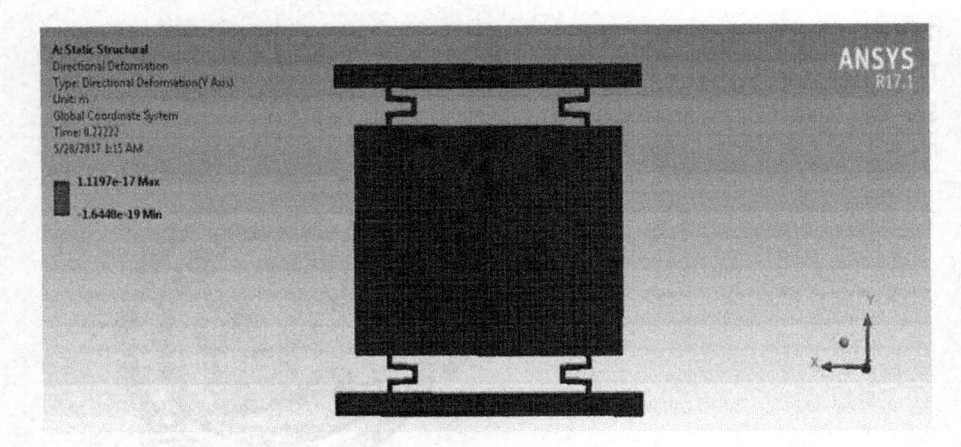

FIGURE 11.14
Y stage static structural analysis (deformation of Y stage).

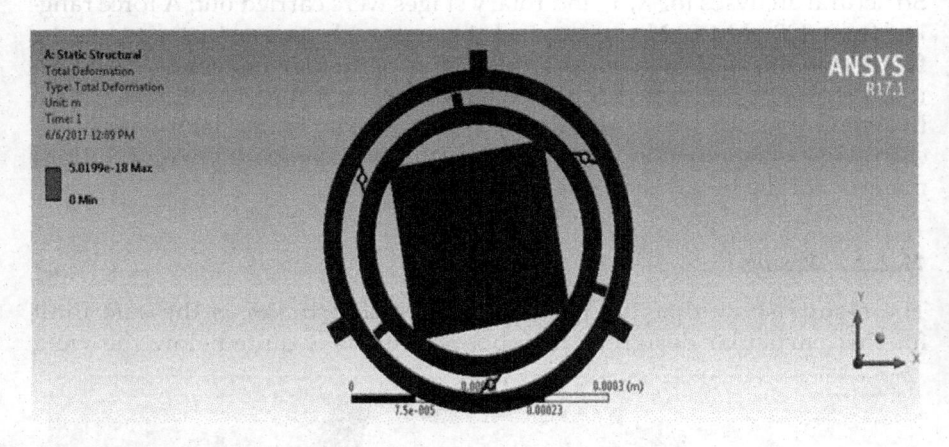

FIGURE 11.15
Rotary stage static structural analysis (deformation of rotary stage).

stress for any material used; however, the maximum deflection can be obtained when gold is used for the structure. The material which shows minimum "stress to yield stress ratio" when the same structure is made from it of similar specifications and subjected to same load is found to be steel (Table 11.3).

The comparison for the maximum deflection obtained when three different materials are used for each stage is shown in Figure 11.16.

The comparison for the stress obtained when three different materials are used for each stage is shown in Figure 11.17.

TABLE 11.3

Results Obtained from Three Stages Designed with Different Materials Subjected to Different Loads

Stage	Material	Force	Yield Strength	Obtained Stress	Deflection (m)
X stage	Gold	1 N	205 MPa	1203.9 GPa	2.2709e–003
		100 μN		120.39 MPa	22.709e–6
Y stage		1 N		1168.3 GPa	2.1724e–003
		100 μN		116.83 MPa	21.72e–6
Rotary stage		1 N × 3		1049.2 GPa	95.283e–003
		100 μN × 3		104.92 MPa	952.83e–6
X stage	Steel	1 N	250 MPa	1123.3 GPa	1.1679e–003
		100 μN		112.33 MPa	11.67e–6
Y stage		1 N		1091.1 GPa	1.12e–003
		100 μN		109.11 MPa	11.2e–006
Rotary stage		1 N × 3		1030.6 GPa	50.199e–003
		100 μN × 3		103.06 MPa	501.99e–6
X stage	Silicon	1 N	180 MPa	1092.2 GPa	1.4423e–003
		100 μN		109.22 MPa	14.42e–6
Y stage		1 N		1061.3 GPa	1.3844e–003
		100 μN		106.13 MPa	13.84e–6
Rotary stage		1 N × 3		1038.2 GPa	62.6e–003
		100 μN × 3		103.8 MPa	626e–6

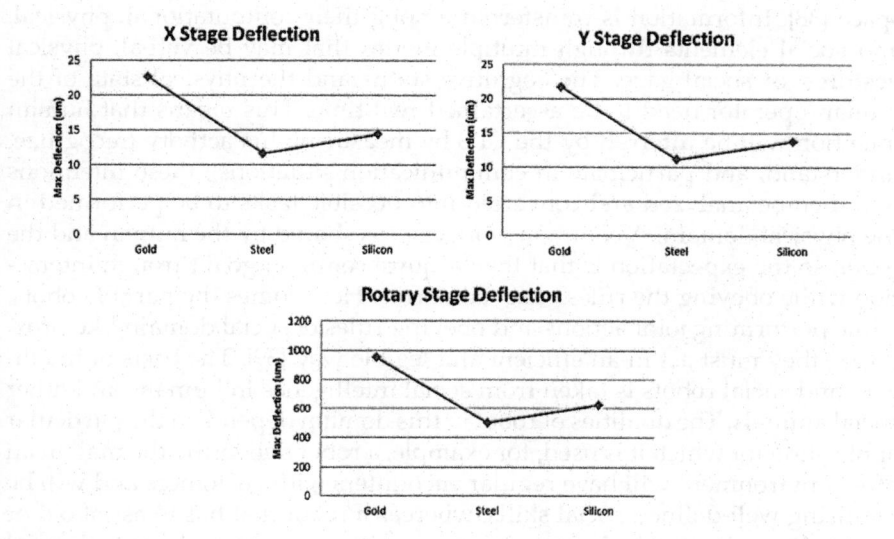

FIGURE 11.16

Comparison of maximum deflection of different stages when 100 μN force is given to each actuator.

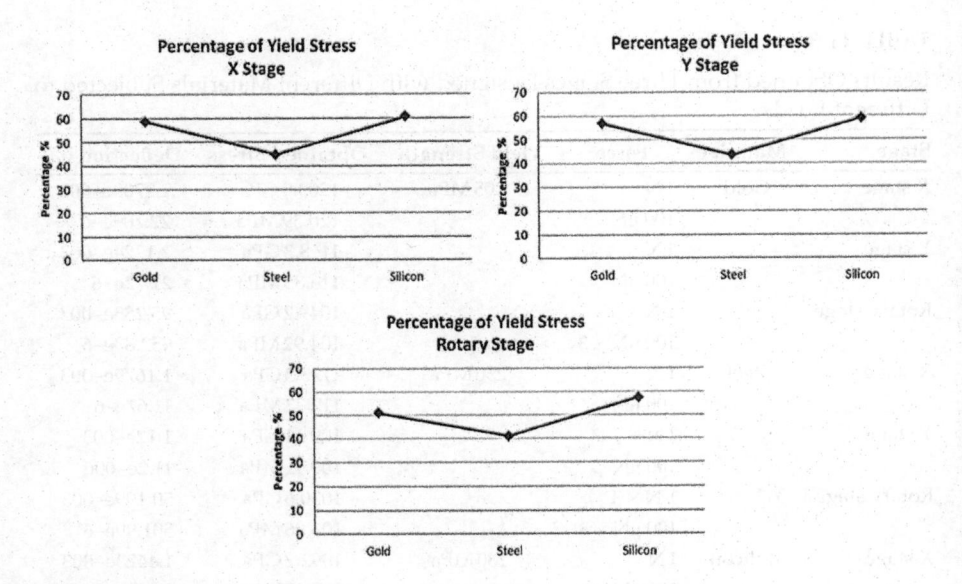

FIGURE 11.17
Comparison of percentage of maximum stress to yield stress for all stages.

11.2.3 Development on Virtual Domain

11.2.3.1 Social Safety of CPS

With increasing HMI, latest CPSs are designed to cater for aspects of social space [38]. Information is transferred among their computational, physical, and social elements through multiple modes that may be verbal, physical gestures, or social gaze. The cognitive status and the physical state of the human operator need to be ascertained real-time. This means that human intention will be inferred by the CPS by measuring his activity (recognize, understand, and participate in communication situations); these intentions will then be analyzed and converted into possible tasks to be performed in the physical domain. A common workspace is shared by the human and the robot, so the expectation is that they acquire common goal through interaction while obeying the rules of social norms. Here comes the part of cobots: while performing joint actions and obeying rules of social domain like proxemics, they must act in an efficient and legible way [39]. The basis of intelligent and social robots is taken from social intelligence in humans and other social animals. The qualities of robot in this domain depend on the particular application for which it is used; for example, a robot delivering the mail in an office environment will have regular encounters with customers and will be requiring well-defined social skills, whereas a robot that has to assist old or disabled people must be in possession of a wide range of qualities and social abilities to make it convenient for humans [40]. Prospects of the social robots in the food industry were presented in [41] that elaborates the roles of robots from food industry to serving robots, keeping in view the social norms like cleanliness to social interaction with humans. The concept has extended to real-time safety system capable of allowing safe HRI. Safety can be classified into two categories: physical safety and psychological safety [42]. The first one is only related to unwanted human–robot contact, whereas the second one is related to HRI that does not cause stress and discomfort for long periods. A conceptual system avoiding both contact and stress is presented in [43]. A real-time safety system was formulated that works at very low separation distance; the system does not require any replacement or modification in robot hardware. The real-time measurement of separation distance thus found can be used for precise robot speed adjustment. A 3D sensor is used in the system which formulates a dynamic safety zone and calculates the safety distance. Collision can be prevented between humans and robots thus making it comfortable for human operators to work stress free and for robots to do their work efficiently; this can be done by leveraging known robot joint angle values and accurate measurements of human positioning in the workspace. A matching idea for human psychological comfort due to the effect of robot motion is presented in [44]. A safe HRC is proposed in [45] for heavy payload industrial robots: an integrated concept is used by combining the concept of security and safety using off the shelf sensors and components of CPS. A defensive strategy to avoid cyber-physical attacks is proposed for safety of CPSs in [46]: the concept includes secured data monitoring at different nodes based on the technique of system reconfiguration and health monitoring. Three categories of robot motion were compared, based on the criteria of human comfort when exposed to particular type of motion. The categories presented were functional, predicable, and legible motions. Overall, the work supports the use of legible motion over predictable motion in collaborative tasks; both are types of functional motion. Functional motion is the one in which the robot reaches the goal without collision, though not efficiently; predictable motion is one that matches collaborator expectation given the goal is known, whereas in legible motion the human infers the goal while the robot is undergoing motion. While the comfort for human operator increased as the system lacks flexibility to encounter any contingency in task, for example in a manufacturing line the robot while performing work on a nut instead found a bolt. Different HRI safety systems are presented in [47–50]. Mainly two types of sensors are used in broader category, one based on vision systems and the other based on proximity/contact. The safety system presented will come into action as soon as the human arm will come into contact or in near vicinity to the robot; however, these systems do not provide the choice to identify the user. Also they do not take into account any foreign element, just for example a pet, if it enters into the work zone. A list of the state-of-the-art existing collaborative robots is presented in [51] showing their capabilities for safe HRC. The list shows that force sensors, torque sensors, and visual/infrared (IR) cameras are used for collision detection. The review identifies that the robots lack particular object/user detection in their workspace during operation. An object classification technique was however used in [52] to identify a human body and some objects available in workspace. The objective is to classify objects in areas of interest of the robot, real-time. However, the system neither can differentiate between other humans than the user nor can detect other objects which are not related to task and nor can modify the role if an object currently is not defined for this particular task. As Industry 4.0 recommends the use of intelligent robots, the concept of comfort to human users can be equally valid for intelligent robots, i.e. both physical and psychological safety. As already discussed, a lot of work has been done for physical safety of both humans and robots, but there is no concept of psychological safety for intelligent robots/systems. Safety cannot be termed in the sense of avoiding collision only; it can also be termed in the sense of avoidance/modification of task when the robot/system is not comfortable. Changing scenarios, diverting from the main task, affect the efficiency of system which must be catered keeping in view the optimization criteria and without compromising safety. Affecting the efficiency means an uncomfortable situation for the system or eventually the intelligent robot. This may be in terms of entrance of unwanted object in the workspace or a changing scenario, may be in terms of wrong feed of parts in manufacturing system. The problem can be addressed using detection of a particular type of object within the work zone and then taking action through predefined logic.

11.2.3.2 Methodology

- A microfactory based on tabletop-size machine tool is proposed.
- The machine tool will be fitted with a three-axis micropositioning stage for milling operations.
- A cobot (UR 3) fitted with microgripper will be placed next to micro-factory for handling microparts.
- A human operator will be present for supervision and control of complete operation within the workspace.
- The cobot will take input from laptop fitted with a machine vision camera.
- An object-detection-based algorithm will be used to detect objects within the workspace and give inputs to cobot for safety and control operations.
- Hazard assessment based on predefined logic may be used to provide social and psychological safety to the system.

11.2.3.3 UR3 Robot Components and Capability

The UR3 is a small tabletop cobot; it can be used for automated workbench tasks of light payload scenarios. It is a compact tabletop robot which can handle payload of 6.6 lbs (3 kg), but its weight is only 24.3 lbs (11 kg); it has a capability of infinite rotation on the end joint and 360° rotation on all wrist joints. The robot system consists of three main parts: the robotic arm, the teach pendant, and the controller box. The controller box contains both digital and analog input and output sockets which can be used for interfacing other components or system components itself. The teach pendant can be used to program the robot as per the requirement of user and can be based on inputs and outputs. The robot can be set up quickly without programming experience using patent technology and can be operated with 3D intuitive visualization. It requires a simple movement of the robotic arm by giving waypoints or from the controls given on the touch pad. UR robots can be set up very quickly; thus they can reduce usual deployment that can take weeks and can be done in hours. The average time calculated is half-day. It can only take an hour to unpack, by an unexperienced operator, and even to program it first time with a simple task. They are lightweight, can save space, are easy to install/relocate, and can be used for multiple applications without changing the layout of the factory. Altering UR3 to new processes is quick and easy, giving the agility to automate almost any manual task; they can even handle small batches or quick changeovers. The programs can be reused for recurrent tasks. No safeguards are required when using UR robot; almost 80% of the thousands all over the world perform with no safeguards, along with humans. It is approved and certified by TÜV (The German Technical Inspection Association).

11.2.3.4 Object-Detection API

The object-detection API is one of the frameworks provided on an open source of TensorFlow or YOLO algorithm. These provide an opportunity to construct models very easily and then train and deploy different models. They had broken the challenge faced by machine vision developers for creating a model which can accurately localize and identify multiple objects in a single image. A tutorial on the object detection is placed in the Appendix at the end of this chapter. The images for the test are shown in Figures 11.18 and 11.19.

The results after the program was run on the test images are shown in Figures 11.20 and 11.21. The algorithm correctly identified the objects in the pictures. Same can be run for real-time video after little modification in the algorithm.

11.2.3.5 Hazard Assessment

The level of interaction and risk play a major role in defining effective HRC. A formal grading to ascertain HRC is introduced in [51]; a concept of risk and hazard assessment was introduced in [45,53] along with HRC assessment. Based on a similar approach, number of hazards are outlined on different

FIGURE 11.18
First image for test.

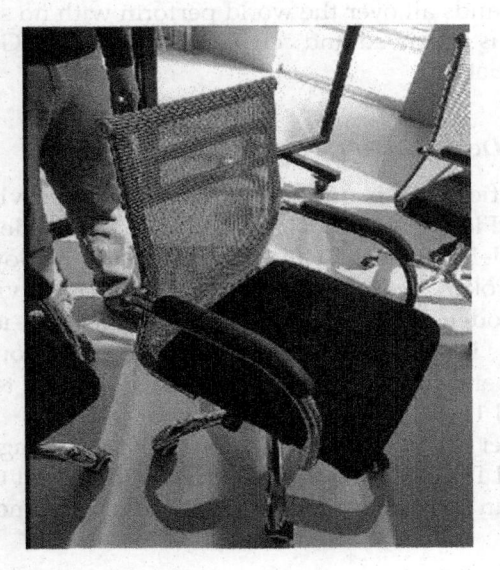

FIGURE 11.19
Second image for test.

criteria, i.e. hazard posed by robot, industrial process, and robot control system. These hazards are then gauged against social space characteristics and graded a particular value. An effective collaborative system can be designed based on the assessment carried out through this process. For this, a chart

FIGURE 11.20
Result of first image showing bottle with 74% accuracy.

FIGURE 11.21
Result of second image showing chair with 84% accuracy.

considering all possible scenarios is developed which will be helpful for risk/hazard assessment. The particular type of hazard will identify which output is affected to which level. This on the other hand can be used as a predefined logic for an automated system by using AI algorithms and will be helpful in providing social and psychological safety to the system. This also can be used as a checklist to design a working cell for HRC. Possible hazards for different categories are stated in a table format where a number is assigned to each; the number can be identified by combining the category number and the hazard number (Table 11.4).

TABLE 11.4

Hazards Posed to CPPS from Robot, Industrial Process, and Control System

(a) Hazards from robot during collaboration

1. Robot characteristics: speed, force, torque, acceleration, momentum, power, etc.
2. Operator in dangerous location of working under heavy payload robot.
3. Hazards from end-effector and work-part protrusions.
4. Sensitivity of the parts of the operator body that can come in contact in case of collision.
5. Mental stress to operator due to robot characteristics (e.g. speed, inertia, etc.)
6. Hazard from trajectory taken by the robot.
7. Physical obstacles against robot operation during collaboration.
8. Hazard from fast worker approach speed and robot's slow reaction time.
9. Hazard from tight safety distance limit in the collaborative workspace.
10. All parts of the robot are not covered using the safety distance approach.

(b) Hazards from the industrial process during collaboration

1. Ergonomic design deficiency.
2. Time duration of collaboration in the process.
3. Transition time from collaborative operation to other operation.
4. Potential hazards from the industrial process (e.g. temperature, loose parts, etc.).
5. Mental stress to operator due to collaborative industrial process.
6. Work material routing during the process.
7. Physical obstacles tackled by worker in order to accomplish process requirement in collaborative workspace.
8. Hazards due to task complexity in collaborative workspace.

(c) Hazards from robot's control system malfunction during collaboration

1. Hazards from biomechanical pressure limits for operator during reasonably foreseeable misuse.
2. Misuse of collaborative system by operator or under a cyber-attack in a connected environment.
3. Physical obstacles in front of active sensors used in the collaborative workspace. (e.g. obstacle in front of camera).
4. Non-provision of transition from collaborative operation to manual system in case of system malfunction.
5. Number of workers involved in the collaborative process.
6. Hazard created due to wrong perception of industrial process completion by the robot.
7. Hazards from obstacles against unobstructed means of exiting the collaborative workspace at any instant.
8. Hazard from visual obstruction for robot in collaborative workspace due to vantage point of operator.

The grading criterion for hazard assessment is stated as follows:

- High influence on output: 3
- Medium influence on output: 2
- Low influence on output: 1
- No influence on output: 0.

A score chart based on the grading criterion by pitching hazard against social space characteristics is given in Table 11.5.

TABLE 11.5

Hazard Assessment Score Chart

			Social Space Characteristics			
		Hazards	Industrial Process Quality	Quality of HRC	Collaborative System Security	Operator Safety	Operator Health
Hazards	Robot	1a	0	3	0	3	3
		2a	0	3	0	2	2
		3a	0	3	0	3	2
		4a	0	2	0	3	2
		5a	0	3	1	2	3
		6a	1	1	0	1	1
		7a	1	2	0	2	1
		8a	0	3	0	3	2
		9a	0	3	0	3	2
		10a	0	3	0	3	2
	Industrial Process	1b	2	2	0	2	2
		2b	0	2	0	2	1
		3b	1	1	0	1	0
		4b	3	3	1	3	3
		5b	2	2	1	2	3
		6b	2	2	0	2	0
		7b	2	2	0	2	0
		8b	1	2	0	2	1
	Robot Control System Malfunction	1c	1	2	0	3	2
		2c	3	3	3	2	0
		3c	3	3	1	3	0
		4c	2	2	1	2	1
		5c	1	1	2	1	0
		6c	2	1	0	2	0
		7c	0	1	0	3	1
		8c	2	2	0	2	0

11.3 Conclusion

Recent developments in research related to smart factory have paved the way for development of new scenarios. Two different approaches in physical and virtual domains are dovetailed and presented in a scenario of microfactory; that is, a CPPS is envisaged incorporating both domains to ensure improvement in product quality, process improvements, physical and social safety, mass customization, and mass production. First a three-axis microstage was presented based on flexure design. The results showed that use of piezoactuators ensured deflection of micromachine stages in micrometer range.

Secondly a virtual domain was considered based on vision system in collaboration with a collaborative robot which ensured satisfactory performance of task keeping, considering the social safety constraints. A new concept of psychological safety of the system was introduced that will provide comfort to the system ensuring optimum utilization. The simultaneous approach that incorporated both the domains collaborated in real time for a smart microfactory has opened a new avenue of research for the particular domain.

11.4 Appendix: Tutorial on Object-Detection API

Object-detection API in the TensorFlow requires the following libraries:

- Tensorflow
- Python-tk
- Pillow 1.0
- Protobuf 3+
- lxml
- tf Slim
- Matplotlib
- Jupyter notebook
- Cython
- Cocoapi.

Tensorflow can be installed by using one of the following commands:

```
#  For CPU
pip  install  tensorflow
# For GPU
pip  install  tensorflow-gpu
```

The remaining libraries can be installed on Ubuntu 16.04 using apt-get:

```
sudo  apt-get  install  protobuf-compiler  python-pil  python-
lxml  python-tk
sudo  pip  install  Cython
sudo pip install jupyter
sudo  pip  install  matplotlib
```

Alternatively, users can install dependencies using pip:

```
sudo  pip  install  Cython
sudo  pip  install  pillow
```

```
sudo  pip  install  lxml
sudo pip install jupyter
sudo  pip  install  matplotlib
```

The Anaconda is another open source which makes it even easier to cater for machine learning and Python data science. There are more than 250 famous data science packages, virtual environment manager for Windows, conda packages, MacOS, and Linux packages. TensorFlow, Scikit-learn, and SciPy are easy to install in Anaconda; it is even easy to upgrade environments and complex data packages. Anaconda 3 includes all the libraries required for object-detection API. The TensorFlow Object-detection API uses Protobufs to configure model and training parameters. Before the framework can be used, the Protobuf libraries must be compiled. Protobuf 3.4 is required for compilation; others don't work. Either add Protobuf in system path or give full path to the protos folder. This should be done by running the following command from the "tensorflow/models/research/" directory:

```
#  From  tensorflow/models/research/
protoc  object_detection/protos/*.proto  --python_out=.
```

Anaconda 3 which is a python environment is used for running all the libraries. After downloading all the libraries and compiling the protos folder in object-detection module, open jupyter notebook in Anaconda prompt. In the prompt, give the path where object-detection folder is present like

E:\Software\tensorflow\model\models-master\research.

Inside it, find the file "object detection tutorial.ipynb" and convert it to ".py" file. After running jupyter notebook, it may happen that Internet Explorer will open, but nothing will happen. A token will appear on screen on command prompt. Copy that token in Word, and place this token in Google Chrome; jupyter notebook will open where file format is to be converted. Then open another Anaconda prompt and run spyder in it. In spyder, open the "object-detection.py" file and run the cells of the program one by one. Detailed description of cells and their purpose is given below: the cell which imports all the libraries is shown in Figure 11.22.

The cell shown in Figure 11.23 imports mataplotlib for images.

The cell shown in Figure 11.24 provides the path and name of the model that are required for object detection.

The cell in Figure 11.25 extracts the model and is not required to run if model file is already downloaded and extracted.

The cell shown in Figure 11.26 loads the model.

This cell in the Figure 11.27 loads the labels.

The cell shown in Figure 11.28 converts the image into array.

This cell shown in Figure 11.29 gives path to the images and declares the image size.

```
import numpy as np
import os
import six.moves.urllib as urllib
import sys
import tarfile
import tensorflow as tf
import zipfile

from collections import defaultdict
from io import StringIO
from matplotlib import pyplot as plt
from PIL import Image

# This is needed since the notebook is stored in the object_detection folde
sys.path.append("..")
from object_detection.utils import ops as utils_ops

if tf.__version__ < '1.4.0':
    raise ImportError('Please upgrade your tensorflow installation to v1.4.'

# ## Env setup
```

FIGURE 11.22
Part of program that imports libraries.

```
# This is needed to display the images.
get_ipython().run_line_magic('matplotlib', 'inline')
```

FIGURE 11.23
Part of program that imports mataplotlib.

```
# In[ ]:

# What model to download.
MODEL_NAME = 'ssd_mobilenet_v1_coco_2017_11_17'
MODEL_FILE = MODEL_NAME + '.tar.gz'
DOWNLOAD_BASE = 'http://download.tensorflow.org/models/object_detection/'

# Path to frozen detection graph. This is the actual model that is used for
PATH_TO_CKPT = MODEL_NAME + '/frozen_inference_graph.pb'

# List of the strings that is used to add correct label for each box.
PATH_TO_LABELS = os.path.join('data', 'mscoco_label_map.pbtxt')

NUM_CLASSES = 90
```

FIGURE 11.24
Part of program that downloads object-detection model.

```
# In[ ]:


opener = urllib.request.URLopener()
opener.retrieve(DOWNLOAD_BASE + MODEL_FILE, MODEL_FILE)
tar_file = tarfile.open(MODEL_FILE)
for file in tar_file.getmembers():
  file_name = os.path.basename(file.name)
  if 'frozen_inference_graph.pb' in file_name:
    tar_file.extract(file, os.getcwd())


# ## Load a (frozen) Tensorflow model into memory.
```

FIGURE 11.25
Part of program that extracts the downloaded model.

```
# In[ ]:


detection_graph = tf.Graph()
with detection_graph.as_default():
  od_graph_def = tf.GraphDef()
  with tf.gfile.GFile(PATH_TO_CKPT, 'rb') as fid:
    serialized_graph = fid.read()
    od_graph_def.ParseFromString(serialized_graph)
    tf.import_graph_def(od_graph_def, name='')
```

FIGURE 11.26
Part of program that loads the model.

```
# In[ ]:


label_map = label_map_util.load_labelmap(PATH_TO_LABELS)
categories = label_map_util.convert_label_map_to_categories(label
category_index = label_map_util.create_category_index(categories)


# ## Helper code
```

FIGURE 11.27
Part of program that loads the labels.

```
# In[ ]:


def load_image_into_numpy_array(image):
    (im_width, im_height) = image.size
    return np.array(image.getdata()).reshape(
        (im_height, im_width, 3)).astype(np.uint8)



# # Detection
```

FIGURE 11.28
Part of program that converts images into array.

```
# In[ ]:

# For the sake of simplicity we will use only 2 images:
# image1.jpg
# image2.jpg
# If you want to test the code with your images, just add path to
PATH_TO_TEST_IMAGES_DIR = 'test_images'
TEST_IMAGE_PATHS = [ os.path.join(PATH_TO_TEST_IMAGES_DIR, 'image

# Size, in inches, of the output images.
IMAGE_SIZE = (12, 8)
```

FIGURE 11.29
Part of program that provides image path and size.

```
# In[ ]:

for image_path in TEST_IMAGE_PATHS:
    image = Image.open(image_path)
    # the array based representation of the image will be used late
    # result image with boxes and labels on it.
    image_np = load_image_into_numpy_array(image)
    # Expand dimensions since the model expects images to have shap
    image_np_expanded = np.expand_dims(image_np, axis=0)
    # Actual detection.
    output_dict = run_inference_for_single_image(image_np, detectio
    # Visualization of the results of a detection.
    vis_util.visualize_boxes_and_labels_on_image_array(
        image_np,
        output_dict['detection_boxes'],
        output_dict['detection_classes'],
        output_dict['detection_scores'],
        category_index,
        instance_masks=output_dict.get('detection_masks'),
        use_normalized_coordinates=True,
        line_thickness=8)
    plt.figure(figsize=IMAGE_SIZE)
    plt.imshow(image_np)
```

FIGURE 11.30
Part of program that contains the main model.

The cell shown in the Figure 11.30 contains the main model; it takes in input images in a loop and converts them into an array, identifies classes, shows results in the form of boxes, classes, and scores.

Place the images in the image path folder and run the program to get results.

References

1. Wertschöpfungsketten. "Industrie 4.0", VDI VDE status report, 2014.
2. Cyber Physical System, available at https://en.wikipedia.org/wiki/Cyber-physical_system.
3. C. R. Rad, O. Hancu, I. A. Takacs, and G. Olteanu. "Smart monitoring of potato crop: a cyber-physical system architecture model in the field of precision agriculture." *Agriculture and Agricultural Science Procedia*, vol. 6, pp. 73–79, 2015.
4. L. A. C. Salazar, "Proportional_Reliability_of_Agent- Oriented_Software_Engineering_for_the_Application_of_Cyber_Physical_Production_Systems", available at www.researchgate.net/publication/320614822.
5. J. Gubbi, R. Buyya, and S. Marusic. Internet of things (IoT): a vision, architectural elements, and future directions. *Future Generation Computing Systems*, vol. 29, no. 7, pp. 1645–1660, 2013.
6. D. Zuhlke. "Smart factory—from vision to reality in factory technologies." In: *Proceeding of the 17th International Federation of Automatic Control World Congress (IFAC)*, South Korea, pp. 82–89, 2008.
7. B. C. Pirvu, C. B. Zamfirescu, and D. Gorecky, "Engineering insights from an anthropocentric cyber-physical system: a case study for an assembly station," *Mechatronics*, vol. 34, pp. 147–159, 2016.
8. A. R. Razali, and Y. Qin. "A review on micro-manufacturing, micro-forming and their key issues." *Procedia Engineering*, vol. 53, pp. 665–672, 2013.
9. Industry 4.0, available at https://en.wikipedia.org/wiki/Industry_4.0.
10. What everyone must know about Industry 4.0, available at www.forbes.com/sites/bernardmarr/2016/06/20/.
11. Industry 4.0, available at www.i-scoop.eu/industry-4-0/.
12. ISOC IoT Overview, available at https://cdn.prod.internetsociety.org/wp-content/uploads/2017/08/.
13. D. Miorandi, S. Sicari, F. D. Pellegrini, and I. Chlamtac. "Internet of things: vision, applications and research challenges." *Ad Hoc Networks*, vol. 10, no. 7, pp. 1497–1516, 2012.
14. Y. Xua, and M. Chena. "Improving Just-in-Time manufacturing operations by using Internet of Things based solutions." *9th International Conference on Digital Enterprise Technology - DET 2016*, Nanjing, China. 2016.
15. A. Radziwona et al. "The smart factory: exploring adaptive and flexible manufacturing solutions." *Procedia Engineering*, vol. 69, pp. 1184–1190, 2014.
16. Smart Factory Connected Manufacturing, available at www2.deloitte.com/insights/us/en/focus/industry–4–0/.
17. S. Wang et al. "Implementing smart factory of Industrie 4.0: An outlook." *International Journal of Distributed Sensor Networks*, vol. 12, p. 3159805, 2016.

18. G. Johannsen, "Human-machine interaction." *Control Systems, Robotics and Automation*, vol. 21, pp. 132–162, 2009.
19. Human Computer Interaction, available at https://en.wikipedia.org/wiki/Human%E2%80%93computer_interaction.
20. Human Machine Interaction, available at www.xsens.com/tags/human-machine-interaction/.
21. J. Edward Colgate, W. Wannasuphoprasit, and M. A. Peshkin. "Cobots: Robots for collaboration with human operators." In *Proceedings of the 1996 ASME International Mechanical Engineering Congress and Exposition*, Atlanta, GA, ASME, 1996.
22. M. Peshkin and J. Edward Colgate. "Cobots." *Industrial Robot: An International Journal*, vol. 26, no. 5, pp. 335–341, 1999.
23. O. Görür, B. Rosman, F. Sivrikaya, and S. Albayrak. "Social cobots: anticipatory decision-making for collaborative robots incorporating unexpected human behaviors." In *Proceedings of the 2018 ACM/IEEE International Conference on Human-Robot Interaction*, Chicago, IL, ACM, pp. 398–406, 2018.
24. J. Chae, S. S. Park, and T. Freiheit. "Investigation of micro-cutting operations." *International Journal of Machine Tools and Manufacture*, vol. 46, no. 3, pp. 313–332, 2006.
25. D. Huo, K. Cheng, and F. Wardle. "Design of a five-axis ultra-precision micro-milling machine—UltraMill. Part 1: holistic design approach, design considerations and specifications." *The International Journal of Advanced Manufacturing Technology*, vol. 47, no. 9, pp. 867–877, 2010.
26. M. Tanaka, Development of desktop machining microfactory. Riken Review N. 34 Focused on Advances on Micro-mechanical Fabrication Techniques, April, 2001.
27. Y. Okazaki, N. Mishima, and K. Ashida. "Microfactory - concept, history, and developments." *Journal of Manufacturing Science and Engineering*, vol. 126, pp. 837–844, 2004.
28. E. Kussul et al. "Development of micromachine tool prototypes for microfactories." *Journal of Micromechanics and Microengineering*, vol. 12, no. 6. pp. 795–812, November 2002.
29. Y.-B. Bang, K.-M. Lee, and S. Oh. "5-axis micro milling machine for machining micro parts." *The International Journal of Advanced Manufacturing Technology*, vol. 25, no. 9–10, pp. 888–894, 2005.
30. D. Huo, K. Cheng, and F. Wardle. "Design of a five-axis ultra-precision micro-milling machine—UltraMill. Part 1: holistic design approach, design considerations and specifications." *The International Journal of Advanced Manufacturing Technology*, vol. 47, no. 9, pp. 867–877, 2010.
31. P. Wang and Q. Xu. "Design of a flexure-based constant-force XY precision positioning stage." *Mechanism and Machine Theory*, vol. 108, pp. 1–13, 2017.
32. H. Wang and X. Zhang. "Input coupling analysis and optimal design of a 3-DOF compliant micro-positioning stage." *Mechanism and Machine Theory*, vol. 43, no. 4, pp. 400–410, 2008.
33. L.-J. Lai, G.-Y. Gu, and L.-M. Zhu. "Design and control of a decoupled two degree of freedom translational parallel micro-positioning stage." *Review of Scientific Instruments*, vol. 83, no. 4, p. 045105, 2012.
34. R. I. Shakoor, S. A. Bazaz, and M. Mubasher Saleem. "Mechanically amplified 3-DoF nonresonant microelectromechanical systems gyroscope fabricated in low cost MetalMUMPs process." *Journal of Mechanical Design*, vol. 133, no. 11, p. 111002, 2011.

35. H. S. Alexander. "Precision machine design." SME, Dearborn, Michigan, 1992.

36. J. J. Allen. *Micro Electro Mechanical System Design*. Boca Raton, FL: CRC Press, 2005.

37. H. Fujita and H. Toshiyoshi. "Micro actuators and their applications." *Microelectronics Journal*, vol. 29, no. 9, pp. 637–640, 1998.

38. H. Zhuge. "Interactive semantics." *Artificial Intelligence*, vol. 174, pp. 190–204, 2010.

39. S. Lemaignan, M. Warnier, E. A. Sisbot, A. Clodic, and R. Alami. "Artificial cognition for social human–robot interaction: an implementation." *Artificial Intelligence*, vol. 247, pp. 45–69, 2017.

40. K. Dautenhahn, "Socially intelligent robots: dimensions of human–robot interaction." *Philosophical Transactions of the Royal Society B: Biological Sciences*, vol. 362, no. 1480, pp. 679–704, 2007.

41. J. Iqbal, Z. H. Khan, and A. Khalid. "Prospects of robotics in food industry." *Food Science and Technology (Campinas) AHEAD*, vol. 37, 159–165, 2017.

42. B. S. McEwen and E. Stellar, "Stress and the individual: mechanisms leading to disease." *Archives of Internal Medicine*, vol. 153, no. 18, pp. 2093–2101, 1993.

43. P. A. Lasota, G. F. Rossano, and J. A. Shah, "Toward safe close-proximity human-robot interaction with standard industrial robots," in *The IEEE International Conference on Automation Science and Engineering*, Taipei City, Taiwan, vol. 2014, January, pp. 339–344, 2014.

44. A. D. Dragan, S. Bauman, J. Forlizzi, and S. S. Srinivasa, "Effects of robot motion on human-robot collaboration." In *Proceedings of the Tenth Annual ACM/IEEE International Conference on Human-Robot Interaction - HRI'15*, vol. 1, New York, NY, pp. 51–58, 2015.

45. A. Khalid, P. Kirisci, Z. Ghrairi, J. Pannek, and K.-D. Thoben. Implementing Safety and Security Concepts for Human-Robot Collaboration in the context of Industry 4.0, *Proceedings of the 39th International MATADOR Conference on Advanced Manufacturing*, Manchester, UK, 2017.

46. A. Khalid, P. Kirisci, Z. H. Khan, Z. Ghrairi, K.-D. Thoben, and J. Pannek. "Security framework for industrial collaborative robotic cyber-physical systems." *Computers in Industry*, vol. 97, pp. 132–145, 2018.

47. J. Krüger, T. K. Lien, and A. Verl. "Cooperation of human and machines in assembly lines." *CIRP Annals - Manufacturing Technology*, vol. 58, no. 2, pp. 628–646, 2009.

48. P. A. Lasota and J. A. Shah. "Analyzing the effects of human-aware motion planning on close-proximity human-robot collaboration," *Human Factors*, vol. 57, no. 1, pp. 21–33, 2015.

49. C. Morato, K. N. Kaipa, B. Zhao, and S. K. Gupta. "Toward safe human robot collaboration by using multiple kinects based real-time human tracking." *Journal of Computing and Information Science in Engineering*, vol. 14, no. 1, p. 011006, 2014.

50. F. Flacco, T. Kröger, A. De Luca, and O. Khatib. "A depth space approach to human-robot collision avoidance." *Proceedings 2006 IEEE International Conference on Robotics and Automation*, Orlando, FL, pp. 338–345, 2012.

51. A. Khalid, P. Kirisci, Z. Ghrairi, K. D. Thoben, and J. Pannek. "A methodology to develop collaborative robotic cyber physical systems for production environments." *Logistics Research*, vol. 9, no. 1, p. 23, 2016.

52. V. Sharma and F. Dittrich, "Efficient real-time pixelwise object class labeling for safe human-robot collaboration in industrial domain," 2006.

53. A. Khalid, P. Kirisci, Z. Ghrairi, J. Pannek, and K.-D. Thoben. "Safety requirements in collaborative human–robot cyber-physical system." In *Dynamics in Logistics*, Springer, Cham, pp. 41–51, 2017.

12

Reverse Engineering the Organizational Processes: A Multiformalism Approach

Abbas K. Zaidi and Edward Huang
George Mason University

CONTENTS

12.1 Introduction 255
12.2 Literature Review 256
    12.2.1 Reverse Engineering 256
    12.2.2 Reverse Engineering Techniques 257
    12.2.3 Reverse Engineering Processes 258
    12.2.4 Multiformalism Modeling 259
12.3 The Proposed Approach 260
12.4 Applications 263
12.5 Conclusions 270
References 271

12.1 Introduction

Modern enterprises employ a host of different technologies to undertake their missions. The advancements in communication and sensor technologies, and availability of affordable computational and storage resources, have offered unprecedented opportunities for enterprises to develop new, faster, and more efficient ways to undertake their traditional and new missions made possible by the convergence of technologies. With the use of these technologies and accessible computing power, the underlying process model, employed by the enterprises, becomes the key enabler of these missions. The performance of an organization, as a result, depends not only upon the systems, algorithms, and resources employed but also on the way these entities are utilized in a process workflow model. An underlying process model that utilizes the resources in an efficient and intelligent manner easily outperforms another that might have access to the same resources but employs an inefficient process model. In large-scale, complex enterprises, e.g., manufacturing, aviation, banking, command and control, etc., these underlying processes are so vital and important to the overall enterprise's yield or performance that their details are closely guarded and are not available outside an organization. In this paper, we present an overview of an approach developed for reverse engineering an enterprise's underlying process model in the form of an executable workflow. The proposed approach can also be used to develop solution workflows for new decision and/or inference problems. The approach is based on the premise that domain knowledge, which includes domain-specific data, concepts, their structure, relationships, and employed algorithms, together with a multiformalism integration platform are means to understanding and recreating an enterprise's process model. A domain ontology with concepts, their properties, and mutual relationships, allows reasoning about data available or generated by an enterprise. A process ontology with relevant algorithms and computational entities allows reasoning about the way these entities can be put together to solve a problem that the individual entities cannot solve on their own. This ontology identifies the possible workflows that can be instantiated with the available data and/or computational entities employed by an enterprise. Finally, a multimodeling integration platform allows implementation of the instantiated workflow incorporating desperate computational entities and execution of the integrated workflow for evaluation (e.g., sensitivity analysis, performance, etc.) purposes.

This chapter is organized as follows: Section 12.2 provides some basic definitions and relevant references on reverse engineering and techniques. It also provides a brief survey of the developed multiformalism approach. Section 12.3 presents the approach as a multiformalism framework and a few use cases for its applications. Section 12.4 provides details of the approach as it was developed and applied on a variety of decision, inference, and process reengineering problems.

12.2 Literature Review

To put the proposed effort in context, research in three distinct areas must be described. First, the relevant definitions of reverse engineering are reviewed. Then, the related prior efforts at developing reverse engineering techniques and processes are discussed. And finally, the proposed work leading up to the development of the proposed approach is overviewed.

12.2.1 Reverse Engineering

Several definitions of reverse engineering are presented in literature. Rekoff (1985) considers reverse engineering as a process of developing a set of specifications from a complex hardware system. The specifications should be defined by the people other than the original designers and be developed for making a clone of the original system. Rekoff proposes a hierarchical structure, which is the decomposition of the physical structure, to characterize complex hardware systems and improve the understanding of the interconnectivity of the physical components.

Chikofsky and Cross (1990) define reverse engineering as a process of analyzing a subject system to identify the system's components and inter-relationships and to create representation of the system in another form or a high level of abstraction. CanforaHarman and Penta (2007) use the same definition of reverse engineering. They present the overview of the field of reverse engineering and highlight three main targets of the future research on reverse engineering: aiding software comprehension, providing support for maintenance and reengineering, and enabling autonomic computing and service-oriented architectures.

Rugaber and Stirewalt (2004) define reverse engineering as the process of comprehending software and producing a model of it at a high abstraction level. Reverse engineering can help better handle foreign code and better understand systems' structure and functions. They propose to use a formal specification or industrial practice of model-driven engineering techniques, such as the Unified Modeling Language (UML) and Object Constraint Language (OCL), as well as automatic code generation for reverse engineering.

12.2.2 Reverse Engineering Techniques

Prior research has developed several reverse engineering techniques. Keller et al. (1999) propose to use design patterns to capture the rationale behind possible design and support studying the trade-offs among alternatives. The complexity of reverse engineering for a complex system can be high. Design patterns enable to narrow down the possible alternatives based on the structural description. They also apply their approach to three case studies of software systems using C++. Shi and Olsson (2006) extend the concept and develop an automated pattern-detection approach from Java source code in specific application domains.

Otto and Wood (1998) introduce a reverse engineering process. The process starts from formulating customers' need, followed by creating a functional model through teardowns. They illustrate the proposed process using an electric wok redesign. Krikhaar (1997) proposes a similar approach by creating reverse architectures consisting of three phases: extraction, abstraction, and presentation. He applied it to a medical system, a communication system, and a consumer electronics system.

Liang et al. (1998) study the genetic network architectures. They propose to use binary models of genetic networks to study input/output patterns of variables in a model and maximize functional inference from large datasets.

They implement a prototype in a C program to analyze incomplete state transition tables to produce the original rules.

Schmidt (2006) focuses on software systems and suggests that model-driven engineering technologies offer an approach to alleviate the complexity of reverse engineering. Model-driven reverse engineering is used to construct models from existing source code. Classical approaches on reverse engineering for software systems usually deal with machine code directly. Model-driven technologies, on the other hand, provide an abstraction, which shields developers from complexities of programming.

Bongard and Lipson (2007) study complex nonlinear dynamic systems under uncertainty. They propose an automated reverse engineering method, which can automatically generate symbolic equations for a nonlinear dynamic system from time series data. The method enables to simplify the equations during modeling. They apply the proposed method to mechanics, ecology, and systems biology.

Krogmann et al. (2010) propose to use genetic search for reverse engineering of performance prediction of components. The genetic algorithm is used to construct a behavior model from collected data, runtime bytecode counts, and bytecode analysis. They apply the proposed approach to a file sharing application and predict its performance.

Müller et al. (2000) review the research on reverse engineering in the fields of computer science, computer engineering, and software engineering. They focus on code and data reverse engineering as well as research strategies for tool development and evaluation. They review various techniques for evaluating and comparing software exploration tools, which are a type of reverse engineering tools.

Although several reverse engineering techniques are proposed in literature, each technique has its own assumptions. This problem arises because these approaches may not work in a different context. What is needed is an approach which will allow to combine different techniques to produce stronger results.

12.2.3 Reverse Engineering Processes

The other related research in literature studies reverse engineering processes. Reverse engineering processes are the processes of analyzing a subject-developed system, identifying its systems components, and defining the interfaces between these components. The subject systems can be a single software/hardware system, a set of software or hardware systems, or a set of both software and hardware systems.

Arnold (1990) defines reverse engineering processes as consisting of two major phases: (i) the phase of identifying system's components and their interrelationships and (ii) the phase of extracting system abstractions and design information. The artifacts of the reverse engineering process are the basis for system comprehension and analysis.

Aiken et al. (1994) study the database systems of United States Department of Defense (DoD) across various divisions. Data stored in various noncombat information systems is not standardized, resulting in the difficulty of exchanging information between divisions. They design a reverse engineering process for identifying functional requirements, domain information, and business rules for database and data structures. They apply their framework shown in Aiken et al. (1993) to reverse engineer legacy information systems in DoD's heterogeneous environment.

Müller (1993) proposes a reverse engineering process to identify subsystem structure. He shows that the top-down decomposition of a subject system can be constructed via bottom-up subsystem composition. Premerlani and Blaha (1993) propose a reverse engineering process by taking a past implementation to extract the essential problem domain. They suggest to use object-oriented models to facilitate the reengineering process.

Although several researchers have addressed reverse engineering processes, the general framework is still not well-defined. Most subject systems are hardware-oriented. The reverse engineering methodology or processes for the systems with business or operational processes need further exploration.

12.2.4 Multiformalism Modeling

The research on multiformalism modeling is based on the observation that computational models, created using different modeling techniques, usually serve different purposes and provide unique insights. While each modeling technique might be capable of answering specific questions, complex problems require multiple models interoperating to complement/supplement each other; we call this multiformalism modeling or multimodeling. This multimodeling approach for solving complex problems is full of syntactic and semantic challenges. Mansoor et al. (2009, 2010), Levis et al. (2010, 2012), and Jbara et al. (2013a–c) discuss issues in creating workflows of model interoperations involving social networks, timed influence nets, organization structures, and geospatial models in a variety of problem domains. Section 4 provides details on some of these efforts. Their approach builds on earlier work by Kappel et al. (2006) and Saeki and Kaiya (2006). AbuJbara (2013) develops a domain-specific model workflow language and a domain ontology and applied these to develop problem-solving workflows combining several different modeling formalisms.

There is a recent advent of several academic and commercial modeling and simulation (M&S) platforms that support multimodeling. The Command and Control Wind Tunnel (C2WT) (Hemingway et al., 2012) developed by Vanderbilt University and the Service Oriented Architecture for Socio-Cultural Systems (SORASCS) (Garlan et al., 2009) developed by Carnegie Mellon University are examples of multimodeling capable platforms developed in academia. While the first provides a federated approach, utilizing the High-Level Architecture

(HLA) standard and the meta-programmable Generic Modeling Environment (GME) (Davis, 2003); the second employs service-oriented architecture techniques in providing model interoperation capabilities. The list of commercially available multimodeling platform is growing as more and more M&S software offer extensions and provisions for interoperating multiple models within a single execution environment. Noted among them is AnyLogic's multimethod simulation modeling tool. It supports agent-based, discrete event, and system dynamics simulation methodologies within a single M&S environment. Another notable application is called ModelCenter, developed by Phoenix Integration, Inc. ModelCenter is a software package that allows integration of multiple analysis models in support of design and optimization of systems. A recent effort by Laskey et al. (2018) presents a multimodeling approach to inference enterprise modeling (MIEM). MIEM combines multiple models to generate multiple predictions of inference enterprise performance. These predictions are combined into an overall estimate of performance with error bounds. The approach is implemented as a Semantic Testbed for Inference Enterprise Modeling (STIEM). The STIEM approach is useful to a broad set of areas that involve multiple analysis models – examples include defense/national security, homeland security, intelligence operations, etc.

12.3 The Proposed Approach

In this section, we describe the developed approach for reverse engineering an enterprise's underlying process model in the form of an executable workflow. It should be noted that the approach is highly domain and problem specific. It requires a substantial amount of domain and problem-solving knowledge in a narrow area to be encoded before a meaningful application can be carried out. The framework supporting the approach can, however, be employed to collect this required domain knowledge during problem-solving episodes to build the required knowledge base. This framework consists of four key components: domain knowledge, situational knowledge, process description module, and an implementation platform. The underlying methodology can be viewed as a two-phase approach. Phase 1 is where the domain data and relevant processes are identified and encoded in the form of data and process ontologies, respectively. This phase also includes the workflow language definition for process description. For a specific domain, this phase might take multiple iterations until an enriched domain knowledge base is developed and a multimodeling workflow language that addresses a domain of interest and is capable of capturing model interoperations is reached. Phase 2 takes place when the workflow language is used to create or reverse engineer process models for specific scenarios/problems. Once a new process model is generated by the approach or domain experts,

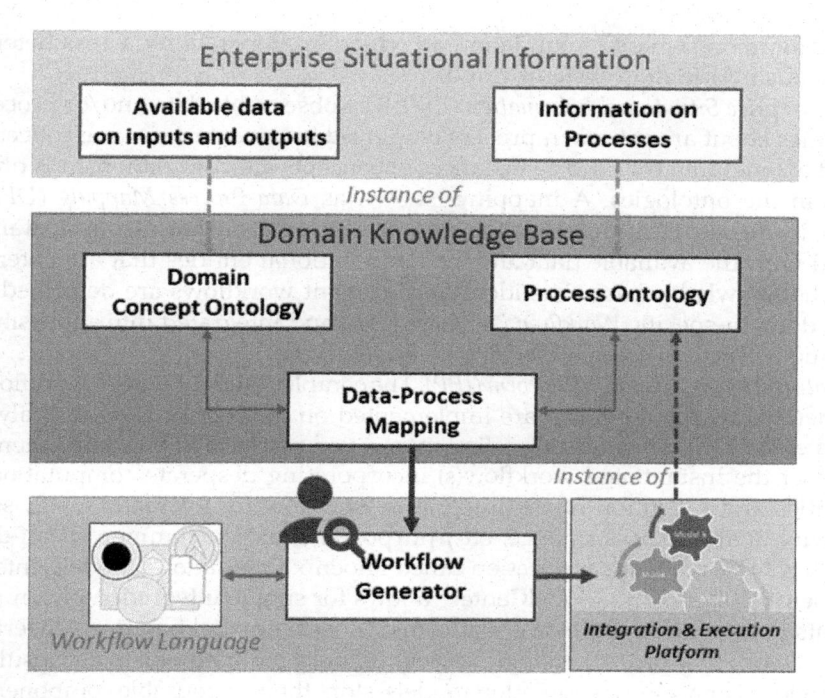

FIGURE 12.1
Framework of the approach.

it is generalized and made part of the domain knowledge base to be used for future problem-solving episodes. Figure 12.1 provides an overview of the framework supporting our proposed approach.

Domain Knowledge Base (DKB): The domain-specific data, concepts, their structure, relationships, and employed algorithms are means to understanding and reverse engineering an enterprise's process model. A *Domain Concept Ontology (DCO)* with domain- and problem-specific concepts, their properties, and mutual relationships allows reasoning about data available or generated by an enterprise. A *Process Ontology (PO)* with relevant algorithms and computational entities allows reasoning about the way these entities can be put together to solve a problem that the individual entities cannot solve on their own.

Gruber defines ontology as "an explicit specification of conceptualization" (1993). Ontologies can transform what is tacit knowledge into explicit knowledge. Gruber also suggests more pragmatic definition and role of ontology in the context of information and computer science (2008). According to the definition, ontology is a set of primitive constructs to represent knowledge of a target domain. This community has widely accepted a consensus on what the primitives are; they are classes, their attributes, relationships between them, and constraints on them. In that sense, ontology plays a role as a common semantic foundation on which interoperability across heterogeneous information systems stands.

Enterprise Situational Information (ESI): The observable data and/or process entities about an unknown process employed by an enterprise are collected and entered into the *DKB* as instances of concepts and/or relationships present in the ontologies. A mapping, shown as *Data-Process Mapping (DPM)* box in Figure 12.1, identifies the possible workflows that can be instantiated with the available data and/or computational entities that are entered in the knowledge base. The identified relevant workflows are described in the domain-specific *Workflow Language (WL)* and integrated into a (possibly) unified structure using a *Workflow Generator (WG)*.

Integration & Execution Platform (IEP): The complete and/or partial workflows generated by the approach are implemented on the IEP for further analysis and evaluation. The multimodeling integration platform allows implementation of the instantiated workflow(s) incorporating desperate computational entities and execution of the integrated workflow for evaluation (e.g., sensitivity analysis, performance, etc.) purposes. It uses a commercial off-the-shelf (COTS) software application called Phoenix Integration's ModelCenter®. Phoenix Integration's ModelCenter® allows for simple integration of components using various software platforms. It uses a model-based engineering framework that provides a wide variety of tools and methods to encapsulate individual analysis or simulation models, store them as reusable components, and create simulation workflows with different data sources. The individual simulation and/or analysis components of a workflow can be developed using any software application, programming language, spreadsheet, analytical model, and databases. ModelCenter's simulation workflows can be easily developed using an editor that allows linking of reusable components (stored in a library) in a "building-block" approach.

The following are two possible scenarios for the employment of the presented approach for reverse engineering problems.

Scenario 1 – From Data to Processes: In this case, the problem and the data used by an enterprise are known; however, the process employed is unknown. The addition of known data (both inputs and output) items in the DCO allows inferences on missing data items and relationships among data entities. The available and inferred information is used by DPM to identify processing templates from PO. The templates are combined into a problem-solving workflow by WG and implemented on IEP for performance studies.

Scenario 2 – From Processes to Data: In this case, the algorithms and computational entities employed by an enterprise to solve a problem are known; however, the data used is not known. The addition of algorithms and instances of computational entities into PO allows inferences on data requirements and any missing processes. If the required data is made available, WG constructs the workflow using the defined multimodeling interoperations. On the other hand, if the data is unavailable, the approach looks for redacted data. The DCO here is added with the available summary statistics to identify computational algorithms that can be used to synthesize the required data for the solution workflow. This additional step will add data synthesis processes to the workflow and implement the solution.

The two solution scenarios, described above, can be combined together in a *Hybrid Scenario* where both data and process information are partially known. We provide an application of the approach on such a scenario in Section 12.4.1. There may also be multiple alternative solution workflows generated by DPM and WG. For such cases, the approach utilizes a human analyst or a fitting criterion to identify the workflow for implementation. The fitting criterion assumes availability of performance data on algorithms and the performance of the process being reverse engineered.

In the following section, we describe two different applications of the presented approach. In the first, a reverse engineering application is illustrated with the help of an Inference Enterprise Modeling (IEM) problem (Laskey et al., 2018). The second application illustrates the approach for designing a new process for evaluating performance of a cyber-physical system.

12.4 Applications

In this section, a cyber-enterprise is used to illustrate the application of the proposed approach for a new design problem. A cyber-domain requires prevention of threats before the actual attack takes place and compromises physical and/or software systems. The proposed approach is relevant for this problem since getting real attack data is usually impractical during the design and operational stages. The proposed approach improves the protection of the systems and services from damage, disruption, and misdirection even if the actual hacking data may not be available.

Lee et al. (2019) describe an Inference Enterprise (IE) as an organizational entity that uses processes which involve data, tools, and people to make mission-focused inferences. In our work on IEM to support reverse engineering of an IE, we primarily focus on the Insider Threat Detection (ITD) problem. ITD systems seek to detect and identify a small proportion of the population that are potential insider threats. Laskey et al. (2018) propose an IEM methodology to understand and improve the operational quality of such enterprises that forecasts and evaluates the performance of the modeled enterprise. Figure 12.2 presents an overview of this IEM

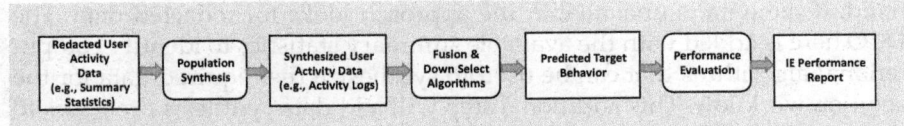

FIGURE 12.2
IEM methodology.

methodology. As part of this effort, the researchers built several IE models and acquired a wealth of knowledge in the domain of modeling ITD systems. Models of an IE are used to understand the operations of an enterprise and to help develop predictions of the effects of introducing changes to the enterprise. An important application of IEM is identifying ways to improve the performance of an IE.

In this application of the proposed approach (in Section 12.3), the researchers developed a comprehensive DCO and PO. These ontologies capture the ITD domain concepts and the IEs employed by organizations to address the ITD problem. The details of these ontologies can be found in Lee et al. (2019). In this application, PO is a collection of predefined workflow templates that are generalizations of past IE workflow models or fragments of models. These templates are represented using Business Process Modeling Notation (BPMN). A comprehensive DCO captures ITD domain concepts, their mutual relationships, and relationships to algorithms, used in ITD systems, in Web Ontology Language (OWL). DPM, in this application, encodes the mapping rules between the concepts in DCO and the workflow templates stored in PO. In this approach, DCO, PO, and DPM are independent assets developed in the domain-modeling phase, i.e., IEM Domain KB. Lee et al. (2019) describe a modular development of this knowledge base and suggest additional components. Figure 12.3 shows all the components of this proposed IEM knowledge base: IEM Problem Specification Ontology describes the properties that characterize an IEM; the IEM Module Ontology represents the manual tasks and software assets used in the solution process to solve previous problems. The IEM Module Ontology also contains information about the data types of the input and output parameters and the physical location (URL or path) of the software assets. The Mapping Box, an extension of DPM, includes the Feature Model Ontology, which defines the mandatory and optional IEM features, as well as any mutually exclusive relationships among IEM features. It also includes the Mapping Rules Ontology that relates problem specifications to IEM features. The Process Template is represented in the Template Ontology that reuses the class constructs from the BPMN Ontology such as Start, End Events, Service Tasks, Gateways, and Sequence Flows (Lee et al., 2019).

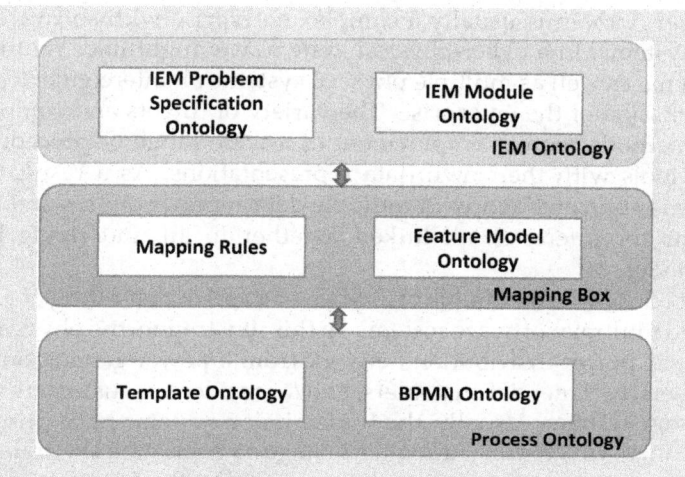

FIGURE 12.3
IEM domain knowledge base components.

For a new or a reverse-engineered instance of an ITD system, the methodology requires an input description of an IE or IEM. A Java application parses the description into classes, objects, and properties and maps them as instances of DCO. This step can also be carried out by a human operator who can directly add the instances to DCO. Standard Description Logic reasoning is used by the Mapping Box to collect relevant templates in PO. The WG ensures the subgraphs retrieved in this step comply with the syntax and some basic consistency rules of BPMN. The output of the WG is then implemented on STIEM by integrating all the required software assets. The execution and analysis capabilities of STIEM allow validation and effectiveness analysis of the proposed workflow for the ITD system under investigation. Figure 12.4 shows an example output of WG of this approach in the form of a proposed solution workflow of an ITD system.

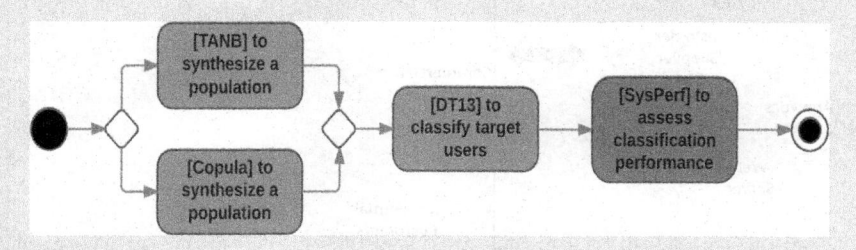

FIGURE 12.4
Example of a solution workflow (Lee et al., 2019).

The cyber-domain is usually a complex enterprise with software and/or physical systems. In a cyber-physical system, we might have multiple software systems as well as multiple physical systems all interconnected to support the mission of the enterprise. The variety of threats and complexity of solution methods require employment of a wide range of procedures and technical tools with their own data representations. As a result, a cyber-enterprise is characterized with multiple data representations and multiple formalisms for inference, all linked together in an underlying business process model.

Figure 12.5 shows an example of a cyber-enterprise, consisting of a physical system and multiple software systems. In this illustration, the physical system is an electric power distribution network from a power generation plant to the end users, commercial customers, and/or residential customers via some transmission systems. Usually, the transmission systems have primary distribution lines, which carry a higher voltage to minimize the transmission losses, and secondary distribution lines supplying the power to customers. The software systems, on the other hand, are deployed at different locations. These software systems, including control program, enterprise resource planning (ERP), database, analysis server, and web servers, are connected using internet or proprietary local (e.g., Ethernet networks) or wide-area networks. All these physical and software systems are interconnected, and the data flow can be either bi-directional or one-way depending upon the procedures of the underlying process model of the system.

Evaluating the performance of a cyber-enterprise requires employment of a wide range of procedures and technical tools. The diagram in Figure 12.6 illustrates an example of an evaluation workflow. The workflow starts from

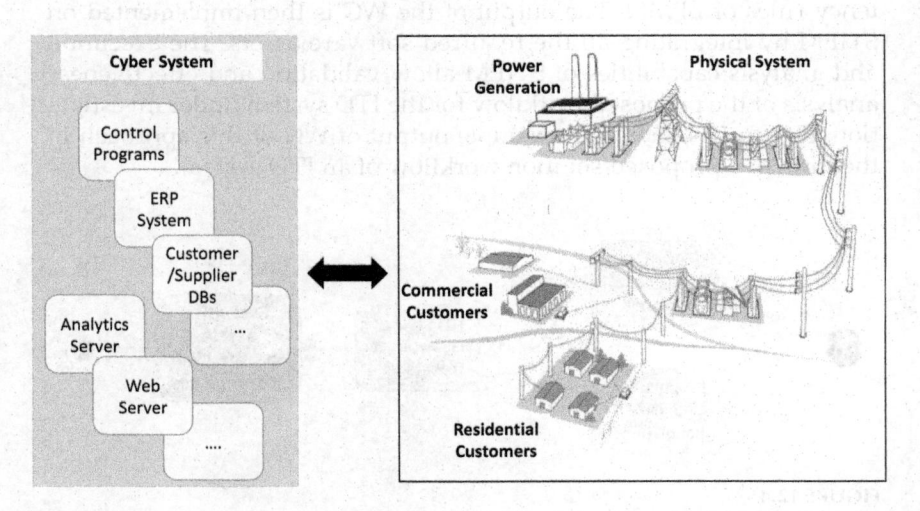

FIGURE 12.5
A cyber-physical enterprise.

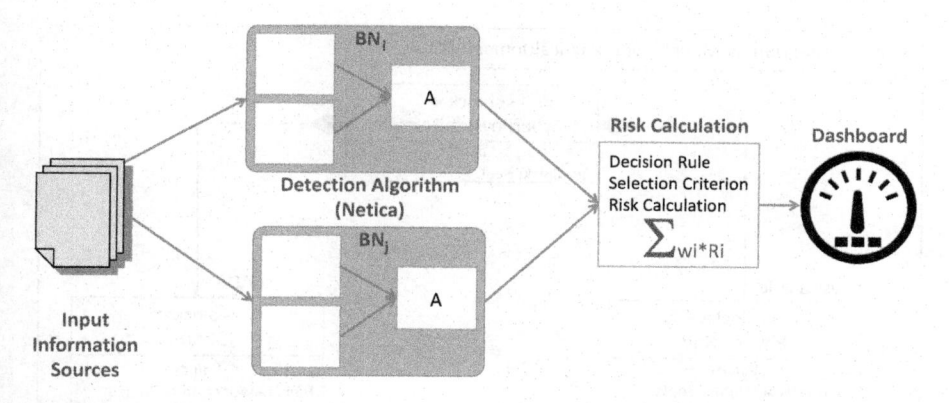

FIGURE 12.6
An example evaluation workflow.

collecting input information sources created by different physical or software systems. Partial data are known, e.g., the structure or layout of the electronic distribution network, and the other unknown data can be estimated by domain experts' subjective judgments or generated by simulations. The unknown data is used to construct detection algorithms using Bayesian networks in order to estimate the performance of the cyber-enterprise given some data is unknown. We create multiple Bayesian network models. One Bayesian model represents the physical distribution network, and the other represents the network of all software systems. In the last step, we compute the total risk based on multiple Bayesian models and display the risk level on a dashboard.

For the development of DKB, we create domain-dependent ontologies for a cyber-domain using constructs from SysML. Several papers discuss the use of UML class diagrams or object diagrams to capture an ontology (Cranefield and Purvis, 1999; Guizzardi et al., 2004). The main advantages cited for using UML as the ontology language are as follows: (i) the large UML user community, and (ii) the standard graphical representations of the UML language provide clearer communication to users than alternative textual representations such as eXtensible Markup Language (XML) files or Rich Text Format (RTF) files, which are usually hard to understand. SysML has the same advantages as UML and also provides modeling capabilities that are intended specifically to support the type of system modeling required in the cyber-domain.

Figure 12.7 shows an example of the domain ontologies captured in a SysML Block Definition Diagram (BDD) representation. We define the cyber-physical domain and its relationships to the physical grid and Supervisory Control and Data Acquisition (SCADA) system. Each block can have its own value properties. Note that the purpose of the example is to illustrate a cyber-enterprise in a SysML BDD. The same approach can be applied to a larger cyber-enterprise or other domains.

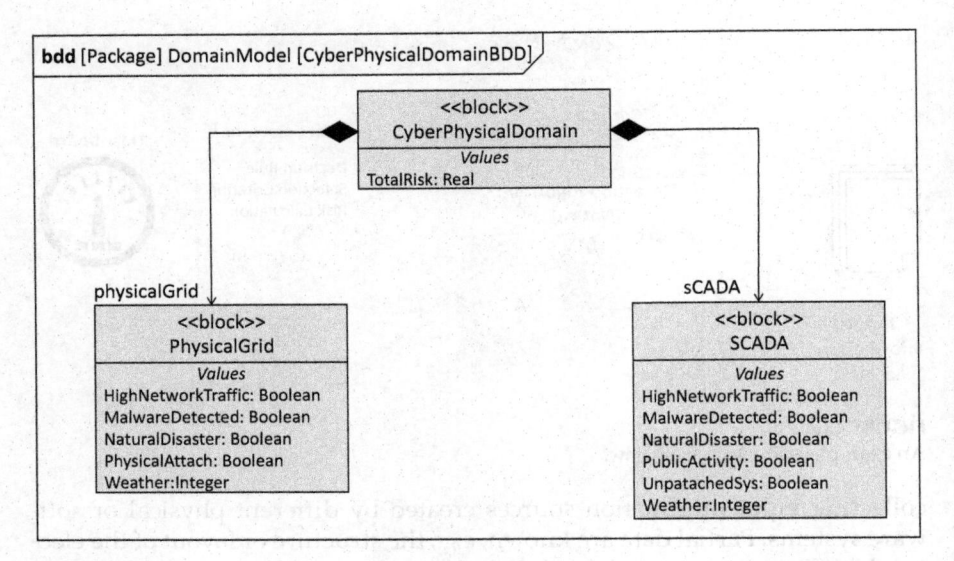

FIGURE 12.7
An example domain ontology.

In this application, the PO contains analysis models and their input and output parameters. Some examples of the analysis models that are contained in the PO are statistical methods, information theoretic methods, Bayesian networks, discrete event models, game theoretic methods, principal component analysis, Markov processes and hidden Markov models, data mining methods (classification/clustering-based), decision or event trees, and rule-based (production/logic-based) methods. These analysis models can be the parts of a more complex evaluation workflow enabling integration of multiple formalisms for the evaluation process or a single model implementing the entire evaluation process. Figure 12.8 shows an example of Bayesian model used in this application.

The rules in the DPM map the data flows between these ontologies, e.g., model domain attribute to the input/output to an analysis model. Figure 12.9 shows an example of the mappings between the domain ontologies and analysis ontologies. The mappings are represented using a SysML parametric diagram. The top two constraint properties represent the domain ontologies. Three constraint properties shown below the domain ontologies are the analysis ontologies. The diagram shows that parameters in a domain ontology will be the inputs to the analysis ontologies. For example, the parameter "HighNetworkTraffic" in a physical network grid is an input to the Bayesian physical network grid model. This flow is shown as a connector drawn between the parameters shown as a small box at the edge of the analysis model.

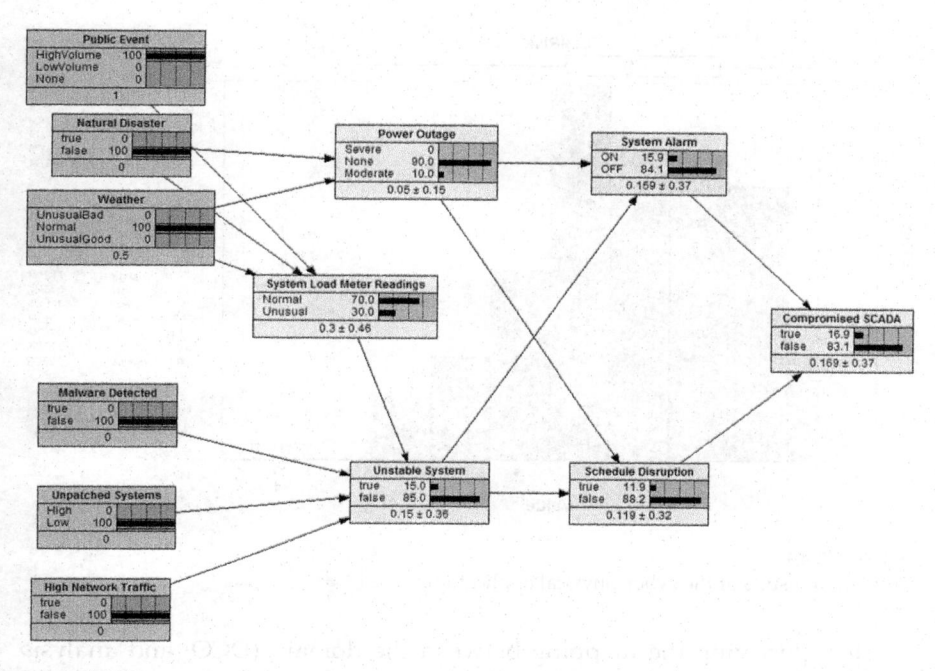

FIGURE 12.8
An example Bayesian model.

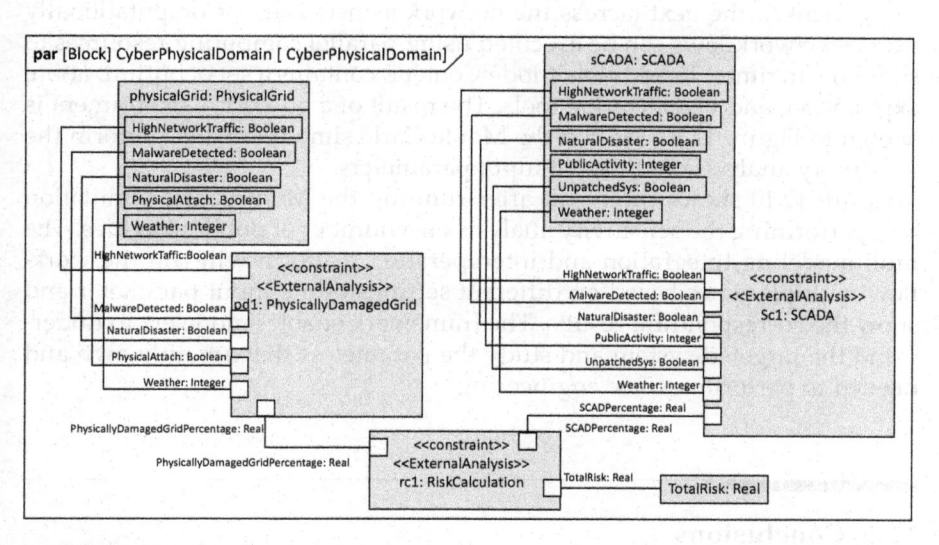

FIGURE 12.9
An example mapping between DCO and PO.

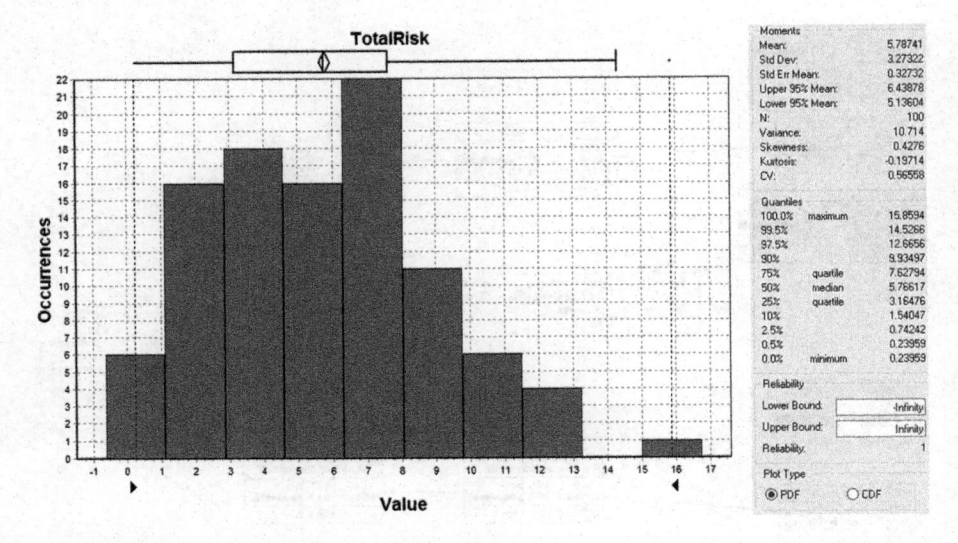

FIGURE 12.10
Simulation results of the cyber-physical application.

After specifying the mapping between the domain (DCO) and analysis (PO) ontologies, the WG generates the workflow that can execute the multi-modeling workflow. Once a workflow has been created, ModelCenter® can automatically execute the simulation workflow as many times as needed. As the workflow is executed, data is automatically transferred from one component to the next (across the network as necessary). Computationally expensive workflows can be executed using parallel computing resources to shorten run times. In addition, ModelCenter® contains a set of optimization, exploration, and visualization tools. The result of a numerical experiment is shown in Figure 12.10. We run the Monte Carlo simulation and perform the sensitivity analysis on different input parameters.

Figure 12.10 shows the result after running the Monte Carlo simulation and performing the sensitivity analysis on a number of public activities. The multimodeling integration and interoperation platform will run the workflow multiple times based on different settings of the input parameter and show the corresponding results. The framework enables modelers to understand the targeted system and study the parameters that are unknown and needed to perform reverse engineering.

12.5 Conclusions

Performing reverse engineering requires high computational resources. An efficient and intelligent reverse engineering process can directly improve the results of reverse engineering using the same computational resources. In this chapter, we propose an approach for reverse engineering an enterprise's underlying process model in the form of an executable workflow. We applied the proposed framework to two applications including IEM and cyber-physical systems evaluation. The proposed framework enables modelers to study the unknown parameters and also improve the understanding of the target systems.

References

P. Aiken, A. Muntz, and R. Richards, "A framework for reverse engineering DoD legacy information systems," in *Reverse Engineering, 1993., Proceedings of Working Conference on*, 1993, pp. 180–191, Baltimore, MD.

P. Aiken, A. Muntz, and R. Richards, "DOD legacy systems: Reverse engineering data requirements," *Communications of the ACM*, vol. 37, pp. 26–42, 1994.

AnyLogic, www.anylogic.com/.

R. Arnold, "Tutorial on software reengineering," in *Conference on Software Maintenance*, 1990, pp. 26–29, San Diego, CA.

J. Bongard and H. Lipson, "Automated reverse engineering of nonlinear dynamical systems," *Proceedings of the National Academy of Sciences*, vol. 104, pp. 9943–9948, 2007.

G. CanforaHarman and M. D. Penta, "New frontiers of reverse engineering," in *2007 Future of Software Engineering*, 2007, pp. 326–341, Minneapolis, MN.

E. J. Chikofsky and J. H. Cross, "Reverse engineering and design recovery: A taxonomy," *IEEE Software*, vol. 7, pp. 13–17, 1990.

S. Cranefield and M. K. Purvis, *UML as an Ontology Modelling Language*, Department of Information Science, University of Otago, Dunedin, 1999.

J. Davis, "GME: the generic modeling environment," in *Companion of the 18th Annual ACM SIGPLAN Conference on Object-Oriented Programming, Systems, Languages, and Applications*, 2003, pp. 82–83, Anaheim, CA.

D. Garlan, K. M. Carley, B. R. Schmerl, M. W. Bigrigg, and O. Celiku, "Using service-oriented architectures for socio-cultural analysis," in *SEKE*, 2009, pp. 370–375, Boston, MA.

T. R. Gruber, "A translation approach to portable ontology specifications," *Knowledge Acquisition*, vol. 5, pp. 199–220, 1993.

T. R. Gruber, Ontology (Computer Science). In: Liu, L. and Özsu, T. (eds.). *Encyclopedia of Database Systems*, http://tomgruber.org/writing/ontology-definition-2007.htm.

G. Guizzardi, G. Wagner, and H. Herre, "On the foundations of UML as an ontology representation language," in *International Conference on Knowledge Engineering and Knowledge Management*, 2004, pp. 47–62, Whittlebury Hall, UK.

G. Hemingway, H. Neema, H. Nine, J. Sztipanovits, and G. Karsai, "Rapid synthesis of high-level architecture-based heterogeneous simulation: a model-based integration approach," *Simulation*, vol. 88, pp. 217–232, 2012.

E. Huang, A. K. Zaidi, and K. B. Laskey, "Inference enterprise multimodeling for insider threat detection systems," in *Disciplinary Convergence in Systems Engineering Research*, Springer, Los Angeles, CA 2018, pp. 175–186.

A. A. Jbara and A. H. Levis, "On defining a domain specific multi-modeling workflow language to address complex multi-modeling activities," in *Systems Conference (SysCon), 2013 IEEE International*, 2013a, pp. 209–214, Orlando, FL.

A. A. Jbara and A. H. Levis, "Semantically correct representation of multi-model interoperations," in *FLAIRS Conference*, 2013b, St. Pete Beach, FL.

A. A. Jbara, A. H. Levis, and A. K. Zaidi, "On using multiple interoperating models to address complex problems," *Procedia Computer Science*, vol. 16, pp. 363–372, 2013c.

G. Kappel, E. Kapsammer, H. Kargl, G. Kramler, T. Reiter, W. Retschitzegger, et al., "Lifting metamodels to ontologies: A step to the semantic integration of modeling languages," in *International Conference on Model Driven Engineering Languages and Systems*, 2006, pp. 528–542, Genova, Italy.

R.K. Keller, R. Schauer, S. Robitaille, and P. Pagé, "Pattern-based reverse-engineering of design components," in *Proceedings of the 21st International Conference on Software Engineering*, 1999, pp. 226–235, Los Angeles, CA.

R. L. Krikhaar, "Reverse architecting approach for complex systems," in *Software Maintenance, 1997. Proceedings, International Conference on*, 1997, pp. 4–11, Bari, Italy.

K. Krogmann, M. Kuperberg, and R. Reussner, "Using genetic search for reverse engineering of parametric behavior models for performance prediction," *IEEE Transactions on Software Engineering*, vol. 36, pp. 865–877, 2010.

K. B. Laskey, A. Zaidi, D. Buede, M. Imran, E. Huang, D. Brown, et al., "Modeling inference enterprises using multiple interoperating models," in *INCOSE International Symposium*, 2018, pp. 1764–1777, Washington, DC.

J. Lee, Matsumoto, S., Zaidi, A. K., and Laskey, K., "Towards automating design and development of inference enterprise models," Submitted to *13th Annual IEEE International Systems Conference*, Orlando, FL, 2019.

A. H. Levis and A. A. Jbara, "Multi-modeling, meta-modeling, and workflow languages," in *Theory and Application of Multi-Formalism Modeling*, IGI Global, 2014, pp. 56–80, Portland, OR.

A. H. Levis, L. W. Wagenhals, and A. K. Zaidi, "Multi-modeling of adversary behaviors," in *Intelligence and Security Informatics (ISI), 2010 IEEE International Conference on*, 2010, pp. 185–189, Vancouver, BC, Canada.

A. H. Levis, A. K. Zaidi, and M. F. Rafi, "Multi-modeling and meta-modeling of human organizations," in *Advances in Design for Cross-Cultural Activities*, CRC Press, Boca Raton, FL, 2012, p. 148.

S. Liang, S. Fuhrman, and R. Somogyi, "Reveal, a general reverse engineering algorithm for inference of genetic network architectures," 1998.

F. Mansoor, A. K. Zaidi, L. Wagenhals, and A. H. Levis, "Meta-modeling the cultural behavior using timed influence nets," in *Social Computing and Behavioral Modeling*, Springer, 2009, Boston, MA, pp. 1–9.

F. Mansoor, A. K. Zaidi, and A. H. Levis, "Meta-model driven construction of timed influence nets," in *Intelligence and Security Informatics (ISI), 2010 IEEE International Conference on*, 2010, pp. 212–217, Vancouver, BC.

L. McGinnis, E. Huang, K. S. Kwon, and V. Ustun, "Ontologies and simulation: a practical approach," *Journal of Simulation*, vol. 5, pp. 190–201, 2011.

H. A. Müller, M. A. Orgun, S. R. Tilley, and J. S. Uhl, "A reverse-engineering approach to subsystem structure identification," *Journal of Software Maintenance: Research and Practice*, vol. 5, pp. 181–204, 1993.

H. A. Müller, J. H. Jahnke, D. B. Smith, M.-A. Storey, S. R. Tilley, and K. Wong, "Reverse engineering: A roadmap," in *Proceedings of the Conference on the Future of Software Engineering*, 2000, pp. 47–60, Limerick, Ireland.

K. N. Otto and K. L. Wood, "Product evolution: a reverse engineering and redesign methodology," *Research in Engineering Design*, vol. 10, pp. 226–243, 1998.

Phoenix Integration, ModelCenter, www.phoenixint.com/modelcenter/integrate. php.

W. J. Premerlani and M. R. Blaha, "An approach for reverse engineering of relational databases," in *Reverse Engineering, 1993., Proceedings of Working Conference on*, 1993, pp. 151–160, Baltimore, MD.

M. G. Rekoff, "On reverse engineering," *IEEE Transactions on Systems, Man, and Cybernetics*, vol. SMC-15, pp. 244–252, 1985.

S. Rugaber and K. Stirewalt, "Model-driven reverse engineering," *IEEE Software*, vol. 21, pp. 45–53, 2004.

M. Saeki and H. Kaiya, "On relationships among models, meta models and ontologies," in *Proceedings of the Proceedings of the 6th OOPSLA Workshop on Domain-Specific Modeling (DSM 2006)*, 2006.

D. C. Schmidt, "Model-driven engineering," *Computer-IEEE Computer Society*, vol. 39, p. 25, 2006.

N. Shi and R. A. Olsson, "Reverse engineering of design patterns from java source code," in *Automated Software Engineering, 2006. ASE'06. 21st IEEE/ACM International Conference on*, 2006, pp. 123–134, Tokyo, Japan.

Part 7

Next-Generation Communication

Industrial systems are being designed to communicate over continents for the sake of product development, for exchange of production data, for customization of products by clients spread worldwide, and for B–C to B–B client service. Next-generation communication networks are going to play an important role in data communication for aforesaid purposes. This part addresses this issue in the subsequent chapter.

13

Next-Generation National Communication Infrastructure (NCI): Emerging Future Technologies—Challenges and Opportunities

Javed I. Khan
Kent State University

CONTENTS

13.1 Introduction 278
    13.1.1 NCI Ecosystem 279
        13.1.1.1 Physical Assets of NCI 279
13.2 New Generation of Mobile Communication 280
    13.2.1 4G to 5G 280
    13.2.2 Technology Components for 5G 284
    13.2.3 5G Architecture 285
    13.2.4 Comments on the Vision 285
    13.2.5 Network Virtualization 287
    13.2.6 5G Players 288
13.3 A Brief Review of Emerging Future Network Technologies 289
13.4 Smart Communities 289
    13.4.1 What it is 289
    13.4.2 Technology and Implied Requirements from NCI and
           NCI Ecosystem 291
    13.4.3 NCI Issues 292
13.5 Internet of Things (IoT) 293
    13.5.1 What is IoT? 293
13.6 Transformation of Industry 295
13.7 Issues, Challenges, and Opportunities for NCI 296
    13.7.1 Geographic Coverage 297
    13.7.2 Expansive NCI Infrastructure 298
    13.7.3 Quality and Consistency of Network 299
    13.7.4 The Explosion of Cloud Services 300
    13.7.5 Spectrum Roadmap 301
    13.7.6 Information Protection, Privacy, and Ethical
           Applications Framework 302

13.7.7 Modernization of Assessment Mechanism.................................304
13.7.8 The Ecosystem for Future Network Innovation........................304
13.8 Conclusions...305
13.9 Acknowledgments...305
References...306

13.1 Introduction

Digital transformation is upon us. Industry 4.0/5.0, 4G LTE/5G, Internet of Things (IoT), artificial intelligence (AI), smart city, and the recent blockchain are initiatives that originated from divergent techno-evolution threads and were advanced by seemingly different communities; yet all are phenomena of this same underlying digital transformation that is unfolding. All the threads of this transformation, irrespective of their origin of drive, however, will require one thing at the core—a capable and robust national cyber/communication infrastructure (NCI).

Are NCIs ready for it? IoT is expected to suddenly bring an estimated 75 billion devices in the next 5 years to the internet. Every major public infrastructure and every private asset (not only every home but also every appliance in them!) will have their pulse on the internet. Smart communities will emerge as a new set of citizen-centric digital services aimed to improve every aspect of community living. Community governments around the world will be an active and forceful political driver in digital ecosystem. AI will see a proliferation of networked applications that will have autonomous decision power on sensitive aspects of our life from personal safety to social justice and equity. Industry 5.0 will bring data-intensive personalization. The blockchain is poised to impact every conceivable business process disrupting every service sector including banking and finance, manufacturing supply chain, and medical and voting management. Cloud infrastructure has already bifurcated world Internet traffic reshaping global IT systems. Today more private traffic flows between data and computing centers than on public Internet.

In this backdrop, the NCI in the majority of developing countries is deeply reflective of the past than being ready for the future. NCIs have evolved mostly around past 20th-century telecommunication industries network infrastructure. Since the 1980s with the emergence of mobile phone, radio infrastructure has been added. Further since the 1990s with the advent of data communication, optical fiber infrastructure of Internet service providers has also been added. The use of radio spectrum is dominated by TV and voice telephony. It is quite apparent that the entire radio and fixed NCI infrastructure and underlying technology have to be gradually redesigned for a much-varied set of data communication applications that have arrived riding the wave of Internet and web. A silent organic transformation of the NCIs are already underway as data centers and cloud computing are increasingly becoming commonplace.

The management of old world NCI is guided by an old regimen of voice spectrum licensing and weak or non-existent data usage or citizen's basic data protection policies. The telecommunication regulatory ecosystem has marginal participation of the players of the new economy if any. To take full advantage of Industry 5.0 or as a digital nation, a country requires a significant drive to update its NCI architecture and bold transformations in the management of the new ecosystem. The impending technological tsunami creates very interesting challenges for any NCIs. This chapter takes a closer look into a set of massively disruptive technologies that are on the horizon including 5G transition. It provides a succinct overview of each with a goal to critically analyze their communication needs and potential implications for a future NCI. It identifies the gaps in current NCIs. This chapter does not suggest any specific design or architecture.

13.1.1 NCI Ecosystem

13.1.1.1 Physical Assets of NCI

So what is an NCI? The physical elements of an NCI have many physical elements including (i) the optical fiber core networks (which are often rings) built by many organizations; (ii) the Point of presence (POP)which allow subscribers to connect to the core network; (iii) land distribution networks, normally spread regionally (often copper or fiber) which connect individual homes and offices from localities to the core; (iv) the cellular base stations—another type of distribution network which collects wireless signals, (v) the spectrum which allows cell phones and end devices to communicate with the base station; (vi) communication satellites which provide another type of wide area distribution network via wireless; (vi) exchange points which bring multiple local networks together and exchange their traffic; and (vii) landing stations where normally a country's networks exit the national boundary and join international peering exchanges.

NCIs have already added newer elements such as (i) gateways, (ii) national data exchanges, and (iii) national DNS servers. There are additionally (i) ID/authentication servers, (ii) firewall and deep packet inspection facilities, (iii) data centers, (iv) supercomputing centers, (v) public cloud computing facilities, (vi) proxy servers, and (vii) content distribution/caching networks.

Next-generation NCI is a rapidly evolving system for the participation of many diverse players. Unlike 4G, where mobile communication and broadband companies played the key roles, the Future Network ecosystem will see a much-expanded ecosystem with major players including a much-wide variety of application and service providers, CKD (content-knowledge-and-data) organizations, as well as public entities as service and infrastructure providers. It will be a national cyber-infrastructure.

13.2 New Generation of Mobile Communication

13.2.1 4G to 5G

Bell's telephony started the journey of fixed telephony in 1878. After a hundred years, finally around 1980s the radio cellular infrastructure was added to the fixed lines starting the age of mobile voice calls. People can make a voice call without using a wire. It is characterized by the addition of a permanent cellular base station network infrastructure launched in the early 1980s. Since then many improvements have taken place. Major stages are characterized as mobile "generation". Each generation is normally characterized by a family of protocols and technologies which were proposed/envisioned and/or implemented to improve the state-of-the-art. Table 13.1 lists some of the influential protocols of the generations.

TABLE 13.1

Major Mobile Standards and Generations

Family	Technology	Year
1G	1G Analog advanced mobile phone system (AMPS)	1980
1G	D-AMPS	1990
2G	GSM circuit switched data (CSD)	1992
2G	GSM high speed circuit switched data (HSCSD)	1993
2G	AMPS cellular digital packet data (CDPD)	1996
2G	GSM general packet radio services (GPRSs)	1998
2G	GSM EDGE	1999
3G	UMTS (universal mobile telecommunications service)-W-CDMA (wideband code division multiple access)	2001
G	UMTS HSDPA (high-speed downlink packet access)	2002
3G	UMTS HSUPA (high-speed uplink packet access)	2005
3G	UMTS TDD (time-division duplexing)	2006
3G	CDMA2000 1xRTT (code-division multiple access)	2000
3G	CDMA2000 EV-DO (evolution-data optimized)	2002
3G	CDMA200 EV-DO REVB (EV-DO revision B)	2010
3G	GSM Edge-Evolution	2006
3G	UMTS HSPA+ (evolved high speed packet access)	2007
3G	Mobile WiMAX	2008
4G	LTE (mimo)	2020
4G	LTE (single)	2011
4G	LTE-Tmobile 16	2016
4G	LTE-New Zealand 16	2016
4G	LTE-Version 18	2018
4G	LTE Advanced (fixed)	2022
4G	LTE Advanced Pro	2022
4G	LTE TDD/LTE FDD (frequency division) (White Space)	2010

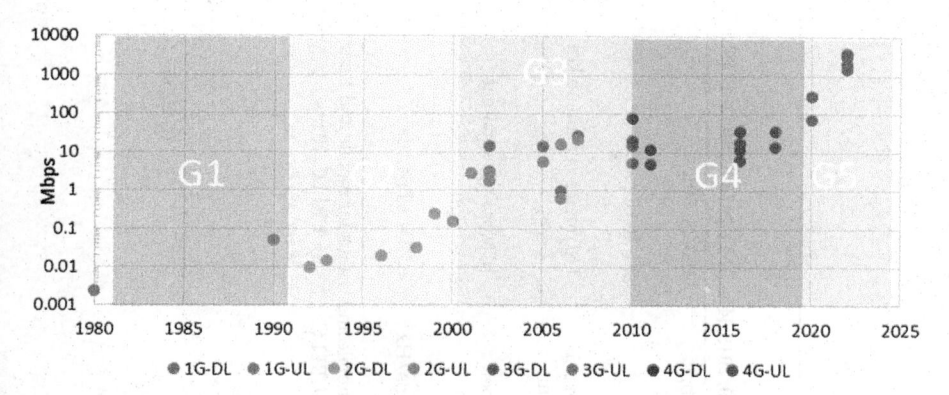

FIGURE 13.1
Bandwidth growth with mobile generations.

Until now the radio bandwidth is used as the key hallmark to distinguish the generations. Figure 13.1 plots the "bandwidth" growth over time for these protocols (each dot represents a technology) representing the mobile generations that emerged during these eras. At later times, the upload and download bandwidths started to be different. The plot shows both upload and download bandwidths. Approximately every generation lasted a decade and improved the bandwidth tenfold.

As of 2018, the 4G long-term evolution (LTE) has been undergoing deployment around the world. After 4G LTE, the next two upgrades, namely 4G LTE Advanced and 4G LTE Advanced Pro (which are still considered 4G technologies), are also on schedule to debut into real world out of the laboratories. In this backdrop, there are also plans underway for debuting 5G overlapped in time. These include 2012–2013 European Union initiative project METIS, International Telecommunication Union Radia Sector (ITU-R MT) for 2020 and beyond, next-generation mobile networks (NGMNs), and companies and national governments like Japan and South Korea.

Now let us look into the state of 5G as it is being shaped. It is a work in progress with the participation of many groups [1–6]. For ease, Table 13.2 provides key aspects of the visions provided by few. One of the earliest vision of 5G has been posted by project METIS [2], funded by EU, a consortium of 29 partners spanning vendors and operators. Its target is given in column 1. Column 3 provides the targets given by Next Generation Mobile Networks Alliance (NGMNA). It a forum of 24 mobile operators, mobile-related companies, including network and handset vendors, and research institutes. The 5G vision of ITU-2020 was published in 2015 which provided two incremental versions international mobile telecommunications (IMT)-Advanced and IMT-2020. In 2017, ITU also published the vision 5G radio interface.

How will 5G be more advanced than 4G? To have a comparative view, some features given by the 5G proponents are compared in Table 13.2. Here are some numbers for 4G LTE Advanced. LTE Advanced offers peak data

TABLE 13.2

Contemporary 5G Visions

Criteria	METIS-2013 [2]	GSMAI-5G, (GSM – Artificial Intelligence) 2014 [1]	NGMN-5G, 2015 [3]	IMT-Advanced, 2015 [4]	IMT-2020, 2015, [4]	IMT-2020 Radio 2017 [5]
Data rate general	100/20 Mbps	1–10 Gbps	Anything above 50 Mbps	10/5 Mbps	100/50 Mbps	100/50 Mbps
Data rate peak	1–10 Gbps	High in metro	1 Gbps	1 Gbps	20 Gbps	20/10 Gbps
Device density	900 Gb/h in stadium	Enormous	7.5 Tbps/square km	0.1 M device per square km	1 M device per square km	1 M device per square km
	300 K/access point		1.5 Tbps/stadium	0.1 Mbps per square m	10 Mbps per square m	10 Mbps per square m
Spectral efficiency				2.25–0.6 bps /Hz	6.75–1.6 bps/Hz	6.75–1.6 bps/Hz
Latency	Less than 5 ms	1 ms end-to-end	1–10 ms	4 ms for mobile and 1 ms for ultra-reliable low latency communication (URLLC)	4 ms for mobile and 1 ms for URLLC	4 ms for mobile and 1 ms for URLLC, 10 ms wake-up
Coverage		100% anytime, anywhere	50 mbps everywhere 12 km above air			

(*Continued*)

TABLE 13.2 (*Continued*)

Contemporary 5G Visions

Criteria	METIS-2013 [2]	GSMAI-5G, (GSM – Artificial Intelligence) 2014 [1]	NGMN-5G, 2015 [3]	IMT-Advanced, 2015 [4]	IMT-2020, 2015, [4]	IMT-2020 Radio 2017 [5]
Availability	99.999%	99.999%	Guaranteed availability for critical services	99.999%	99.999%	99.999%
Energy	1 decade battery	90% reduction from 4G	3 days for smartphone, 15 years for sensors	Support for high sleep rate and long sleep duration	Support for high sleep rate and long sleep duration	Support for high sleep rate and long sleep duration, 10-year battery for IoT
Mobility				Quality of service (QOS) to hold up to 120–350 km/h	QOS to hold up to 120–500 km/h	QOS to hold up to 120–500 km/h
Bandwidth aggregation					Up to 1 GHz	100 MHz

download/upload rate of 1/0.5 Gbps; transmission bandwidth is expected to be wider than 70 MHz in download (DL) and 40 MHz in upload (UL). The Cell edge user throughput will be two times higher than that in LTE; average user throughput is three times higher than that in 4G LTE. An ambitious goals is that it also envisions increase in spectral capacity to be three times higher than that in LTE through peak spectrum efficiency in DL about 30 bps/Hz and in UL: 15 bps/Hz. User plane latency will be less than 5 ms. In each way in the Radio Access Network (RAN), taking into account 30% retransmissions (FFS), Control Plane Latency from Idle (with IP address allocated) to fully connected will be less than 50 ms. The protocols must have support for scalable bandwidth and spectrum aggregation. 5G must remain backward compatible with 4G LTE and 3G partnership project (PP) legacy systems.

13.2.2 Technology Components for 5G

Below we present the novel technological aspect of the 5G, which is to make use of new spectrum about 30 GHz.

New Spectrum of Millimeter Waves: 5G enthusiasts envision that new spectrum needs to be opened up to support the huge data need by the increasing Internet users and devices. Lower frequencies have already been allocated to 2G/3G and 4G; the new swath has to come from the unused spectrum (total 252 GHz unused) in the range 30–300 GHz (whereas the wavelengths are in mm range), and this will be one of the key aspects of 5G. However, mm waves cannot easily travel through physical obstacles such as buildings and walls. They are almost always absorbed by trees, foliage, and rain. They are unlikely to travel long distances. So, they need to be used in relatively close distance communications using another new technology called **small cells** [1]. The short range is a major problem with mm waves. Thus almost obviously, it will require the deployment of the small radio cells. Thus, it is envisioned that miniature base stations be densely placed in service areas of today's macro base station (MBS). Small cells can be femtocells (10–20 m), picocells (200 m), and microcells (2 km) [7,8]. Network densification using small cells enhances the coverage and services of a base station. It also has the advantage that small cells are easier to mount on smaller towers or poles and building tops or walls. Antennas can potentially be embedded into smart windows or decorative tiles.

Massive MIMO (mMIMO): 4G LTE already uses a multiple antenna called multiple-input and multiple-output (MIMO) (IEEE 802.11n-2009) [7,8]. MIMO uses a maximum of four transmitters and four receivers. Since millimeter wave can work with much smaller antenna by packing potentially hundreds of antennas or massive MIMOs (mMIMO). By exploiting short-range and reflective properties of millimeter waves, same frequency can be reused by different antennas in different directions. The key issue of using a large number of densely packed antennas in mMIMO is interference.

This needs to be mitigated by carefully placing antennas, using volumetric spacing, and spatial modulation.

Beamforming: Another technology which can be associated with millimeter wave is beamforming [9,10]. It is a process of focusing the radio waves and guiding them to the specific receiver antenna intended. This helps more data to reach a targeted device instead of radiating out into the atmosphere. It improves Signal to Noise Ratio (SNR) by constructive and canceling interference. A perfect beam can be formed using multiple transmitting antennae. The 3D beamforming technology can electronically adjust the pattern of radiation beam in both azimuth/horizontal planes and elevation (tilting antenna) to provide more degrees of freedom in adjusting to various directional scenarios.

13.2.3 5G Architecture

To understand the impact of NCI, however, it is very important to understand the deployment architecture that will be required. This is explained in Figure 13.2. The 5G cellular architecture is heterogeneous, so it must include macrocells, microcells, small cells, and relays. Figure 13.2 shows a proposed 5G architecture in [10]. It explains how the above technology pieces, mMIMO network, and mobile and static small-cell networks fit into the 5G cellular network architecture to support various legacy and emerging applications and use cases such as Device to Device (D2D) communication, IoT, Internet of Vehicular (IoV) Networks, etc. It is a multitier architecture with static and mobile small cells and Cognitive Radio Network (CRN) based communications. The 5G architecture also aims to use cloud-based RANs (C-RANs). The idea of C-RAN was first provided by a China Mobile Research Institute. The basic idea behind any C-RAN is to execute most of the functions (control layer) of an MBS in the cloud.

13.2.4 Comments on the Vision

It will be difficult to foretell how 5G will actually play out. It seems that the characterization of 5G in terms of further advancement in peak data rate, average data rate, availability, and latency may not be enough to make it a "new" generation. The millimeter wave may not be able to contribute to the radio communication beyond a few 100 m. The average bandwidth target of 100 Mbps is very close to what even 4G LTE is already offering today (30–50 Mbps). About ten orders of magnitude increase have been normally expected in previous generational changes.

An interesting question is, "Will there really be a tenfold increase in user bandwidth demand in the next 10 years?" There are reasons to believe that user demand of bandwidth from mobile telephony may not increase as seen in previous generations and the new 5G might be right to set the limit to 100 mbps. However, this might be anyway within the reach of 4G. Since most

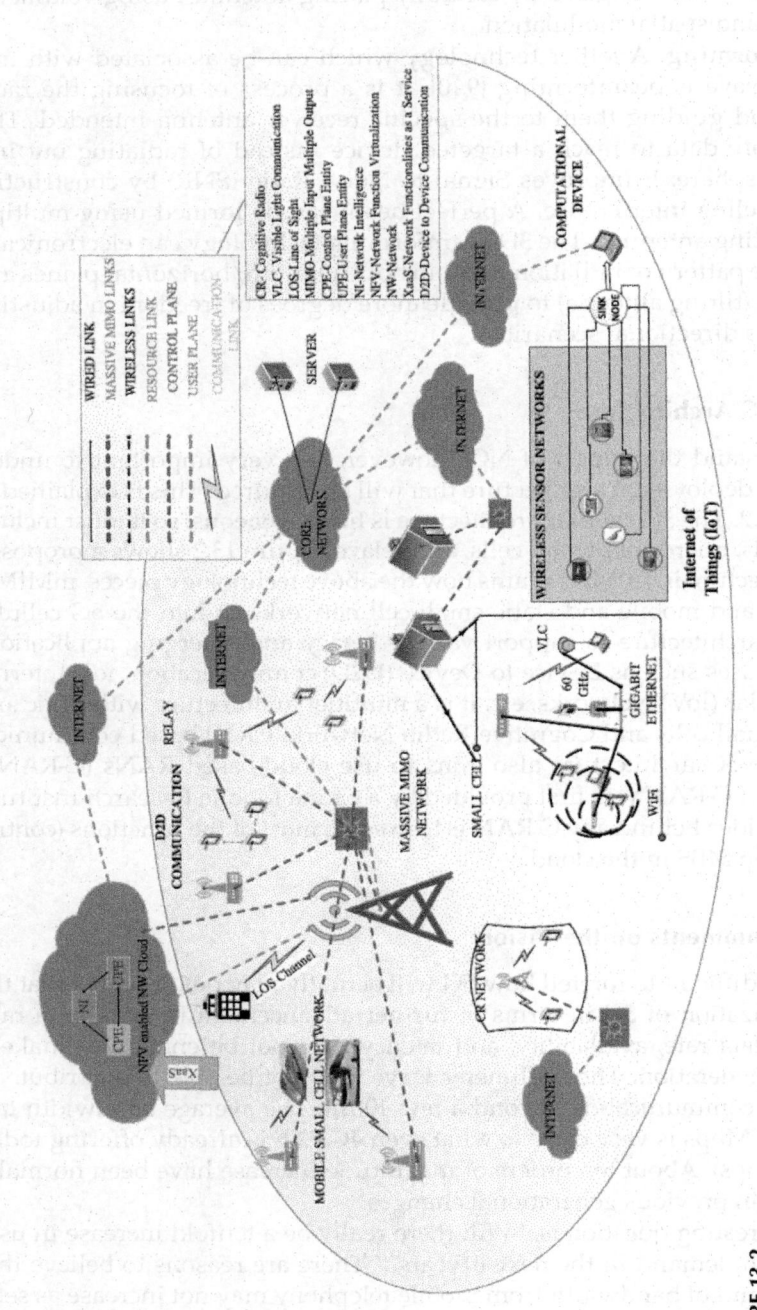

FIGURE 13.2

5G architecture.

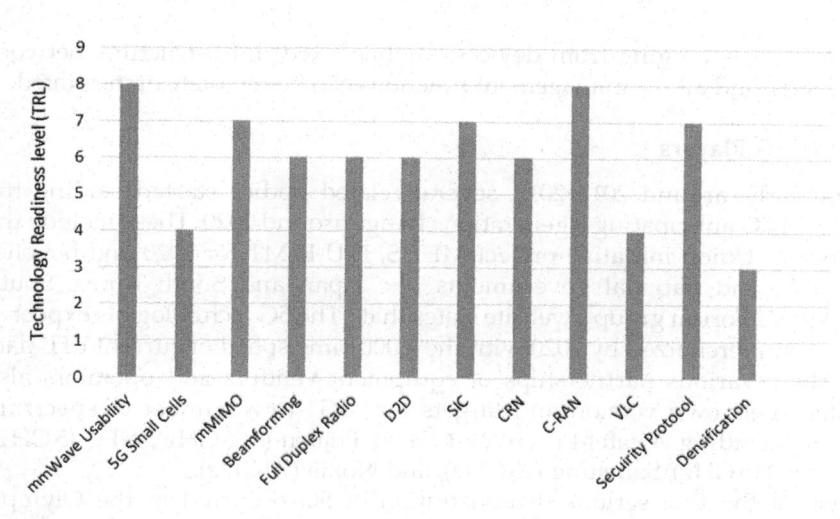

FIGURE 13.3
Readiness levels of technological components of 5G.

countries are still rolling out 4G, when will 5G be ready? Figure 13.3 shows an assessment of the technology readiness levels of various technology pieces of 5G as of 2018 [11,12].

The deployment of 5G has also brought into the limelight some other compelling issues in current NCI growth such as the rising cost of technology, its relentless upgrades, the expanding gap between what a specification claims and the user experience, lack of consistent networking, geospatial coverage, and growing concern about security and privacy. Indeed, technological advances in some of these are so urgently needed; technology improvement in few of these may end up defining the truly next-generation mobile communication than the current version of 5G itself.

13.2.5 Network Virtualization

Indeed there are promising technologies which can address the infrastructure cost issue. As evident to make use of millimeter wave, a rather large number of new cells/towers have to be built, which will be a major issue for any country.

Recent focus[1] has shifted to another important technology: network flow virtualization (NFV). Out of many potential uses, software-defined networking (SDN) leverages the structural separation of hardware and software. The idea is to reuse the underlying hardware and costly resources and, with software, create the appearance of many "virtual" networks running on those. NFV has the potential of reducing the cost of infrastructure by enabling easy sharing of real hardware. As such, the latest 5G enthusiasts see that 5G architecture will also use a new generation device capable of SDN and NFV technology where

---

[1] Initially there was no specific plan of adopting SDN in first of the 5G visions.

many aspects ranging from devices, (mobile/fixed) infrastructure, network functions, and all the management functions can be remotely orchestrated.

13.2.6 5G Players

Quite early, around 2012–2013, several related bodies started leading the vision of 5G anticipating a generation change around 2020. These include the European Union initiative project METIS, ITU-R MT for 2020 and beyond, NGMNs, and national governments like Japan and South Korea. South Korea's 5G Forum group's website states that "The 5G technology is expected to be commercialized by 2020 with the 1,000-time speed of current LTE data transfer". Various partnerships of equipment vendors and operators also posted their own version and targets for "5G" in a number of spectrum bands, including Alcatel-Lucent (3–6 GHz), Fujitsu (3–6 GHz), NEC (5 GHz), Ericsson (15 GHz), Samsung (28 GHz), and Nokia (70 GHz).

One of the first serious demonstration of 5G occurred in the Olympic Winter Games in Pyeongchang in February 2018. Here Intel, Samsung, and Korea Telecom (KT) partnered to complete the world's first broad-scale 5G network. Spectators enjoyed the Olympic Winter Games more vividly than ever before, as 5G allowed viewers to enjoy high-resolution media streaming at gigabit speeds with very low latency. The Olympic Ice Arena was fitted with about a hundred highly advanced cameras to capture the action of the performers in the ice rink to unprecedented visual details. The idea is to capture the visuals better than what even live spectators can see. Over a 5G network nearby, edge servers collected the captured videos and transmitted those large streams in real time to the data center of KT's Olympics. Here powerful video servers fitted with Intel scalable processors produced time-sliced views of the athletes in motion and then streamed them back to user phones and tablets. Viewers could use interactive time-slice technology to switch between the best angles to view the actions of the athletes. There was also "Synch View", which lets one stream two screens of action to a user's phone at the same time. The live streaming was available nearly live along with the 5G network capable of handling up to 250,000 devices.

Player	Type of 5G	Year
South Korea—KT	Commercial 5G Service	2019
Japan NTT	Commercial 5G Service	2020
China Mobile	10K 5G Base Stations	2020
US—Verizon	Commercial Fixed-5G Service (3–5 cities)	2018
US—ATT	Commercial 5G Service (12 cities)	2018
US—T-Mobile	Commercial Mobile and NB-IoT 5G at 600 and 200 MHz US-wide	2019
Europe	Commercial 5G Service (1 city in country)	2020
Europe	Commercial 5G in all major cities and highways	2015
UAE—Etisalat	Commercial 5G Service	2020

Japan's telecommunication industry is to show off their 5G in 2020 Olympics. The table above shows other recent or planned commercial deployments around the world. There is already a race to deploy 5G. Players want to have firm role in defining the 5G standard as it is being framed. They are also competing to gain the first-mover advantage. Operators worldwide are announcing ambitious targets for commercial 5G launches.

13.3 A Brief Review of Emerging Future Network Technologies

Now let us look into a set of massively disruptive technologies on the horizon. Besides mobile telephony, at the dawn of 2020, there is also an overwhelming set of new groups of applications rising much beyond mobile telephony. These all require communication, and all are going to require a share of the national communication infrastructure, with far-reaching significance for national development. Now we present the overview of each with a goal to analyze their communication needs and potential implications for a future NCI.

13.4 Smart Communities

13.4.1 What it is

US National Science Foundation (NSF) states: "Cities and communities in the US and around the world are entering a new era of transformational change, in which their inhabitants and the surrounding built and natural environments are increasingly connected by smart technologies, leading to new opportunities for innovation, improved services, and enhanced quality of life" [13]. There are also other definitions. But the essence is the enhancement of a citizen's quality of life.

There are various forms of communities. However, in conventional sense they mean geographically defined entities, such as towns, cities, or incorporated rural areas, consisting of various populations, with a governance structure. There are a broad range of applications. Each community provides various services such as roads and highways, traffic management, public safety, law enforcement, health care, recreation, entertainment, social services, utilities. For the larger community at a national level, it can mean clean air to national roads and highway management to energy optimization. Many services such as trash collection, maintenance of roads, etc. are often directly organized by the cities. However, many services such as transportation and utility are often provided by a third party closely operating with the community administration. All these services can be improved with the deployment of smart technologies. In the context of emerging IoT and big data, smart communities can often be considered as one more user group constituency.

Here are some sample smart city projects. One of the notable smart cities is *Songdo*, South Korea that is now under construction from scratch with the goal of green living; smart transportation; and state-of-the-art waste, water, and power management; and state-of-the-art education. Everything in this city is planned to be wired, and numerous computers will be embedded into the buildings and streets [14]. Constant streams of *data* will be processed by army of computers with marginal human intervention. Residents of Songdo will be able to control many aspects of their living including their place's lighting, heating, air conditioning, more with a push of button or voice control. They can join in instant video conferencing with their neighbors and attend remote classes. Embedded sensors everywhere will watch and analyze information on things like traffic flow, route of public transportation, weather quality, and consumption of energy. This kind of information can be converted into alerts that tell citizens when a bus will arrive or notify the authorities when a crime is taking place [15]. The water pipes are designed to prevent drinkable water from being wasted in showers and toilets.

In the city of *Santander*, Spain, with 180,000 inhabitants, a smartphone app is connected to 10,000 sensors that enable a multitude of services like parking search, environmental monitoring, digital city agenda, and more. Other examples of large-scale deployments underway include the Sino-Singapore, Guangzhou Knowledge City; work on improving air and water quality; reducing noise pollution; increasing transportation efficiency in San Jose, California; and smart traffic management in western Singapore and Columbus, Ohio. French company, *Sigfox*, commenced building an ultra-*narrowband* wireless data network in the *San Francisco Bay Area* in 2014, the first business to achieve such a deployment in the US It subsequently announced it would set up a total of 4,000 *base stations* to cover a total of 30 cities in the US by the end of 2016, making it the largest IoT network coverage provider in the country thus far.

Another example of a large deployment is the one completed by New York Waterways (NYWW) in New York City to connect all the city's vessels and be able to monitor them live 24/7. The NYWW network is currently providing coverage on the Hudson River, East River, and Upper New York Bay. With the wireless network in place, NY Waterway is able to take control of its fleet and passengers in a way that was not previously possible.

Among the notable government initiatives, the Government of India has awarded 100 smart cities distributed among the states based on urban population and the number of statutory towns in the various states with technical know-how to be provided by 50 or so consulting firms and handholding agencies [16]. New applications can include security, energy and fleet management, digital signage, public Wi-Fi, paperless ticketing, and others.

13.4.2 Technology and Implied Requirements from NCI and NCI Ecosystem

Smart city and smart community applications and services will also require new communication technology over its geographical locality. These may overlap with IoT and 5G, but there will be significant difference in the details. The following are few of the problems being investigated.

City Block Millimeter-Wave Cell: Independent of telecommunications industry, the smart community leaders are also exploring millimeter wave. This is to enable open R&D and systems testing at the millimeter-wave bands between 20 GHz and 200 GHz, with a target of 100 Gbps in data rates for small-cell networks that cover a few city blocks, specifically for the cities to deploy. The communication use cases are different from voice/video telephony applications. Block communication technology is needed for highly secured intranetworking of city and infrastructure sensors and non-connection-oriented traffic.

Dynamic Spectrum (DS) [17]: The idea is to find ways where new applications from the cities and communities can coexist in the bands without destroying the quality of service of legacy services. Unfortunately, most of the lower frequencies have already been allocated/licensed to applications/parties such as TV or voice telephony often for very long terms. Currently, there are is no good motivation for sharing them and hence no good technology. The current model of licensing ties spectrum bands to applications (such as TV, voice, voice over IP (VOIP)). This approach particularly limits the new applications needed by the cities. The DS is aimed to develop the technology pieces. DS focuses on the spectral bands that are sub-6 GHz (and thus has long reachability required by the cities) and aims to identify spectral opportunities in existing networks and establish usage models for novel spectrum-driven applications to be used by communities while also studying co-existence and protection issues (Table 13.3).

Mobility-at-Scale (MaS): Currently wireless technology such as the 802.11 family is designed to handle much less than 100 or so end devices and with limited definition of mobility. It is not clear if these can scale up and cope with heterogeneous mobility. In city-scale deployment, wireless network devices need to work at a much larger scale (perhaps of the order of 1,000 devices) and at various speeds such as 10 mph walking to 25–80 mph driving scenarios. A new MaS technology will be needed to address larger issues

TABLE 13.3

US Unlicensed TV Band Spectrum Estimated Payload Bandwidth Based on 4 bits per Hz Payload Coding Rate [SC3]

North American White-Space Band	Frequency Range Available	Bandwidth	TV Channel	Estimated Payload (Mbps)
VHF-High Band	2, 5, 6	54–60, 76–88	18	72
VHF-Low Band	7–13	174–216	42	168
UHF	14–35, 39–51	470–602, 620–698	210	840

with network mobility from the transport to media access control (MAC) layers, including evaluation of large-scale, dense, heterogeneous wireless networks and issues such as connection management, load balancing, and mobility management in roads and vehicular communication. Dense block communication technology is also needed for smart vehicle-to-vehicle and vehicle-to-road-side-things communication.

White-Space Communication (WSC): Another frequency-sharing technology under investigation is White-Space communication (WSC). The aim of WSC [18] is to foster increased dynamic utilization of the unused/underused TV frequencies by open users. Many parts of the world have dozens of open TV channels, especially in rural areas. White-Space spectrum increases as one moves away from urban areas. Development of White-Space technology can give cities and communities the much-required way to communicate in wide natural areas than 2.4–8 GHz unlicensed bands. Google provides a spectrum application programming interface (API) for smart radios to quickly find and use such White Space in USA (Table 13.2). On September 23, 2010, with the change from analog to digital TV, the US FCC opened up 270 MHz of prime VHF (54–60, 76–88, and 174–216 MHz) and UHF (470–602, 620–698 MHz) spectrum for unlicensed secondary use. Novel White-Space-based wireless networks have been demonstrated to provide up to 1 Gbps connectivity to remote locations via long-range wireless mesh connections that can reach 10 miles and withstand vegetation, hills, forest, or weather-related obstructions much better. WSC could help bridge the digital divide that exists between rural areas and cities.

Transparent Networking and Network Metrology: Smart community networking groups are also feeling the need to have more network transparency, so advanced sharing is possible. To enable sharing of spectrum, in-band protocols can be placed to monitor in real time the current usage as well as network quality. Similarly, real-time monitoring becomes increasingly important to support dense networking. Smart community networking groups are also exploring infrastructure-wide network metrology to advance capabilities to measure and monitor wireless network performance. Network transparency is also critical for improving the security, reliability, and performance besides the yield of wireless networks. Currently, there is hardly any transparency in NCI.

13.4.3 NCI Issues

The elements and pieces of technology for smart communities are deceptively similar to IoT, where sensors, massive data collection, and cloud-based analysis infrastructure are key. However, there are interesting differences.

Compared to business or home applications, smart community applications span across larger outdoor geographical areas such as a city or neighborhood with a spread bigger than a set of building. Thus, the availability of outdoor communication infrastructure is critical. Yet city applications may not need longer-distance communication as generally required for public telephony.

The user issues regarding privacy, safety, and transparency inclusion have to be designed quite differently than a solution for a corporate client or a private individual. For many smart city infrastructure applications, security and safety will be a major concern. Localized communication, rather than communication through wide-reach or public infrastructure such as the Internet, might be more desirable.

Cities have a strong case for being direct receivers of millimeter-wave spectrum licenses. This will require a much bigger chunk of spectrum to be opened up for shared-license or unlicensed use. A new regulatory paradigm (beyond spectrum licensing) is likely to emerge for ensuring public safety and privacy of the devices and applications.

13.5 Internet of Things (IoT)

13.5.1 What is IoT?

The IoT is the network of smart physical devices embedded with electronics and computer code which are capable of connecting and exchanging data. The physical devices can be a smart version of anything such as vehicles, home appliances, home thermostat, doors, lights, sensors, actuators, household tables, beds, watches, weight machines, smart wearables, biochips, environmental pathogen detectors—the possibilities are limitless.

Experts estimate that the IoT will consist of about 30 billion objects by 2020. It is also estimated that the global market value of IoT will reach $7.1 trillion by 2020. The vision of IoT was inspired by radio frequency identification (RFID) technology developed at MIT. The IoT consists of myriads of devices to be sensed and also controlled remotely across the network infrastructure. This creates opportunity for more direct integration of the physical world into computer-based systems and results in deeply integrated cyber-physical systems.

The global IoT market will grow from $157B in 2016 to $457B by 2020, attaining a Compound Annual Growth Rate (CAGR) of 28.5%. According to GrowthEnabler and MarketsandMarkets analysis, the global IoT market share will be dominated by three sub-sectors—smart cities (26%), industrial IoT (24%), and connected health (20%)—followed by smart homes (14%), connected cars (7%), smart utilities (4%), and wearables (3%).

The IoT device's communication characteristics are very different from classical voice telephony. IoT will require communication infrastructure to support more densely packed devices. Though an individual device may not require particularly large bandwidth, still it can create substantially high communication density. The devices will normally communicate in a tree- and hub-based topology. Many devices will be required to communicate with the low order of magnitude power.

IoT will require both short-range and long-range communication. For short-range communication (20–30 m), there is already a wide range of choices available such as Bluetooth, Z-Wave, ZigBee, Wi-Fi, etc. and their derivatives. The technical issues which need to be solved are secured communication and accommodation of a large number of devices. However, the main challenge lies where monitoring and control devices are in distant areas, where traditional infrastructure is limited, and where they send a small amount of data at regular interval. Areas where physical reach is difficult indeed make a more compelling case for automated or robotic-presence IoT devices. Both power and communication have been brought to these devices.

For power provisioning, there are ongoing researches which employ all three strategies, i.e., building a device which can sense and communicate with extremely low power, improving battery technology which can last for years, and harnessing of ambient energy.

For long-range wireless, based on authority/licensing mode of the spectrum, communication options in this domain can be classified as (1) Licensed: (i) Extended Coverage GSM for IoT (EC-GSM-IoT), (ii) LTE-M, and (iii) Narrowband IoT and (2) Unlicensed: LoRa.

EC-GSM-IoT: It is optimized for GSM networks. It is deployable over commonly used spectrum bands for GSM in 800–900 MHz and 1,800–1,900 MHz with coverage up to 20 db. EC-GSM IoT spec was ratified in 3G PP release 13. It uses half-duplex-frequency-division multiplexed channel in 200 KHz frequency. EC-GSM has a data rate of 70–240 kbps. It is designed to operate with existing GSM base-station hardware. Software update in GSM core network alone is sufficient for EC-GSM-IoT. It enables a combined capacity of up to 50,000 devices per cell on a single transceiver. 3G PP eDRX power-saving mode improves idle-mode behavior by allowing the use of a number of inactivity timers. eDRX can enable almost a 10-year battery life cycle for many use cases. Orange telecom has operational EC-GSM-IoT-based network for environmental monitoring in Europe.

LTE-M (LTE Cat M1): LTE-M is the IoT-optimized version of LTE. LTE-M uses the same spectrum (1.4 MHz frequency) and base station of LTE. It can also enable true TCP/IP data sessions. But it is more power efficient than LTE. Some use cases may enable 10 years of a lifetime with a single charge. Its power-saving mode (PSM) and extended discontinuous repetition cycle (eDRX) allow it to go idle without rejoining the network after wake-up. LTE-M has impressive data rate ranging from 375 kbps to 1 Mbps. In the USA, Verizon, AT&T, and other major LTE operators have operational LTE-M-based solutions in different countries.

Narrowband IoT (LTE Cat-NB1, LTE Cat-M2): Narrowband IoT (NB-IoT) was standardized in 3G PP's Release 13. It operates based on a direct sequence spread spectrum (DSSS) modulation technique. It is designed to be tightly integrated and to interwork with LTE. NB-IoT can be deployed as a self-contained carrier with only 200 kHz bandwidth. A standalone

200 kHz NB-IoT carrier can support more than 200,000 subscribers. It can also operate on both in-band and guard band in the spectrum. This makes NB-IoT a hot choice for telecom operators. It has better power efficiency than LTE-M but has a lower data rate than LTE-M. Its data rate ranges from 20 to 300 kbps.

LoRa: LoRa is an unlicensed spectrum long-range wireless communication standard for IoT. A LoRa device range varies from a few kilometers in dense urban areas up to 15–30 km in rural areas. LoRa devices usually produce power within 10–25 mW range. They enable around 10+ years of battery life. LoRa devices have a data rate of 0.3–50 kbps. These devices are most suitable for on/off type applications with only a few messages per hour. They require fewer base stations than 3G/4G. Currently, LoRa is being widely used in various fields. About 500+ member organizations across the world have joined LoRa alliance.

The scale of innovation associated with IoT will require communication technology to be opened up. Very few companies can participate in today's fixed spectrum allocation model. IoT, thus, will require a much bigger chunk of both short-range and long-range spectrum bands to be opened up for shared-license or unlicensed use. While the IoT provider companies will be expected to meet various safety and privacy standards. A new IoT regulatory paradigm (beyond spectrum licensing) is likely to emerge.

13.6 Transformation of Industry

Industry 4.0 [19] refers to the ongoing trend of automation and data exchange in manufacturing technologies. The term "Industry 4.0" originates from a project in the high-tech strategy of the German government to promote the computerization of manufacturing. It was popularized in 2012 when in one of the world's largest industrial fair Hannover Fair, the Working Group on Industry 4.0 presented a set of Industry 4.0 implementation recommendations to the German federal government. Its cornerstone is the interconnection of cyber-physical systems with increased adoption of the IoT, cloud computing, data analytics, and AI.

The term "4.0" refers to three past major turning points of the industrial revolution. *Mechanization* powered by water and steam (by inventions of James Watt 1800–1900), *mass production* by assembly line and electricity (by inventions of Ford and Edison 1900–1950), and *automation* by the adoption of industrial electronics, computers, and robotics (1950–2000) are the first three stages. Industry 4.0 is commonly referred to as the fourth industrial revolution that industries are currently going through by interconnecting their industrial production system with cyber-physical systems. Essentially the fourth evolution is **networking**.

More recently, the concept of Industry 5.0 has emerged pointing to product personalization with the use of AI and communication. There has been technology demonstration in Education and Apparel industry.

Yet, another emerging technology that may upend many aspects of the industry is the blockchain. The advances in blockchain technology promise to improve the efficiency of almost every conceivable service industry. It promises to replace tedious and costly middlemen/experts (ledger-workforce) who are now maintaining, verifying, correcting paper-based ledgers and contracts manually. These tasks can be replaced by automation in the form of machine-verifiable trust network and tamper-proof smart contracts.

Many start-ups are already competing from developing new models for commodity trading, banking, and voting. The blockchain will also facilitate business-to-business digital transactions for various industry sectors across the supply chain. This is likely to require the next level of cloud networking and cloud exchanges [20].

The eventual benefits of this stage are a dramatic improvement in the quality of services and products and the sophistication of products and access to products. Like previous evolutions, these benefits will come with an overall increase in cost efficiency. The networked industry will dramatically improve the management of the supply chain to the management of the product's lifecycle, improve the reliability of production, and improve the safety and working condition of the workforce. In each sector, it will bring full production ecosystem closer. In a national scenario, it is also expected to reorganize the global competitiveness landscape. Both old and emerging economies will compete in a new playing field with opportunities as well as pitfalls.

Yet many networking and IT challenges exist. These include (i) IT security issues, which are greatly aggravated by the inherent need to open up those previously closed production shops; (ii) reliability and stability needed for critical machine-to-machine communication (M2M), including very short and stable latency times; (iii) need to maintain the integrity of production processes; (iv) data security and need to protect industrial know-how; (v) unclear legal and data security regimens, lack of regulation, and standards and forms of certifications; and (vi) new business models.

13.7 Issues, Challenges, and Opportunities for NCI

As per the latest 2017–18 ITU surveys, mobile penetration is more than 100% in more than 100 countries [21]. The fixed broadband connection is more than 40% in about 10 countries [22]. Given the advent of 4G LTE, it is expected that mobile subscribers are in easy reach to be considered as Internet connected. The smartphone penetration is estimated to be 10%–80% in the top 50 countries.

For example, the UK is the top (80%) followed by other countries such as Malaysia (58%), India (27%), Myanmar (21%), Bangladesh (16%), and Pakistan (13%). With such population saturation, particularly countries with 75%+ penetration, what should be the next goal for NCI? Here we discuss some opportunities.

13.7.1 Geographic Coverage

As countries stride to make their NCI Future Network ready, a national roadblock for many will be insufficient geographical coverage. Let us consider the case of Malaysia. It has excellent (133%) mobile penetration and is ranked 33rd in the world [21]. Yet it does not have coverage over a vast swath of its national territories with major void even in central Malaysian peninsula. For a long time, the investment focus was on covering the population centers along the peninsular coastlines. As of 2018, roughly over ~72% of its national ground territory seemingly has been near-zero coverage. Malaysian researchers pursuing scientific projects (coastal plantations, soil erosion containment, environment monitoring, etc.) reported lack of basic radio connectivity in interior areas even with recent long-range IoT communicators [23].

With IoT devices will come a wave of IoT remote-sensing applications, and lack of radio connectivity over vast national territories in many countries will preclude all such applications. The experience of the researchers is a harbinger of national-scale problems that will come with IoT aspirations.

Hard-to-reach places include the vast uncovered zones of inland waterways, mountains, forested highlands, desserts, coastal archipelago areas, and vast economic zones extended in the sea. These areas have low population density but are often rich in natural resources and economic, industrial, climactic, and geo-strategic importance.

Expansion of geo-coverage will require more determined national approach which leaves it up- to market force. The geo-applications, at least in initial years, will require low-power communication and not that much bandwidth. These are physically difficult places, and thus, ground infrastructure will always be expensive to reach and maintain. How can countries encourage the expansion? Several approaches can be explored.

Lower frequencies can reach relatively longer distances with lower power. The regulatory and development bodies (RDBs) may want to accelerate the reallocation (or at least dual use) of lower-frequency spectrums vacating older technology for modern IoT-capable communication technologies. The vacating effort should be highly prioritized in hard-to-reach zones. The low-density 1G and 2G subscribers in small town/village enclaves can be moved to other frequencies (with 4G LTE) with relatively low investment compared to the cost of losing in emerging IoT services and applications to come.

Some military and non-commercial low-frequency spectrums can also be eased up for national geo-territory asset protection and monitoring applications and for university research projects aimed at similar goals.

There are also few promising sky-based technologies near commercialization which can be explored where geo-coverage is severely low (less than 60%). RDBs in cooperation with national universities may set up testbeds in each of these zones for new technologies. One such testbed is Alphabet Inc.'s Loon (by Google). Alphabet is already experimenting with several countries to provide mobile connectivity from the sky with no ground infrastructure.

There is new ultra-low-frequency (ULF) communication technology in development that can reach continental-scale long distances. RDBs can pioneer technology testbeds with collaboration with the national universities for national interest for it also.

Not to mention, the focus on geo-coverage, if addressed, will also mitigate the age-old rural telecommunication access problem as a bonus.

13.7.2 Expansive NCI Infrastructure

Yet another formidable challenge many countries will face is the expansion of NCI infrastructure. NCI needs fibers, radio access points, and towers for them; the towers then need to have backhaul fiber connections; the backhaul fibers need ducts. The infrastructure cost is already high and is rising in most countries. To deploy infrastructure, telecommunication companies face roadblocks such as the high regulatory cost of site acquisition, fiber trenching, fiber leasing, spectrum acquisition, the monopoly by the state-backed company, slow land approval, and high approval fees by state authorities. Over time, land and asset owners have become increasingly savvy in demanding high fees and rents. There is also a concern about aesthetics and the environment. How will countries manage this great expansion?

A problem case at hand is the 5G mobile technology itself. 5G will require multiple ranges of cells such as femtocells (10–20 m), picocells (200 m), and microcells (2 km). An estimated 4–5 times more new types of radio access points will be required by any previous mobile generation. These radio access points will also need backhaul fibers to reach. These fibers will need ducts to get access through national, rural, local roads, and highways and private and public lands. These also need towers, lands for the tower, rooftops for the building, small and medium structures to keep the equipment, heating and cooling systems, power for them as well as equipment, and manpower for maintenance and safety. How will these be provisioned?

In a few countries, there is already regulatory impetus for encouraging infrastructure sharing, particularly for the expansion of rural broadband coverage. Even in those countries, there are practical problems in the field. There is a lack of interest in leasing by the owners. The related information such as existence, location, ownership, pricing, etc. are often kept quasi-hidden.

To make the matter more complex, the future applications are almost certain to see changing roles of the various parties in the infrastructure marketplace. The digital transformation of communities will turn townships, municipalities, or even large property owners potentially into some form of digital service providers themselves. Explosion of IoT and asset management applications will make them more affianced customers of communication services. It is quite possible that some larger property owners will maintain much-sophisticated communication infrastructure themselves, much in the line they own 802.11 wireless access points. These bodies are expected to make use case for short-range (such as 5G) spectrum ownership. Negotiation positions of the parties will significantly change.

RDBs, thus, side by side with other government authorities, need to start engaging these future technology prospective players such as property owners (land grant universities, state governments, public schools, power facilities, etc.), communities, new property developers and owners, and city planners to identify the solution. A much comprehensive resource-sharing model with governmental weight for encouraging sharing of civil infrastructure is needed.

One first step should be to foster trading of excess resources—beginning from civil infrastructure (tower, co-location facilities, fiver duct, right of access) to telecommunication capacity (dark fiber strands, spectrum, lambda, storage, bulk bandwidth, etc.). Even telecommunication equipment (such as radio equipment, expensive edge switch and routers) can be shared.

New technologies like NFV and SDN are coming to make the sharing automatic. It is also important to bring transparency and efficiency into this market by encouraging timely and accurate exchange of "product" information (what, when, whom to contact, etc.). Notably, this transparency is particularly important for the newcomers.

Advance provisioning can be the key to win over the infrastructure challenge. Power-line-borne fiber is an excellent example where, when planned in advance (fiber is added along with the neutral wire), a fiber network can be obtained almost free (by 1%–2%). Countries can move fast to develop design standards for roads and buildings to include NCI items to drastically reduce the overall national capital investment.

Given the overwhelming national strategic importance of NCI in all aspects of national development, countries may have to rethink undertaking national-scale NCI projects. In particular, optimization/topology regularization of NCI national backbone and international connectivity.

Every sector of government and public services is undergoing digital transformation and will need major network services anyway. Worldwide Research and Education Networks (RENs) created by many countries public higher education sectors are excellent examples of how impactful national public investment can be made.

13.7.3 Quality and Consistency of Network

By 2018, most countries had done an excellent job of providing a cell phone in the hand of every citizen. The new frontier of the challenge is the quality and consistency of network [24].

At a very basic level, bandwidth, network delay, and connection establishment time are three core parameters that characterize the quality of network services. For most mobile users, there is no good tool to understand even basic-level quality of service.

Humans are mobile by nature; a person moves around his home, workplace, and marketplace in daily and weekly patterns. In some countries, people also move between urban home and rural home in weekly, monthly and seasonal patterns. Network service may be adequate in specific commercial, residential, and office zones; yet it often falls off with individual mobility. Disconnection during handover, end-to-end fluctuation of bandwidth, uniformity of coverage during mobility across are the indicators of the problem. The dead zones, tower-to-tower handover, potential lack of technical (2G/3G/4G switchover) or business-level synchronization between provider's networks, and opportunistic oversubscription frequently make user experience quite poor across countries.

In the absence of any meaningful data, anecdotal experience seems to reinforce that end-to-end user experience is not great even in the best places in a country. In most countries, there are frequent dead zones in national highways. Too many times, the handover fails even within the city during travel. Both during 2017 and 2018, bandwidth reports suggests worsening quality even for 4G LTE traffic. Poor user experience is endemic; many experts opine that the quality and consistency of NCI is now a bigger problem than speed. In most countries, there is no real regulatory demand for consistent network.

Quality and consistency not only matter to end-users, but to software as well. Cyber-application developer assesses the capability and consistency of the network and assesses up to what level a network can sustain the application and services. The quality and consistency of the underlying telecommunication service are of high quality only in limited places such as urban centers but spotty in most zones—high-performance-application market stalls. Thus, quality and consistency of the network will be increasingly critical for technologically sophisticated and advanced applications to flourish.

Experts from opined regulatory bodies, national planners, and industry as a whole need to focus more on improving the quality of the delivered communication services rather than being fixated on speed. The industry is well equipped to handle the latter and much less equipped to handle the former. Also importantly, consumers should be given capable tools to know about the network quality parameters of the services they receive to make informed choices between providers. Regulatory bodies should encourage national providers to implement the required technical and business measures to increase the quality and consistency of the NCI.

13.7.4 The Explosion of Cloud Services

Many of the developing countries are making an excellent stride in expanding mobile and Internet connections following the steps of advanced countries.

However, many among them have fallen behind in the area of public cloud, data centers, and content delivery networking (CDN) infrastructure. These are emerging components of modern NCI infrastructures in advanced countries.

In the developing countries, larger companies, including telecommunication companies have built large data centers; yet most are being used only for internal data needs. Data centers and cloud will play a big role in the flourishing of the digital transformation of the country. A country must provide access to a healthy and modern cloud infrastructure as businesses will undergo a rapid digital transformation, and these depend on cloud services.

Thanks to global cloud providers in advanced countries, such as USA and UK, global cloud traffic has already surpassed classical Internet traffic. In the coming years, the cloud-service-related traffic is expected to exponentially grow to signify the importance of cloud services.

It is a key enabler for digitization of all other business and commerce sectors. Businesses around the world are embracing deeper cloud services. While many started with older CDN and infrastructure as a service (IAAS) (storage) models, more are increasingly opting for deeper software as a service (SAAS) (storage and software) and platform as a service (PAAS) (processing, software, and storage) models.

The strength of cloud will be essential for Industry 4.0 sectors, as companies will seek to lower the IT capital expenditure for their data-intensive business processes and will require fast, responsive, and reliable IT platforms to deliver their own services to their customers. Countries, however, should look for means to encourage and invigorate their own national public cloud industry besides the global providers.

The blockchain will also facilitate business-to-business digital transactions for various industry sectors across the supply chain. This is likely to require the next level of cloud networking and cloud exchange technology.

Regulatory bodies may start by (i) assessing the current cloud market and usage by the national companies and (ii) encourage the establishment of an indigenous public cloud infrastructure In the meantime, regulatory bodies should work for commissioning National Cloud Exchange and related protocols into its NCI to ensure seamless high-performance interconnection to cloud and fog systems worldwide for its businesses.

13.7.5 Spectrum Roadmap

Though there is much discussion about 5G, many countries have not given specific guidance on 5G spectrum allocation. There are deeper issues countries' spectrum management authorities need to tackle, and thus, this time the spectrum allocations may not be just another auction cycle.

Overall, it is clear with the emergence of IoT, the smart communities will require much nimble spectrum usage and licensing paradigm—a significant shift from old large providers with focus on commercial spectrum-licensing paradigm.

There is a strong case for dynamic spectrum technology focusing on the spectral bands that are sub-6 GHz (and thus having long reachability required by the cities) and aim to identify spectral opportunities in existing networks and establish usage models for novel spectrum-driven applications to be used by smart communities while also studying co-existence and protection issues. Both 4G and 5G technologies enable spectrum-band aggregation. Rapid facilitation of Software-Defined-Radio (SDR) will also allow equipment to be versatile and flexible.

There is also a strong technical case for flexible-width spectrum bands so that low to very high data rates can be supported. Band aggregation or concatenation capabilities (added in 4G and 5G) can solve some of the width problems. These, however, require adjacent bands and thus a flexible spectrum license policy.

Smart city, combined with 5G projects, will require extensive research and development of systems and testing grounds at the millimeter-wave bands between 20 and 200 GHz, with a target of 100 Gbps in data rates for small-cell networks that cover a few city blocks, for the city to deploy. These and university groups might need more clear research usage in wider spectrums.

In a greater scheme, this is a reflection of a larger problem. The legacy model of licensing ties spectrum bands to applications (such as voice, VOIP). This approach, however, limits new applications. The idea is to find ways where new applications from the cities and communities can coexist in the bands without destroying the quality of service of legacy services.

Overall, the spectrum license should make bands more shareable (an example is White-Space technology) and more leasable/sub-leasable and include more use-it-or-lose-it policies.

Reallocation of extremely valuable low-frequency long-range bands should be accelerated. Emerging applications need them more urgently.

New spectrum-allocation policy should create space for smaller but major innovative players such as universities, start-ups, smart communities, etc.

13.7.6 Information Protection, Privacy, and Ethical Applications Framework

About 120 countries in the world currently have enacted some form of citizen data privacy law [25,26]. Another 40 countries are in the second category where the laws are in the process of the formulation (Pakistan, India, Thailand, Indonesia, etc.). The third group has no initiative (Bangladesh, China, etc.).

The data privacy laws confer a set of rights (subject to qualifications) in relation to citizens' personal data. The examples are right to access personal data, the right to correct personal data, the right to withdraw consent to process personal data, the right to prevent processing likely to cause damage and distress, the right to prevent processing for direct marketing, etc.

The laws are important essential requirements for the growth of next-generation data-driven services (DDSs). Yet, countries have to deal with additional issues to accelerate proper emergence of advanced DDSs.

The immediate challenge after the regulations are enacted is the implementation framework. How can citizens exercise their digital privacy rights conferred by the act? It requires company business processes so these rights can be exercised. For example, a telecommunication company A may collect individual location data to optimize service. Customers have options to limit archival duration (say 7 or 30 days) and rights to limit the purpose of use or opt out. These rights and options can be exercised only if certain data archival and storage practices are in place. The mechanisms for letting customers change or withdraw their privacy choices, checking and correcting accuracy, knowing/notifying a client in time (to minimize damage) if his/her personal data has been illegally breached, etc. are non-existent or very weak at best. Implementation of digital rights and options have a significant cost impact on businesses. It is quite possible that company A uses cloud company C for data processing and D from backup archival. The industry as a whole needs investment and time to implement the required technical and business processes. Regulatory agencies need to start working with all the provider companies on the details to create this implementation framework so personalized data-intensive services can flourish with appropriate protection for everyone.

The emergence of IoT and Community Computing will require revisiting the legal framework beyond humans as well. One case at hand is IoT. These applications by design will be intensely invasive and collect myriads of very detailed data related to every aspect of dwelling (both private and public places) and use them for machine control. Misuse of infrastructure data can be lethal (such as power data, security video, when the front doors are open, etc.).

Most personal data protection laws do not adequately cover infrastructure data. For example, operational data of a house may not directly contain information about the owner of the house; this may or may not be covered by privacy laws, and yet it can have a direct implication on the owner's safety.

Infrastructure data protection policies and legal safeguards are imperative to foster safe rollout of these systems. Regulatory bodies should work closely with related international and local bodies for infrastructure data management and protection policy and framework to accommodate safe rollout of IoT and machine-to-machine applications.

Thirdly, the emergence of AI will require a third regimen of application-specific standards and regulations. These potent systems will make decisions without the human in the loop. These require vetting of approved application areas and use cases, protection against ethical and unethical use, adherence to purpose, avoiding unfair systemic bias, careful safety testing before mass launching, etc.

Few companies like Google and IBM are volunteering the development of their own AI principles [26] anticipating backlash. The emergence of software-controlled world calls for a whole new software, system, application certification paradigm, and national regulatory safeguard to enact Fair Information Practices principles, which will need to delve much deeper than the current privacy laws we are seeing.

13.7.7 Modernization of Assessment Mechanism

A good ecosystem requires modern assessment arms that can collect extensive data from service providers. In many countries, regulatory bodies rely on an antiquated reporting system that collects data such as traditional 1G, 2G, 3G, and 4G maps; locations of radio equipment/towers; etc. To improve the policy and regulatory efficacy, regulatory bodies should redesign its overall assessment framework and expand the metrics.

Many monitoring areas will need immediate attention: (i) network quality, (ii) consistency or end-to-end service, (iii) data networking, (iv) infrastructure sharing, and (v) data protection and security. The old paper-based batch reporting of quality assessment and annual-book-styled reporting may not be scalable. It should move to a live and continuous online assessment reporting system.

Many data elements such as bandwidth, delay link health, etc. from providers' not otherwise classified (NOC) can automatically and continuously stream into this the automated assessment system. Standards and measures should be there to ensure interchangeable format, authenticity, and accuracy of reporting. RBDs can also deploy live measurement points in the NCI deploying telemetric tools and adding intelligent analytical and visualization tools. Automation can significantly reduce the cost and overhead of reporting and monitoring and increase transparency.

To create a healthy ecosystem at a national scale, it is important to create a robust mechanism for consumer awareness. Consumer awareness is a multifaceted complex process. A potential flagship project is to lead an initiative to have a Live Label so end subscribers are informed about the type, quality, and health of the communication service provided to them. The closest analogy is the classical food labels adopted by advanced countries.

13.7.8 The Ecosystem for Future Network Innovation

In the current global telecommunication ecosystem, the countries which lead in the research and development (R&D) of telecommunication technology are unsurprisingly few. The vast majority of aspiring countries are mere receivers of technology. The advent of digital transformation creates a fertile ground for innovation. It will open up the opportunity for many countries to break into this innovation ecosystem. A national presence in R&D is also vital for national security and global competence.

Countries should actively promote innovation forum to connect university researchers, engineers, units related to national telecommunication industry's future network services (FNS), smart community/municipality technology leaders, and potential IoT service companies besides international initiatives.

One particular point to note is that this ecosystem has to go beyond traditional telecommunication players. The new national NCIs will need to be readied for many diverse applications beyond telephony. One particular new group to engage in NCI stakeholders group are the smart community initiative leaders who are often far from the technology fray; yet their voices and needs have to be heard.

Countries can further promote setting up a specific set of national testbeds and innovation demonstration center(s) to gain immersive experience into the newest. RDBs can take initiative to encourage international/national companies/vendors to set up product/service demonstration of their newest technology as they plan rollout in their country or in the region. They can also demonstrate flagship NCI-related projects from the universities and local start-ups with a goal to arrange early exposure to policymakers, university researchers, and local innovators, to the latest for next-step ideation and adoption.

13.8 Conclusions

Digital transformation is upon us. Initiatives, such as Industry 4.0/5.0, 4G LTE/5G, IoT, AI, smart city, and blockchain, though originated from divergent techno-evolution threads, are all phenomena of the underlying digital transformation. Certainly, these are not all. More such initiatives are to unfold. All the threads of this transformation, irrespective of their origin of the drive, however, will require one thing at the core—a capable and robust NCI. The aspirations of the 21st-century developing countries are different from that of the 20th-century developing countries. It is not merely poverty elimination but to be a fully developed country along all dimensions—economically, politically, socially, spiritually, psychologically, and culturally. These are anchored on the prosperity and well-being of citizens where there will be equal access to high-quality health care and lifelong education but go way beyond into the lifestyle. Neighborhoods and communities are to be a safer and peaceful place. These should occur irrespective of the geographic location of the community or the socio-economic background of the citizen. NCI will be at the heart of almost all national development efforts. The task at hand is daunting. These goals will be unattainable without a capable NCI.

13.9 Acknowledgments

The article is part of a broader study including extensive field survey and industry and academia discussion that was undertaken in 2017–2018 to identify the evolution of various components of future network and its potential impact on NCIs and sponsored by Fulbright program of US Department of State and hosted by Malaysian Communications and Multimedia Commission (MCMC), Government of Malaysia, and the Malaysian-American

Commission on Educational Exchange (MACEE). More than 50 experts from the industry and academia provided their valuable insights in this future study. The author would like to acknowledge their contributions and support. The author would like to acknowledge the special contribution of several researchers including Badruzzaman Mat Nor, Norzailah Mohd Yusoff, Amjad Hossain, Jeanne Tan, Iftekharul Islam, and Debobrata Das for their ideation in this research area.

References

1. D. Warren and C. Dewar, Understanding 5G: Perspectives on Future Technological Advancements in Mobile, GSMA Intelligence, December 2014.
2. A. Osseiran, F. Boccardi, V. Braun, K. Kusume, P. Marsch, M. Maternia, O. Queseth, M. Schellmann, H. Schotten, "Scenarios for 5G Mobile and Wireless Communications: The Vision of The METIS Project," *IEEE Communications Magazine* 52(5): 26–35, 2014-05-01.
3. NGMN Alliance, 5G White Paper, February 2015.
4. IMT Vision – Framework and overall objectives of the future development of IMT for 2020 and beyond, Telecommunications Union. Retrieved 2017-02-22, Recommendation ITU-R M.2083-0, September 2015.
5. Draft New Report ITU - Minimum Requirements Related to Technical Performance for IMT-2020 Radio Interface(s), 2017-02-23.
6. Mobile Economy 2017, GSMA Intelligence, 2018.
7. M. Agiwal, A. Roy and N. Saxena, "Next Generation 5G Wireless Networks: A Comprehensive Survey," *IEEE Communications Surveys & Tutorials* 18(3): 1617–1655, third quarter 2016.
8. N. Panwar, S. Sharma, and A. K. Singh, "A Survey on 5G: The Next Generation of Mobile Communication," *Physical Communication* 18: 64–84, Part 2, March 2016, ISSN 1874-4907. doi:10.1016/j.phycom.2015.10.006.
9. S. M. Razavizadeh, M. Ahn, and I. Lee, "Three-Dimensional Beamforming: A New Enabling Technology for 5G Wireless Networks," *IEEE Signal Processing Magazine* 31(6): 94–101, November 2014.
10. A. Gupta and R. K. Jha, "A Survey of 5G Network: Architecture and Emerging Technologies," *IEEE Access* 3: 1206–1232, 2015. doi:10.1109/ACCESS.2015.2461602.
11. V. Jamier and C. Aucher, "Demystifying Technology Readiness Levels for Complex Technologies," Leitat, Barcelona, April 24, 2018.
12. Technology Readiness Levels, NASA Document, REF: www.nasa.gov/directorates/heo/scan/engineering/technology/txt_accordion1.html.
13. National Science Foundation Solicitation 18-520, Smart and Connected Communities (S&CC), November 28, 2017.
14. "Built-from-Scratch Songdo Starts Coming to Life," Korea JoongAng Daily. Retrieved 2019-01.
15. "Could Songdo be the World's Smartest City?" World Finance. Retrieved 2019-01-05.

16. Smart Cities Initiative India, http://smartcities.gov.in/content/innerpage/handholding-support.php.
17. Towards Dynamic Regulation of Radio Spectrum: Technical Dream or Economic Nightmare?, Alcatel-Lucent.com. Retrieved 2015-12-22.
18. Delivering Wide Area Broadband Services in Wilderness and Rural Terrain, by Metric Systems, Aug 13, 2015, www.metricsystems.com/delivering-wide-area-broadband-services-in-wilderness/.
19. Cincinnati to be Industry 4.0 Demonstration City. Imscenter.net. Retrieved 2016-07-30.
20. Application Performance: A Framework for Cloud Enablement, White Paper, 2018, www.equinix.com/resources/whitepapers/application-performance-a-framework-for-cloud-enablement/.
21. Mobile-Cellular Subscriptions 2000–2017, International Telecommunication Union official website.
22. Fixed-Broadband Subscriptions 2000–2017, International Telecommunication Union official website.
23. J. I. Khan, Assessing the Readiness of Future Network in Malaysia- Survey Analysis, MACEE/MCMC Fulbright Senior Specialist Report, April 2018. Available at: https://ssrn.com/abstract=1951416.
24. W. Webb, The 5G Myth: And Why Consistent connectivity is a better future, ISBN-13: 978-1540465818.
25. Banisar, National Comprehensive Data Protection/Privacy Laws and Bills 2018, September 4, 2018. Available at SSRN: https://ssrn.com/abstract=1951416 or http://dx.doi.org/10.2139/ssrn.1951416.
26. Data Policy in the Fourth Industrial Revolution: Insights on Personal Data, November 2018, World Economic Forum Report, Switzerland. Available at: https://weforum.ent.box.com/s/cd1m02qcx2yihq58b9l1r31nvgmalo48.

Index

A

Absolute humidity (AH), 58
AC servo motor, 171
Active sensors, 41
Actuation mechanism, 227, 229
Actuators, 30, 41–42, 56
   components of, 42
   electric, 43
   electrostatic, 226
   hydraulic, 42–43
   lateral comb, 225
   lateral thermal, 225–226
   mechanical, 43
   microactuation, 225–226
   micromachine, 221
   omnidirectional, 46
   piezoelectric, 46, 47, 158, 222, 226
   pneumatic, 43
   selection of, 47
   smart soft, *see* Smart soft actuators
   thermal, 225
   transverse comb, 225
Adaptive control system, 115
ADC, *see* Analog to digital converter
   (ADC)
Advance microprocessors, 4
Advance provisioning, 299
AH, *see* Absolute humidity (AH)
Alphabet Inc., 298
Aluminum (Al) alloys, 83
AMD microprocessors, 16–18
Amorphous alloys, 80
Analog to digital converter (ADC),
   26–27
Analysis models, in PO, 268
Anatase, 61–62
Application programming interface
   (API), 292
Arduino, 4, 25, 26
Arduino MKR, 25, 26
Area, blob analysis block, 129

Arithmetic logic unit (ALU)
   in microcontrollers, 20
   in microprocessors, 4
ARM processor, 23
Artificial intelligence (AI), 219, 220, 278,
   303
Artificial neural networks, 92, 93, 117
Atmel AVR microcontroller, 23
Automated Bayesian inspection
   method, 117
Automated reverse engineering
   method, 258
Automated vision inspection system, 117
Automation, 295, 304

B

Back-end data-sharing model, 218
Backpropagation neural network (BPN),
   92
Band aggregation, 302
Bayesian network models, 267, 269
Beamforming, 285
Bias errors, 114
Biological sensors, 36
Blob analysis, 128–130
Blob count, 130
Block communication technology, 291,
   292, 296, 301
Bounding box, 129
BPMN, *see* Business Process Modeling
   Notation (BPMN)
BPN, *see* Backpropagation neural
   network (BPN)
Bridge-type displacement amplifier,
   156–157
Brookite, 61–62
B-spline, 146–147
Bulk forming processes, 167
Business Process Modeling Notation
   (BPMN), 264

C

Calibration process, 120–125
Camera calibration, 120–121
    application of, 122
    extrinsic parameters, 121–122
    intrinsic parameters, 121
    process, 122–125
Capacitive sensors, 39
CDN, *see* Content delivery networking
    (CDN)
Cellular base stations, 279
Centroid, blob analysis block, 129
Chemical sensors, 36
CISC microcontroller, *see* Complex
    instruction set computer
    (CISC) microcontroller
City block millimeter-wave cell, 291
Classical differential geometry, 138, 139,
    149
Closed-loop manufacturing method
    benefits of, 111
    in-process measurement, 113–114
    in-situ measurement, 112–113
    loops, 113–114
    remote measurement, 112
Closed surfaces, 141
Cloud-based RANs (C-RANs), 285
Cloud infrastructure, 278
Cloud services, 300–301
CNC machines, *see* Computer numerical
    control (CNC) machines
CNC servo press, *see* Computer
    numerical control (CNC) servo
    press
Cobots, 219–220, 226, 240, 246
    social, 220
    UR3, 240–241
Coefficient of thermal expansion (CTE),
    228
Collaborative robot, *see* Cobots
CoM, *see* Computer-on-Module (CoM)
Command and Control Wind Tunnel
    (C2WT), 259
Communication models, 218
Communication satellites, 279
Community computing, 303
Complex instruction set computer
    (CISC) microcontroller, 21

Compliant mechanism, 155, 158, 223
Computer aided design (CAD) data, 119
Computer numerical control (CNC)
    machines, 57
Computer numerical control (CNC)
    servo press, 172–173
Computer-on-Module (CoM), 4
Computer vision, 92, 122, 126
Consumer awareness, 304
Contact measurement, 109
Content delivery networking (CDN),
    301
Convex hull computation, 137–138, 149
    algorithm for, 138
    of ellipsoid and ellipse, 148–149
    genus-0 surface and curve
        implicit surface, 145–146
        parametric surface, 144–145
    implementation examples, 147–149
    initial value computation, 143–144
    mathematical fundamentals,
        138–141
    numerical integration, 146–147
    partial convex hull surface, 148
    of surfaces, 141–144
    of two ellipsoids, 147–148
Cost matrix, 199
Coupling effect, nested bottom
    platform, 161
CPPS, *see* Cyber-physical production
    system (CPPS)
CPS, *see* Cyber-physical system (CPS)
Crank mechanism, 169
C-RANs, *see* Cloud-based RANs
    (C-RANs)
Crater wear, 94
CTE, *see* Coefficient of thermal
    expansion (CTE)
Cutting tool errors, 110
C2WT, *see* Command and Control Wind
    Tunnel (C2WT)
Cyber-application developer, 300
Cyber-enterprise, 263
    evaluation workflow, 266–267
    physical and software systems, 266
Cyber-physical production system
    (CPPS), 216, 245
Cyber-physical system (CPS), 216, 219
    social safety of, 238–240

D

Data acquisition mission of spacecraft,
        MDP coding, 198
    cost matrix, 199
    index to state conversion, 203
    policy calculation function, 206
    policy evaluation and plots, 207–208
    reward function, 198–199
    state to index conversion, 203
    state transition function, 199–202
    trajectory generation, 206–207
    transition probability function,
        202–203
    value calculation function, 205–206
    value iteration main file, 203–205
Data centers, 301
Data-driven services (DDSs), 302
Data privacy laws, 302–303
Data-Process Mapping (DPM), 262–264,
        268
Data security, 296
DCO, *see* Domain Concept Ontology
        (DCO)
DC servo motor, 171
Decision epoch, 184
Decision-making problems, 183
Degrees of freedom (DOF), 154, *see
        also* Multi-DOF micro-feed
        mechanism
    micropositioning stage, 222
Department of Defense (DoD) database
        systems, 259
Depth of field, 116
Design-for-manufacturability loop, 114
Design patterns, 257
Detection limit, sensors, 36–37
Developable surfaces, 140–145, 147, 148
Device-to-cloud model, 218
Device-to-device model, 218
Device-to-gateway model, 218
Digital image processing, 92, 125–126
Digital privacy rights, 302, 303
Digital supply network, 219
Digital transformation, 278, 301, 304, 305
Dimensional inspection, 113
Directrix curve, 141
Displacement performance analysis, 164
Displacement sensors, 40

Distortion coefficients, 122–123
DKB, *see* Domain Knowledge Base
        (DKB)
Domain Concept Ontology (DCO),
        261–265
    and PO, mapping, 269, 270
Domain Knowledge Base (DKB),
        261–262, 267
Domain ontologies, 267–268
DPM, *see* Data-Process Mapping (DPM)
DS, *see* Dynamic spectrum (DS)
Durability of sensors, 59
Dynamical tool wear prediction
        method, 93, 103
    accuracy, 95
    idea of, 94
    LS-SVM-based model, 96
    machine vision system, 95–96
    MIKRON UCP710 five-axis milling
        center, 93, 96
        average tool wear model, 98, 100,
            102
        design of experiments, 97–98
        dynamical prediction
            effectiveness, 101–102
        experimental data, 100
        experimental setup, 96–97
        maximum tool wear model, 98,
            99, 101
        stainless steel 0Cr17Ni12Mo2,
            96–97
    process flow, 94–95
Dynamic errors, 110
Dynamic response of sensors, 59, 63–64
Dynamic spectrum (DS), 291, 302

E

ECA, *see* Electrochemical actuator (ECA)
Eccentricity, blob analysis block, 130
Eccentric presses, 171, 172
EC-GSM-IoT, *see* Extended Coverage
        GSM for IoT (EC-GSM-IoT)
8-bit microcontroller, 20
8031 microcontroller, 23
8052 microcontroller, 23
Electric actuators, 43
Electrically erasable programmable read-
        only memory (EEPROM), 23

Electrical sensors, 37
Electric polymerization, 46
Electrochemical actuator (ECA), 44
Electrochemical sensors, 40
Electrochemical transducers, 33
Electromagnetic induction, 40
Electromechanical transducers, 32–33
Electronic-type sensing mechanism, 68
Electrostatic actuators, 226
Electrostatic discharge (ESD), 57–58
Embedded memory microcontrollers, 21
Emerging future network technologies, 289
Enterprise Situational Information (ESI), 262
Error(s), 110, 113
    bias, 114
    manufacturing, 106, 109–111
    prediction, 94, 95
    system, 109–111
ESD, *see* Electrostatic discharge (ESD)
ESI, *see* Enterprise Situational Information (ESI)
Euler–Bernouli beam equation, 223
Euler–Bernouli beam model, 225
Extended Coverage GSM for IoT (EC-GSM-IoT), 294
eXtensible Markup Language (XML) files, 267
External memory microcontrollers, 21
Extrinsic camera parameters, 121–122
    visualizing, 124

F

Fabrication
    actuator, 44, 46
    techniques, 225
Fair Information Practices principles, 303
FEA, *see* Finite element analysis (FEA)
FEM analysis, 230
    modal analysis, 232–235
    structural analysis, 235, 236
FETs, *see* Field effect transistors (FETs)
Field effect transistors (FETs), 40
Field of view (FOV), 116
Finite element analysis (FEA), 164
    ANSYS software, 159

mesh generation, 159
micro-feed platform, 163
nested bottom platform, 159–161
parallel top platform, 162–163
of servo press, 173–175
Five-axis micromilling machine based on PC control system, 222
5-DOF micro-feed platform, 155
5G, 281
    architecture, 285, 286
    commercialization of, 288
    high-resolution media streaming, 288
    network virtualization, 287–288
    players, 288–289
    spectrum allocation, 301
    technology components
        beamforming, 285
        massive MIMO, 284–285
        millimeter waves, 284
        readiness levels of, 287
    visions, 281–283
Flank wear, 94, 96
Flexure hinges, 154
Flexures, 223–225
Force flow diagram, 177
4G long-term evolution (4G LTE), 281
FOV, *see* Field of view (FOV)
FRE, *see* Functional reverse engineering (FRE)
Functional motion, robots, 239
Functional reverse engineering (FRE), 176–177
Future Network ecosystem, 279
Future network services (FNS), 304
Fuzzy logic, 92, 93, 115

G

Gabor filter, 117
Gauss curvature of surface, 139, 145
Gear drive servo press, 171
Genetic algorithm, reverse engineering, 258
Genetic network architectures, 257–258
Genus-0 surface and curve, convex hull of
    implicit surface, 145–146
    parametric surface, 144–145

Geometric measurement, 119
Global cloud traffic, 301
Google, 303
Grotthuss mechanism, 69

H

Harvard architecture microcontroller, 21–22
Hazard assessment, 240–245
Hidden Markov model (HMM), 93
High-level image processing, 126
High-power diode laser, 83
High-torque slow-angular-speed servo motors, 171
Histograms, 127
Human–machine interaction (HMI), 219
Human–robot collaboration (HRC), 220, 239
Human vision system, 118
Humidity sensors, 57
    cost and power consumption, 60
    dynamic response, 59
    hysteresis, 60
    parameters, 58
    recovery time, 59
    response time, 59
    selectivity, 59
    sensitivity, 59
    stability, 59
    TiO_2-based sensors, *see* TiO_2-based humidity sensors
Hydraulic actuators, 42–43
Hydraulic presses, 167–168
Hysteresis
    humidity sensors, 60
    TiO_2-based sensors, 64, 65

I

IAAS, *see* Infrastructure as a service (IAAS)
IBM, 303
    PowerPC processors, 19
IEP, *see* Integration & Execution Platform (IEP)
ILC procedure, *see* Iterative learning control (ILC) procedure
Image acquisition, 126–128

Image enhancement, 126
Image processing, 126
Image representation, 125–126
Image segmentation, 126, 127
Image transformation, 125–126
Implicit surfaces
    convex hull of, 145–146, 148
    gradient of, 140, 142
"indexlist" function, 194–195
Index to state function file, 187–188
Inductive sensors, 40
Industrial transformation, 295–296
Industry 4.0, 112, 216–217, 240, 295
Industry 5.0, 278, 279, 296
Inference Enterprise (IE), 263–264
Inference Enterprise Modeling (IEM)
    knowledge base, 264–265
    methodology, 263–264
Information protection, 302–303
Infrastructure as a service (IAAS), 301
In-process measurements, 110, 113–114, 119–120
In-process tool wear estimation, 96
Insider Threat Detection (ITD), 263–265
In-situ measurements, 112–113
Integrated vision system, 117
Integration & Execution Platform (IEP), 262–263
Intel 4002, 22
Intel 4003, 22
Intel 4004, 22
Intel 8051, 21, 23
Intelligent robot, 238, 240
Intel microprocessors, 8–15
Interactive time-slice technology, 288
Intermediate closed loop, 114–118
Internet Architecture Board, 218
Internet of Things (IoT), 216–218, 299, 303
    communication infrastructure, 293–294
    definition of, 293
    EC-GSM-IoT, 294
    global market value of, 293
    long-range communication, 294–295
    LoRa, 295
    LTE-M, 294
    NB-IoT, 294–295
    short-range communication, 294

Intrinsic camera parameters, 121
Inverse piezoelectric effect, 47
Ionic-type sensing mechanism, 68
I/O ports, 6
IoT, *see* Internet of Things (IoT)
Iterative learning control (ILC)
    procedure, 172
IT security issues, 296

J

Japan's telecommunication industry, 289
Joule effect, 47
Joule heating, 226

K

Kalman filter, 93
K-means clustering, 127

L

Label, blob analysis block, 129
Labeling of object, 127–128
Land distribution networks, 279
Laser cladding process, 79–80
    application materials, 83
    experimental setup, 80, 81
    experimental work, 85
    industrial applicability, 83–84
    process parameters, 82, 84
    remanufacturing, 84–85
    replacement method, 82
    schematic diagram of, 81
    synchronous feeding, 82
Laser hardening process, 83–84
Laser power, 82
Laser surface carburizing and nitriding,
    82
Lateral comb actuators, 225
Lateral thermal actuators, 225–226
LBMM, *see* Lithographybased
    micromanufacturing (LBMM)
Least squares support vector machines
    (LS-SVMs), 93, *see also* LSVM-
    based tool wear model
Legible motion, robots, 239
Lens distortion, 125
Light dependent resistors (LDRs), 39

Linear sensors, 36
Line defects, 67
Lithographic methods, 221
Lithographybased micromanufacturing
    (LBMM), 220
Long-term stability of sensors, 59
Loops, CLM systems, 113–114
LoRa, 295
Low-level image processing, 126
LSVM-based tool wear model, 94, 96,
    98, 103
LTE-M, 294

M

Machine accuracy, 110
Machine-to-machine communication
    (M2M), 296
Machine vision, 92, 93, 103
    online tool wear measurement, 95–96
Machine vision system (MVS), 114–118,
    120
Magnesium (Mg) alloys, 83
Magnetic actuators, 47
Magnetostrictive effect, 47
Major axis length, 130
Manufacturing errors, 106
Manufacturing process, 106–107
    calibration process, 120–125
    closed-loop manufacturing method
        benefits of, 111
        in-process measurement, 113–114
        in-situ measurement, 112–113
        loops, 113–114
        remote measurement, 112
    integrating intelligence, 131
    machine vision system, 114–118
    measurement processes, 107–109
    quality control in, 119
    real-time intelligent monitoring
        system, 118
    system error and variation, 109–111
Markov decision process (MDP) coding,
    183, 212–213
    case studies
        data acquisition mission of
            spacecraft, 198–208
        SIR model of epidemic infection,
            208–212

elements of, 184
MATLAB for, 184
  basic assumptions, 185
  components of, 184–185
  index to state function file,
    187–188
  policy calculation function file,
    193
  reward function file, 185–186
  state to index function file,
    186–187
  state transition function file,
    188–189
  transition probability function
    file, 189–191
  value calculation function file,
    192–193
  value iteration main file, 191–192
optimal policy analysis
  "indexlist" function, 194–195
  trajectory generation, 195–197
time representation, 184
value and policy iteration, 183
MaS, *see* Mobility-at-scale (MaS)
Massive MIMO (mMIMO), 284–285
Mass production, 295
MATLAB coding, for MDP, 184, 212–213
  assumptions, 185
  case studies, 198–212
  components of, 184–185
  index to state function file, 187–188
  policy calculation function file, 193
  reward function file, 185–186
  state to index function file, 186–187
  state transition function file, 188–189
  transition probability function file,
    189–191
  value calculation function file,
    192–193
  value iteration main file, 191–192
MCS-51 microcontroller, 22
MDP coding, *see* Markov decision
  process (MDP) coding
Measurement uncertainty, 108–109
Mechanical actuators, 43
Mechanical micro-machining, 153–154
Mechanical presses, 168
Mechanization, 295
Mega AVR microcontrollers, 23

MEMS, *see* Microelectromechanical
  systems (MEMS)
MEMS-based active sensors, 47
Mesh generation, 159
Metal forming presses, 167
  hydraulic presses, 167–168
  mechanical presses, 168
  reverse engineering, 175–177
  servo presses
    AC/DC motors, 171
    advantages of, 170
    computer numerical control,
      172–173
    drive systems, 170
    features, 169
    finite element analysis, 173–175
    forming technique, 173
    gear drive, 171
    ILC procedure for, 172
    load pulsation mode in, 173
    mechanical eccentric press *vs.*,
      171, 172
    multipoint, 172
    3D servo press, 171
    working mechanisms, 169
Metal-oxide (MOX)-based gas sensor,
  58, *see also* Humidity sensors
METIS project, 281, 288
Metrology
  manufacturing process, 107–109
  network, 292
Microactuation, 225–226
Microcontrollers, 4, 27
  applications, 6
  block diagram of, 5
  classification of, 20–22
  future trends, 26–27
  history of, 23, 24
  internal data bus of, 20
  microprocessors *vs.*, 5–7
  modern age, 25–26
  processing power and memory, 6
Microelectromechanical systems
  (MEMS), 220
Microfactory, 221–222
Micro-feed platform, FEA model, 163
Micromachine actuators, 221
Micromanufacturing techniques,
  220–221

Micromechanical machining, 221
Micromechatronics, 225
Microprocessors, 4
  AMD, 16–18
  applications, 5–6
  block diagram of, 5
  history of, 22, 24
  IBM Power PC, 19
  Intel, 8–15
  vs. microcontrollers, 5–7
  Mostech, 19
  Motorola, 16
  processing power and memory, 6
  SPARC, 19
  Zilog, 18–19
Microstage design, 227–229
Microsystems, 220
Mid-level image processing, 126
Millimeter (mm) waves, 284
Miniature machine tools, 221
Minor axis length, 130
mMIMO, *see* Massive MIMO (mMIMO)
Mobile communication, 279
Mobile generation
  bandwidth growth with, 281
  standards and, 280
Mobility-at-scale (MaS), 291, 292
Modal analysis, 232–235
ModelCenter®, 260, 270
Model-driven reverse engineering, 258
Modern image processing techniques,
    92
Modernization of assessment
    mechanism, 304
Moment of inertia, 230
Monte Carlo simulation, 270
Mostech microprocessor, 19
Motorola microprocessors, 16
Multi-DOF micro-feed mechanism, 154,
    164
  design process, 154
    branched chain, 156
    compliant mechanism, 155
    nested bottom platform, 156–157
    parallel top platform, 157–158
  displacement performance analysis,
    164
  finite element analysis
    ANSYS software, 159

  mesh generation, 159
  micro-feed platform, 163
  nested bottom platform, 159–161
  parallel top platform, 162–163
  model of, 155
Multiformalism modeling, 259–260
Multifunctional smart sensor system, 58
Multimodeling approach to inference
    enterprise modeling (MIEM),
    260
Multimodeling integration platform,
    256, 262
Multiple-input and multiple-output
    (MIMO), 284, *see also* Massive
    MIMO (mMIMO)
Multiple linear regression model, 92
Multipoint servo presses, 172
MVS, *see* Machine vision system (MVS)

N

Narrowband IoT (NB-IoT), 294–295
National Cloud Exchange, 301
National Communication Infrastructure
    (NCI), 278–279
  geographic coverage, 297–298
  information protection, privacy,
    and ethical applications
    framework, 302–303
  infrastructure challenges, 298–299
  issues, 292–293
  modern assessment, 304
  national strategic importance of, 299
  physical assets of, 279
  quality and consistency, 299–300
  smart community applications,
    291–292
  stakeholders group, 304–305
NB-IoT, *see* Narrowband IoT (NB-IoT)
NCI, *see* National Communication
    Infrastructure (NCI)
Negative radial distortion, 122
Nested bottom platform, 156, 164
  coupling effects, 161
  design process, 156–157
  finite element analysis, 159–161
  model of, 156
Network flow virtualization (NFV), 287,
    299

Networking, 295
  challenges, 296
  software-defined, 287, 299
  transparent, 292
Network metrology, 292
Network transparency, 292
Network virtualization, 287–288
Neural network model, 92, 115
New York Waterways (NYWW)
    network, 290
Next Generation Mobile Networks
    Alliance (NGMNA), 281
NFV, *see* Network flow virtualization
    (NFV)
NGMNA, *see* Next Generation Mobile
    Networks Alliance (NGMNA)
Nickel titanium (NiTi) alloys, 44, 45
NLBMM, *see* Non-lithography-
    based micromanufacturing
    (NLBMM)
Non-contact measurement, 109
Non-linear sensors, 36
Non-lithography-based
    micromanufacturing
    (NLBMM), 220
Numerical integration techniques,
    146–147
NYWW network, *see* New York
    Waterways (NYWW) network

O

Object classification technique, 240
Object-detection API, 241–243
Object-detection-based algorithm, 240
Ohmic heating, 226
OMM, *see* On-machine measurement
    (OMM)
Omnidirectional actuators, 46
One-dimensional (1-D) machine vision,
    116
Online tool wear estimation system,
    92–93, 95–96
On-machine measurement (OMM), 110,
    113
Ontologies, 261, 264
Optical actuators, 33
Optical fiber core networks, 279
Optical sensors, 40–41

Optimal policy analysis, 194–198
Optoelectronic transducers, 31–32
Optomechanical transducers, 33
Otsu's segmentation technique, 127

P

PAAS, *see* Platform as a service (PAAS)
Parallel face flexures, 223
Parallel top platform, 155–156, 164
    design process, 157–158
    finite element analysis, 162–163
    model of, 158
    3-PSS/S configuration, 157
Parametric surfaces, 144–145
Passive sensors, 37–38, 47
    capacitive sensors, 39
    electrochemical sensors, 40
    inductive sensors, 40
    optical sensors, 40–41
    resistive sensors, 38, 39
    subcategories, 38
Perimeter, blob analysis block, 130
Peripheral interface controller (PIC), 23
Phase changing materials, 44
Phoenix Integration's ModelCenter®, 262
Photo-etching, 220
Physical safety, 239
Piezoactuators, 154, 157, 227, 245
Piezoelectric actuators, 46, 47, 158, 222,
    226
Piezoelectric crystals, 226
Piezoelectric effect, 47
Piezoelectric transducers, 33
Pin joints, 223
Plane defects, 67
Platform as a service (PAAS), 301
Pneumatic actuators, 43
PO, *see* Process Ontology (PO)
Point defects, 67
Point of presence (POP), 279
Policy calculation function file, 193
Positive radial distortion, 122
Post-processing measurement, 112, 119
Power-line-borne fiber, 299
PowerPC processors, 19
Precision, 108
Precision machine tools, 221
Predictable motion, robots, 239

Press technology, 167, *see also* Metal forming presses
Pressure sensors, 40
Princeton architecture microcontroller, 22
Privacy laws, 302–303
Process control, 109, 110
Process control loop, 114
Processes improvement loop, 113
Process intermittent measurement, 113
Process Ontology (PO), 256, 261, 262, 264
    analysis models in, 268
    DCO and, 269, 270
Process qualification, 110
Projection errors, 124
Psychological safety, 239
Psychological safety of system, 226, 246
Pycom, 26
Pyroelectric transducers, 33
Pyrometers, 84

Q

Qualitative measurement, 108
Quality control
    and manufacturing process, 119
    procedures, 118, 131
Quasi-static errors, 110

R

Radial distortion coefficients, 122–123
Radio bandwidth, 281
Radio frequency identification (RFID), 293
Radio Frequency Identification (RFID), 218
Raspberry Pi, 4, 5, 26
RE, *see* Reverse engineering (RE)
Real-time inspection system, 113, 119
Real-time intelligent monitoring system (RIMS), 118
Real-time safety system, 239
Real-time tool wear estimation, 96
Reciprocal piezoelectric effect, 47
Recovery time, 35, 37
    humidity sensors, 59
    TiO_2-based sensors, 63, 64

Reduced instruction set computer (RISC) microcontroller, 21, 23
Regulatory and development bodies (RDBs), 297–301, 304, 305
Relative humidity (RH), 57
    mathematical formula, 58
    sensor response *vs.*, 64
Remanufacturing process, 84–85
Remote measurements, 112
RENs, *see* Research and Education Networks (RENs)
Repeatability of measurement, 108
Reproducibility of measurement, 108
Reproducibility of sensors, 37
Reprojection error, 124
Research and development (R&D), 304
Research and Education Networks (RENs), 299
Resistive heating, 226
Resistive sensors, 38, 39
Resolution, 116
Resource-sharing model, 299
Response time, 35, 37
    humidity sensors, 59
    TiO_2-based sensors, 63, 64
Reverse engineering (RE), 256, 270–271
    advanced techniques, 176
    applications, 263–270
    definitions of, 256–257
    engineering design, 175–176
    functional, 176–177
    phases, 258
    processes, 258–259
    proposed approach
        Domain Knowledge Base, 261–262
        Enterprise Situational Information, 262
        framework, 261
        Integration & Execution Platform, 262
        scenarios for, 262–263
        two-phase approach, 260–261
    techniques, 257–258
Reward function file, 185–186
RFID, *see* Radio Frequency Identification (RFID)
RH, *see* Relative humidity (RH)
Rich Text Format (RTF) files, 267

RIMS, *see* Real-time intelligent
    monitoring system (RIMS)
RISC microcontroller, *see* Reduced
    instruction set computer
    (RISC) microcontroller
Robot(s)
    collaborative, *see* Cobots
    intelligent, 238, 240
    motion, 239
    social, 238
    system, 241
Ruled surfaces, 140–141
Rutile, 61–62

S

SAAS, *see* Software as a service (SAAS)
Safety, 238–240
Santander, smart city project in, 290
SBC, *see* Single-board computer (SBC)
Scanning velocity, 82, 85
SDN, *see* Software-defined networking
    (SDN)
SDR, *see* Software-Defined-Radio
    (SDR)
Security issues, 296
Segmentation, image, 126, 127
Self-organizing map (SOM), 93
Self-powered sensors, 41
Semantic Testbed for Inference
    Enterprise Modeling (STIEM),
    260
Sensitivity *(S)*
    definition of, 36
    humidity sensors, 59
    TiO$_2$-based sensors, 65
Sensor(s), 30, 33–34, *see also* Humidity
    sensors
    active, 41
    characteristics, 34, 35
        detection range and limit, 36–37
        linearity and calibration, 34–36
        recovery time, 37
        response time, 37
        sensitivity, 36
        specificity, 36
        stability and reproducibility, 37
    in CNC machines, 57
    functions, 56

passive, 37–38
    capacitive sensors, 39
    electrochemical sensors, 40
    inductive sensors, 40
    optical sensors, 40–41
    resistive sensors, 38, 39
    subcategories, 38
    selection of, 47
    self-powered, 41
    system, 56–58
    technology, 55–56
    working principle, 56
Sensor response, definition of, 64
Service Oriented Architecture for
        Socio-Cultural Systems
        (SORASCS), 259
Servo motors, 171
Servo presses
    AC/DC motors, 171
    advantages of, 170
    computer numerical control, 172–173
    drive systems, 170
    features, 169
    finite element analysis, 173–175
    forming technique, 173
    gear drive, 171
    ILC procedure for, 172
    load pulsation mode in, 173
    mechanical eccentric press *vs.*,
        171, 172
    multipoint, 172
    3D servo press, 171
    working mechanisms, 169
SH, *see* Specific humidity (SH)
Sheet metal armor, 167
Short-term stability of sensors, 59
Sigfox, 290
Single-axis flexure-hinge structure, 158
Single-board computer (SBC), 5
SiP, *see* System-in-a-Package (SiP)
16-bit microcontrollers, 20
64-bit microcontrollers, 20–21
64-bit microprocessors, 7
Small cells, 284
Smart city, 302
Smart communities, 278, 289–290
    NCI issues, 292–293
    spectrum usage, 301
    technology requirements, 291–292

Smart community networking groups, 292
Smart factories, 218–219
    physical domain development
        material properties, 228, 229
        mathematical formulation, 230–232
        meshing of structure, 231, 232
        methodology, 230
        microstage design, 227–229
        modal analysis, 232–235
        results, 235–238
        structural analysis, 235, 236
    scenario details, 226–227
    virtual domain development
        hazard assessment, 241–245
        methodology, 240
        object-detection API, 241–243
        social safety of CPS, 238–240
        UR3 cobot, 240–241
Smart materials, 43, 44
Smart objects, 217
Smart radios, 292
Smart sensors, 57
Smart soft actuators
    electrochemical, 44
    fabrication of, 44
    magnetic, 47
    mechanical design concept, 45
    piezoelectric, 46, 47
    smart materials, 43
    SMA/SMP, 44–46
Smart watches, 58
SMA–SMP smart soft actuator, 44–46
SoC, *see* System on Chip (SoC)
Social-cobots, 220
Social robots, 238
Software as a service (SAAS), 301
Software-defined networking (SDN), 287, 299
Software-Defined-Radio (SDR), 302
Software systems, 258
Solar cells, 41
Solid-state laser, 83
SoM, *see* System on Module (SoM)
Songdo, smart city project in, 290
Specific humidity (SH), 58
Specificity of sensors, 36

Spectrum-allocation policy, 302
Spectrum license, 302
Stability of sensors, 35, 37, 59
    TiO_2-based humidity sensors, 63
Stainless steel 0Cr17Ni12Mo2, 96–97
State to index function file, 186–187
State transition function file, 188–189
Statistical learning system, 92
STIEM, *see* Semantic Testbed for Inference Enterprise Modeling (STIEM)
Strain sensors, 39
Stroke-based control and screw mechanism, 168–169
Structural analysis, 235, 236
Sun Scalable Processor Architecture (SPARC), 19
Supervisory Control and Data Acquisition (SCADA) system, 267
Support vector machine (SVM), 92, *see also* Least squares support vector machines (LS-SVMs)
Support vector method, 93
Surface parameterization, 138, 139
Surfaces, convex hull computation of, 141–144
Surface treatment technologies, 79, *see also* Laser cladding process
Susceptible-Infected-Recovered (SIR) model, MDP coding
    index to state conversion, 210
    policy evaluation, 211
    reward/cost function, 208
    reward evaluation and plots, 211–212
    state to index conversion, 210
    state transition function, 208–209
    transition probability function, 209
    value calculation, 211
    value iteration main file, 210–211
SVR model, 93
SysML Block Definition Diagram (BDD), 267
System error, 109–111
System-in-a-Package (SiP), 4
System on Chip (SoC), 4
System on Module (SoM), 5

T

Tangential distortion, 123
Telecommunication regulatory
     ecosystem, 279
Temperature measurement meter, 84
Temperature sensors, 38, 39
Texture filters, 127
Thermal actuators, 225
Thermal oxidation process of titanium,
     60, 61
32-bit microcontrollers, 20
32-bit microprocessors, 7
3D beamforming technology, 285
3D connectivity, 128
3D sensor, 239
Three-dimensional (3-D) machine
     vision, 116
3D servo presses, 171
Threshold, selection of, 127
Tiny AVR microcontrollers, 23
TiO_2-based humidity sensors
     humidity sensing properties
          dynamic response, 63–64
          hysteresis characteristics, 64, 65
          response and recovery times,
               63, 64
          sensitivity, 65
          sensor response, 64
          stability of, 63
     sensing mechanism, 65–69
     sensor performance, 69, 70
     thermal oxidation process of
          titanium, 60, 61
          HRTEM image, 63
          SEM micrographs, 61, 62
          TEM images, 62, 63
          XRD diffractograms, 61, 62
Titanium
     in laser cladding process, 83
     thermal oxidation process of, 60, 61
Tooling accuracy, 110
Tool wear
     crater wear, 94
     definition of, 91
     flank wear, 94, 96
Tool wear estimation, 91–93, *see
          also* Dynamical tool wear
          prediction method

real-time mode and in-process
     mode, 96
Tool wear model, 94–95
     LS-SVM-based model, 96
Trajectory generation, 195–197
Transducers, 30–31, 40, 56
     electrochemical, 33
     electromechanical, 32–33
     optoelectronic, 31–32
     optomechanical, 33
     selection of, 47
     types of, 32
Transition probability function file,
     189–191
Transparent networking, 292
Transverse comb actuators, 225
Triboelectric transducers, 33
Trueness, 108
2-D connectivity, 128
Two-dimensional (2-D) machine vision,
     116

U

Ultra-low-frequency (ULF)
     communication, 298
Uncertainty evaluation, 107–108
Unified Modelling Language (UML), 267
Unit normal vector, 139, 140, 142
UR3 cobot, 240–241

V

Value calculation function file, 192–193
Value iteration main file, 191–192
Von-Neumann architecture
     microcontroller, 22

W

Water molecule, 58
Watershed segmentation, 127
WG, *see* Workflow Generator (WG)
White-Space communication (WSC), 292
Wi-Fi, in microcontrollers, 27
Wireless sensor networks (WSNs), 218
Wireless technology, 291
WL, *see* Workflow Language (WL)
Workflow Generator (WG), 262, 265, 270

Workflow Language (WL), 262
Working distance (WD), 116
Workpiece fixture errors, 110
WSC, *see* White-Space communication
(WSC)
WSNs, *see* Wireless sensor networks
(WSNs)

X

XMega AVR microcontrollers, 23

Z

Zilog microprocessors, 18–19